Springer-Verlag France
26, rue des Carmes, 75005 Paris, France

K+ Channels in Cardiovascular Medicine

From basic science to clinical practice

Denis Escande MD – Nick Standen PhD (Eds)

Springer-Verlag Paris
Berlin Heidelberg New-York
London Tokyo Hong-Kong
Barcelona Budapest

Denis Escande
Laboratoire de Physiologie Cellulaire, URA CNRS 1121, Bât 443, Université Paris XI, 91405 Orsay, France

Nick Standen
Ion Channel Group, Department of Physiology, University of Leicester, Leicester LE1 7RH, UK

All translation, reproduction and adaption rights reserved for all countries.

The law of March 11, 1957 forbids copies or reproductions intended for collective use. Any representation, partial or integral reproduction made by any process whatsoever without the consent of the author or his executors is illicit and constitutes a fraud dealt with by Articles 425 and following of the Penal Code.

© Springer-Verlag France, Paris, 1993
Printed in France

The use for registered names, trademarks, etc. in this publication does not imply, even in the absence of a specific statement, that such names are exempt from the relevant prospective laws and regulations are therefore free for general use.
Product Liability: The publisher can give no guarantee for information about drug dosage and application there of contained in this book. In every individual case the respective user must check its accuracy by consulting other pharmaceutical literature

ISBN 2-287-00406-8 Springer-Verlag Paris Berlin Heidelberg Londres New York Tokyo Hong Kong Barcelone Budapest
ISBN 3-540-59598-8 Springer-Verlag Berlin Heidelberg London New York Paris Tokyo Hong Kong Barcelona Budapest
ISBN 0-387-59598-8 Springer-Verlag New York Berlin Heidelberg London Paris Tokyo Hong Kong Barcelona Budapest
2918 / 3917 - 543210 / Printed on acid-free paper.

Table of contents

List of authors, ...	VII
Preface, ...	XI
Foreword, E Corabœuf	1

1. **Introduction to the properties and functions of potassium channels,**
   NB Standen 3
2. **The molecular structure of potassium channels,**
   O Pongs and W Stühmer 25

Physiology

The heart muscle

3. **The role of potassium channels in maintaining resting potential in normal and anoxic cardiac muscle,** A Noma and H Matsuda 41
4. **Potassium channels and the regulation of refractoriness and heart rate,**
   E Carmeliet 53
5. **Potassium channels and cardiac contraction,**
   MR Boyett, SM Harrison and E White 69

The vasculature

6. **The role of potassium channels in the regulation of peripheral resistance,**
   MT Nelson 95
7. **The role of potassium channels in the regulation of coronary blood flow,**
   J Daut 107
8. **The regulation of blood flow to skeletal muscle during exercise,**
   F Karim and D Cotterrell 123
9. **Potassium channels in the pulmonary vasculature,**
   RZ Kozlowski and PCG Nye 141

Pharmacology

10. **Sulphonylureas: a receptor and a potassium channel?,**
    MLJ Ashford 161
11. **Potassium channel blockers in the heart,**
    I Baró and D Escande 175

12. **Pharmacological profile of potassium channel openers in the vasculature,**
    I Cavero and JM Guillon 193
13. **Potassium channel openers in the heart,**
    D Escande and I Cavero 225

From basic science to the clinic

14. **Prolonging cardiac repolarization as an evolving antiarrhythmic principle,** BN Singh, R Ahmed and L Sen 247
15. **Adenosine, ATP-sensitive potassium channels and myocardial preconditioning,** K Mullane 273
16. **Potassium channel openers in hypertension; a clinician's perspective,** RJ Kovacs 285
17. **Potassium channel openers in coronary heart disease,** A Berdeaux and JM Lablanche 293
18. **The potential of potassium channel openers in peripheral vascular disease,** NS Cook, JR Fozard and RP Hof 305

Index 321

List of authors

R Ahmed
Division of Cardiology, Veterans Administration Affairs Medical Center, California, the Department of Medicine, and the UCLA School of Medicine, California, USA

MLJ Ashford
Department of Pharmacology, University of Cambridge, Tennis Court Road, Cambridge CB2 1QJ, UK

I Baró
Laboratoire de Physiologie Cellulaire, URA CNRS 1121, Bât 443, Université Paris XI, 91405 Orsay, France

A Berdeaux
Faculté de Médecine Paris XI, Département de Pharmacologie, 63 rue Gabriel Péri, 94276 Le Kremlin Bicêtre, France

MR Boyett
Department of Physiology, University of Leeds, Leeds LS2 9JT, UK

E Carmeliet
Laboratory of Physiology, Campus Gasthuisberg, University of Leuven, B-3000 Leuven, Belgium

I Cavero
Rhône-Poulenc Rorer, Département de Biologie, CRVA, 13 quai Jules Guesde, 94403 Vitry-sur-Seine, France

NS Cook
Preclinical Research, Sandoz Pharma Ltd., CH-4002 Basel, Switzerland

D Cotterrell
Department of Physiology, University of Leeds, Leeds LS2 9NQ, UK

J Daut
Physiologisches Institut der Technischen Universität München, Biedersteiner Str. 29, D-8000 München 40, Germany

D Escande
Laboratoire de Physiologie Cellulaire, URA CNRS 1121, Bât 443, Université Paris XI, 91405 Orsay, France

JR Fozard
Preclinical Research, Sandoz Pharma Ltd., CH-4002 Basel, Switzerland

JM Guillon
Centre de Recherche Vitry-Alfortville, Rhône-Poulenc Rorer, 13 quai J. Guesde, 94403 Vitry-sur-Seine, France

SM Harrison
Department of Physiology, University of Leeds, Leeds LS2 9JT, UK

RP Hof
Preclinical Research, Sandoz Pharma Ltd., CH-4002 Basel, Switzerland

F Karim
Department of Physiology, University of Leeds, Leeds LS2 9NQ, UK

RJ Kovacs
Department of Medical Research, Methodist Hospital of Indiana, Indianapolis, USA

RZ Kozlowski
University Laboratory of Physiology, Parks Rd, Oxford OX1 3PT, UK

JM Lablanche
Faculté de Médecine de Lille, Service de Cardiologie B et Hémodynamique, Hôpital Cardiologique, Boulevard du Professeur J. Leclercq, 59037 Lille, France

H Matsuda
Department of Physiology, Faculty of Medicine, Kyushu University, Fukuoka 812, Japan

K Mullane
Gensia Pharmaceuticals Inc., 11025 Roselle Street, San Diego, CA 92121, USA

MT Nelson
Department of Pharmacology, Vermont Center for Vascular Research, University of Vermont, 55A South Park Drive, Colchester, VT 05466, USA

A Noma
Department of Physiology, Faculty of Medicine, Kyoto University, Kyoto 606, Japan

PCG Nye
University Laboratory of Physiology, Parks Rd, Oxford OX1 3PT, UK

O Pongs
Zentrum für Molekulare Neurobiologie, 2000 Hamburg, Germany and Max-Planck-Institut für Biophysikalische Chemie, Göttingen, Germany

List of authors

L Sen
Division of Cardiology, Veterans Administration Affairs Medical Center, California, the Department of Medicine, and the UCLA School of Medicine, California, USA

BN Singh
Division of Cardiology, Veterans Administration Affairs Medical Center, California, the Department of Medicine, and the UCLA School of Medicine, California, USA

NB Standen
Ion Channel Group, Department of Physiology, University of Leicester, Leicester LE1 7RH, UK

W Stühmer
Zentrum für Molekulare Neurobiologie, 2000 Hamburg, Germany and Max-Planck-Institut für Biophysikalische Chemie, Göttingen, Germany

E White
Department of Physiology, University of Leeds, Leeds LS2 9JT, UK

Preface

In recent years, knowledge of the properties, functions, and diversity of K^+ channels has spread from the domain of the basic scientist to a point where it has direct importance for the clinician. The most basic functions of the cardiovascular system involve K^+ channels: they have been shown to play important roles in the regulation of cardiac output through contractility and heart rate, and in the regulation of peripheral resistance and of blood pressure and distribution of blood flow through the control of the diameter of blood vessels.

The enormous increase in our knowledge of K^+ channels has come from the application of many techniques, with patch clamp and recently molecular biology being particularly important. When they were first described, K^+ channels were thought to be regulated only by the electrical difference across the cell membrane, and so to have functions restricted to electrically excitable cells. We now know that K^+ channels form a diverse group of proteins, with varied and complex functions, and subject to modulation by a wide variety of hormones, neurotransmitters, and intracellular regulators. In some cases we have information at every level from the molecular structure of the channel, through the detailed behaviour of the channel protein, to its function at the level of the cell and of the whole organ. At the same time, the pharmacology of K^+ channels is developing fast. K^+ channels are much more diverse than those selective for other ions. This diversity reflects the many roles they play, and offers many opportunities for selective intervention with drugs, so that it is likely that agents acting on K^+ channels will come to dominate developments in cardiovascular pharmacology in the 1990s.

In this book, we have sought to produce an up-to-date guidebook to the properties, functions, and pharmacology of K^+ channels in the cardiovascular system. We have assembled chapters from a panel of experts ranging from basic scientists working on the fundamental structure and behaviour of K^+ channels to practising clinicians. Throughout we have aimed to make the book intelligible to those in clinical practice, both in medicine and in paramedical professions. The book is organized on the basis of the main physiological functions of K^+ channels, moving on to their pharmacology and pathophysiology. The first two chapters provide background information on what K^+ channels are and what they do. These are followed by major sections of the functions of K^+ channels in the heart and vasculature and on their pharmacology in the cardiovascular system. The final section considers the clinical relevance of K^+ channels in situations where some of their part in cardiovascular pathology is already understood. Each chapter is self-contained so that the reader should be able to dip into the book as they wish, referring to other sections for related information, or of course to follow it from beginning to end.

A book like this involves the time and commitment of many contributors, and we should like to thank our co-authors for their help and enthusiasm. We hope that this book will facilitate clinical practice in aspects of cardiovascular medicine and will also stimulate research in this area. If these hopes are achieved to some degree, our efforts will have been worthwhile.

Denis Escande, MD
Professor of Physiology
University of Paris XI
Orsay, France

Nick Standen, PhD
Professor of Physiology
University of Leicester
Leicester, UK

Foreword

"Potassium ions are the richness of cells and potassium permeability is synonymous with rest and stability". In an austere lecture room at the old Sorbonne, Professor AM Monnier explained to his physiology students that, in agreement with Berstein's theory, the nerve resting membrane was essentially permeable to potassium ions. That was in 1946 and for us, his students, it was the first clear and understandable statement indicating that potassium ions play a key role in living phenomena. In contrast, exactly what occurred during the action potential was much more obscure. The illuminating ionic theory of Hodgkin and Huxley was still in preparation.

Three years later, during a short stay in Cambridge, at a time when the sodium hypothesis had been clearly stated, I was lucky enough to work with Silvio Weidmann and to participate in the recording of the first transmembrane action potentials. Very soon, S Weidmann demonstrated that the brief initial phase of the Purkinje fibre was a sodium spike as in nerve or muscle, but the existence of the long lasting plateau was harder to interpret and obviously was of a much more complicated nature. As a result of resistance measurements, S Weidmann was able to write in 1957 that in contrast with what occurs in nerve, the potassium permeability "in cardiac muscle does not rise to any considerable extent as a consequence of depolarization". This statement and his previous demonstration that the turtle action potential can be shortened by K^+-rich perfusion are what really started the potassium story in cardiac tissues. This led our group in Poitiers in 1958, and then OF Hutter and D Noble in 1960 and E Carmeliet in 1961 to show that the potassium permeability in fact decreases during the action potential. In 1965, this phenomenon, known as the inward-going rectification, and its potassium sensitivity were analyzed in detail by D Noble.

The existence of the inward-going rectification of the background potassium current, i_{K1}, appeared clearly as a highly efficient means for cardiac cells to spare inward currents during depolarization, i. e. to develop their long lasting action potential plateau at a low energy cost, but it remained to be understood how such a maintained plateau finally repolarized. Obviously, the existence of some delayed repolarizing current still had to be demonstrated. This was done by O Rougier, G Vassort and R Stämpfli and, more particularly, by D Noble and R Tsien who, in 1969, showed the existence in Purkinje fibres of two delayed potassium currents, i_{x1} and i_{x2}, recently described as i_{Kr} and i_{Ks} in guinea pig ventricular cells.

Several years earlier, KA Deck and W Trautwein, using their new technique of artificially shortened Purkinje fibres had observed a large transient outward current which was initially considered as a chloride current, then as a potassium current. In 1982, I was able with E Carmeliet, during a short stay in Leuven, to show that there exist in Purkinje fibres not one but two transient outward currents both supposedly carried by potassium ions. We proposed that these currents be labelled i_{lo}, for long lasting outward current, and i_{bo}, for brief outward current. We were possibly wrong to assume that both i_{lo} and i_{bo} (also labelled i_{to1} and i_{to2}) were potassium currents because recently i_{bo} has been described by AC Zygmunt and WR Gibbons as a chloride current in the rabbit heart.

As early as 1953, ASV Burgen and KG Terroux, and three years later OF Hutter and W Trautwein performed experiments indicating that acetylcholine increases the potassium permeability of sino-atrial and atrial cells. Was this permeability one of those already known in cardiac tissues? Apparently not, as shown by A Noma and W Trautwein in 1978 and E Carmeliet and K Mubagwa in 1986. Such experiments clearly proved that background potassium currents different from i_{K1} could participate in the modulation of both resting and action potentials of cardiac cells under normal conditions and actively stimulated the search for potassium currents activated by other ligands.

By that time, the explosion of the patch clamp technique had changed the face of the world in physiology as in many other fields of biology and in 1981 G Trube, B Sakmann and W Trautwein reported the first observation of the inward rectifier K^+ channel in cardiac cells. The following year R Coronado and R Latorre recorded the activity of several cardiac potassium channels reconstituted in planar lipid bilayer membranes, whereas in 1983 A Noma discovered the ATP-regulated K^+ channel in atrial and ventricular cell membranes. But we are now directly into the subject...

This book offers the remarkable advantage of covering the field of potassium channels from molecular structure to phenomena of direct and essential pathophysiological and clinical importance such as myocardial preconditioning and vascular diseases. Well balanced, a large section is given over to the role of K^+ channels in the different areas of the vasculature. The pharmacology of K^+ channel openers, a field of increasing interest and vast potentialities, is also extensively covered. Clearly, such a book will be of invaluable help to all those who need to gain a sound insight into, as well as a general overview of, the role of K^+ channels in cardiovascular physiology, pharmacology and medicine.

Edouard Corabœuf, PhD
Professor of Physiology
University of Paris XI

CHAPTER 1
Introduction to the properties and functions of potassium channels

NB Standen

The aim of this chapter is to provide a background for understanding the roles of potassium channels considered elsewhere in this book. We shall consider what ion channels are and what they do, and introduce some of the concepts needed to understand the way they work. Since the description and classification of channels is bound up with the electrophysiological methods used for their study, we shall also briefly consider some of these methods. The chapter will also introduce the diversity of potassium channels, and give an overview of their functions. There is only space here for a brief consideration of channel behaviour; much further detail may be found, for example in Hille [1].

Ion channels are membrane spanning proteins

Cell membranes have a basic lipid bilayer structure that is very impermeable to ions. Special proteins have evolved that provide pathways for ions to cross cell membranes, and so make the membrane permeable to certain physiological ions. These ion channels span the cell membrane, and have an aqueous pore running through the protein molecule that provides the pathway for ion movement across the membrane (fig. 1). The term channel is sometimes used to refer just to the pore itself, but more often for the whole protein molecule. Here, we shall refer to the whole protein as a channel.

Because channels make the cell membrane permeable to a number of different ions, and because this permeability can be regulated by the opening and closing of channels, they underlie many properties and functions of the cell membrane. Thus channels enable cells to set up membrane potentials, and allow currents to flow that change these membrane potentials, so underlying electrical signalling by the cell membrane. They also provide the means for the transduction of electrical signals into other cellular processes; a mechanism that involves calcium flow into the cell. In secretory and absorptive tissues channels provide pathways for ion flow, and it is likely that the variety of their known functions will increase as new cell types are studied. Channels also form targets for many transmitter substances, both extracellular transmitters involved in communication between cells, and intracellular messengers. They are also targets for an ever increasing number of drugs aimed at affecting a wide range of cellular functions.

There are many different types of ion channel, but they share a number of characteristic properties as follows:

Channels select between ions

The pore of an ion channel permits certain ions to pass through, but excludes others. Usually this selectivity is just for one physiological ion species, providing a basis for the

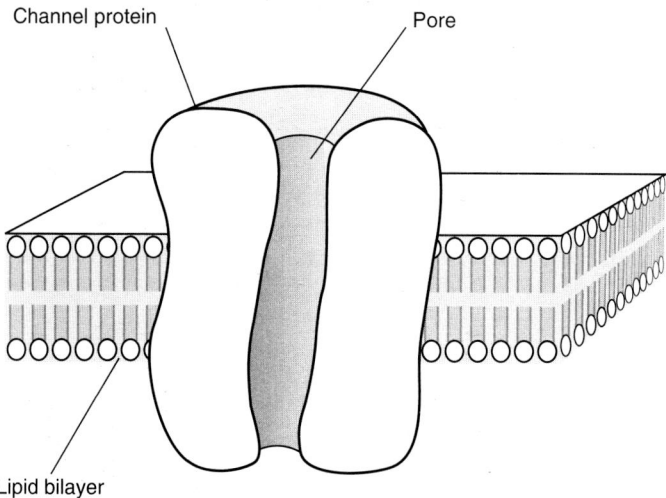

Fig. 1. Diagrammatic view of an ion channel protein. The protein molecule, shown here cut open to reveal the pore, spans the lipid bilayer of the cell membrane. When it is open, the pore provides an aqueous pathway across the membrane through which ions can pass

initial classification of channels into groups on the basis of their permeant ion. Thus we can distinguish sodium, potassium, calcium and chloride channels. Channels permeable to cations but not anions have also been described, and are usually called non-selective cation channels. If channels are classified according to their permeability in this way, several types of channel permeable to a given ion but differing in other properties can be distinguished. As we shall see, potassium channels are particularly diverse, with several families of K^+ channels having been described. At the molecular level, selectivity depends on the detailed structure of the pore, and therefore on the interactions between the pore walls and ions as they pass through. The details of these processes, and of the relation between channel structure and function in general, are beginning to be investigated using a combination of molecular biological and electrophysiological methods.

The channel pore is opened and closed by conformational changes in the protein

In general, the pore of an ion channel is not permanently open, but can be opened and closed by a change in the conformation of the protein. The process is called **gating**, and can occur in response to a number of factors. Channels can be divided on the basis of their gating into two broad categories; **voltage-gated channels** and **ligand-gated channels** (fig. 2). An increase in the chance that a channel will be open is often referred to as **activation**, and so it is also common to speak, for example, of voltage activated channels, or channels activated by acetylcholine. Voltage-gated channels respond primarily to changes in the membrane potential, that is the electrical potential difference across the cell membrane, and play a major role in cellular electrical activity, for example in the action potential. Ligand-gated channels respond primarily to chemical agents, either

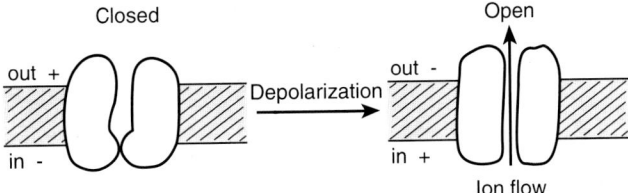
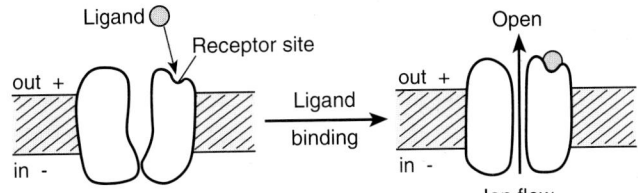

Fig. 2. Voltage-gated and ligand-gated channels

transmitters acting from outside the cell or intracellular messengers. Such channels enable cells to respond to transmitters with changes in ion permeability and so in electrical activity. Neither category is exclusive; the activity of voltage-gated channels is often modulated by neurotransmitters, while many ligand-gated channels show some sensitivity to membrane potential.

Voltage-gating involves a charged part of the channel protein, the voltage sensor, that is able to move in the voltage field across the membrane to open and close the pore, although the details of the mechanism at the level of movement of the protein molecule are not yet known. Ligand-gating involves a receptor site (or sites) on the protein, either extracellular or intracellular, for the particular chemical involved. Sometimes the receptor is not on the channel protein itself, but on a separate receptor protein. In this case the receptor often communicates with the channel by means of a GTP-binding protein.

Ion flow through channels is passive

The flow of ions through a channel protein is passive; that is to say ions flow downhill and metabolic energy is not consumed. Because ions are electrically charged the hill in question involves a consideration of both the concentration and voltage differences across the membrane. This electrochemical gradient determines the direction and influences the rate of ion flow through the channel. It will be further discussed below when we consider ionic equilibrium potentials. Because channels cannot move ions uphill, they do not establish the differences in ion concentration that occur across cell membranes, but rather use the energy stored in those concentration gradients for cellular signalling. These gradients are maintained by other membrane transport proteins that are able to move ions up their electrochemical gradient provided an appropriate energy source is available, for example the sodium/potassium pump, an ATPase that transports Na^+ out of and K^+ into cells.

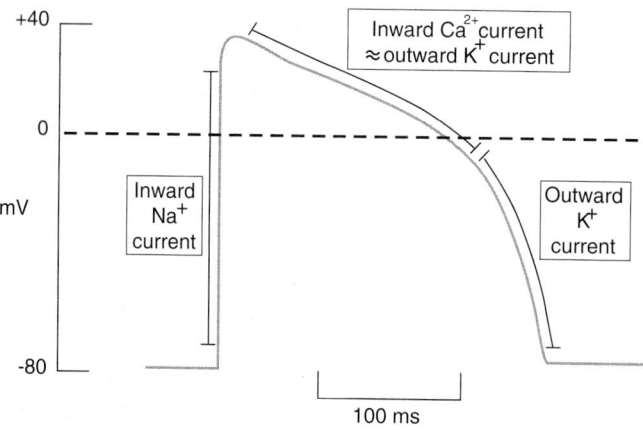

Fig. 3. Ion currents in the ventricular action potential. The action potential involves several different sorts of channel. Na$^+$ channels open to give an inward Na$^+$ current that causes the rapid initial depolarization. Both Ca^{2+} and K$^+$ channels are open during the plateau, while K$^+$ channels dominate the final repolarization

Ion flow through the pore of an open ion channel is typically fast, in the order of millions or tens of millions of ions per second. This carries a significant electrical current through the cell membrane; a current that may give rise to changes in the membrane potential. The current can be measured quite readily using the methods of electrophysiology, and in particular voltage- and patch-clamp which will be considered below.

Different channels have different functions

The diversity of ion channels reflects a diversity of function between different channel types. Different cells express a subset of the range of available channels and so acquire their characteristic membrane properties. Much of this book will be concerned with the diverse roles of different types of K$^+$ channel and with its important consequence of allowing for a selective pharmacology between channels.

As a simple example of the roles of different types of channel in a cellular process we can consider the action potential of ventricular cardiac muscle as shown in fig. 3. At rest, the cardiac muscle cell has a membrane potential of about -80 mV. In the action potential there is a rapid depolarization to about +40 mV followed by a plateau phase lasting over a hundred milliseconds in which the potential is above 0 mV, and finally a repolarization which returns the membrane potential to its resting value. The duration of the action potential plays a critical role in controlling the duration and force of ventricular systole. The resting membrane potential is primarily maintained by those K$^+$ channels which are open in the cell membrane at rest. The initial rapid depolarization is caused by the opening of Na$^+$ channels leading to a flow of Na$^+$ ions into the cell, so making the cell interior more positive. During the plateau phase the membrane potential is changing quite slowly, so that there is little net current flow across the membrane. This is because inward and outward currents more or less balance one another. The inward current is predominantly carried by Ca^{2+} entering the cell through Ca^{2+} channels, while the outward current is carried by K^+ ions flowing through K^+ channels. Lastly, the repolarization phase is caused by an outflow of K^+ ions from the cell, again through K^+ channels.

Cell membrane potentials

The voltages that occur across cell membranes, called membrane potentials, are by convention expressed as the voltage of the inside of the cell relative to the extracellular fluid, and are generally measured in millivolts (mV). At rest, animal cells have negative membrane potentials, resting potentials, ranging from about -30 mV to -90 mV, depending on the type of cell. A reduction in the membrane potential, such that the inside of the cell become less negative, is called a **depolarization**, while an increase in membrane potential to a value more negative than the usual resting potential is called a **hyperpolarization.** In electrically excitable cells the initial positive going upstroke of the action potential is thus a depolarization, while the return of the membrane potential to its resting value, the downstroke, is called a **repolarization**. A flow of positive ions, for example Na^+, from the extracellular medium to the inside of the cell will move the membrane potential in a positive direction, in other words it will cause a depolarization, while an outflow of positive ions will cause a hyperpolarization (or repolarization). Conversely an outflow of negatively charged ions like Cl^- will depolarize the cell.

The number of channel protein molecules in the membrane of a cell is often relatively small. This is so for the tissues considered in this book, where the overall membrane permeability to ions is quite low, especially so for smooth muscle. Often there may be only a few thousand, or even a few hundred, channels of a given type in the cell membrane.

Ionic equilibrium potentials are important for the understanding of channel function

Knowing the equilibrium potential for an ion in a particular cell provides important information about the way in which a channel permeable to that ion will affect the membrane potential of the cell. This is because it enables us to calculate both the direction and size of the electrochemical gradient for the ion and so to predict the direction and rate of ion flow through an open channel. In simple terms, opening a channel which is selectively permeable to a particular ion will tend to move the membrane potential towards the equilibrium potential for that ion.

As an example, we will consider the equilibrium potential for potassium ions. K^+ is present inside cells at high concentrations (around 140 mM), while its concentration in extracellular fluid is much lower, about 5 mM. So the concentration gradient tends to cause K^+ to flow out of the cell. As the positively charged K^+ ions cross the cell membrane an electrical gradient will be set up, so that the inside of the cell becomes negative with respect to the outside. This electrical gradient will oppose the outflow of K^+. For given

Table 1. Typical equilibrium potentials for physiologically important ions at 37°C

	Extracellular concentration	Intracellular concentration	Equilibrium potential
K^+	5 mM	140 mM	-88 mV
Na^+	140 mM	15 mM	+59 mV
Ca^{2+}	2 mM	0.1 µM	+131 mV
Cl^-	130 mM	15 mM	-57 mV

concentrations of K^+ inside and outside the cell, there will be an electrical potential difference (voltage) across the membrane at which the diffusional and electrical forces exactly balance. This potential is known as the potassium equilibrium potential, often called E_K, and is the potential where there will be no net movement of K^+ across the membrane. The number of ions which have to move to set up the electrical gradient is tiny; far too small to have any significant effect on the intracellular or extracellular concentration of K^+. Exactly the same argument may be applied to any other species of ion, so that there will also be equilibrium potentials for Na^+, Cl^-, Ca^{2+} and so on. These equilibrium potentials can be calculated as shown below, and Table 1 gives their values for typical extracellular and intracellular ionic concentrations. Of course, the value of the equilibrium potential for an ion will change if either the extracellular or intracellular concentration of the ion changes. Such changes can be important for example in ischaemia, when K^+ accumulation in extracellular spaces can lead to E_K becoming more positive.

Calculating the K^+ equilibrium potential

It is quite simple to calculate the equilibrium potential for K^+ (or any other ion). The expression needed is known as the **Nernst equation** and is:

$$E_K = RT/ZF \ln ([K^+]_o / [K^+]_i)$$

where E_K is the potassium equilibrium potential, $[K^+]_o$ and $[K^+]_i$ are the extracellular and intracellular concentrations of K^+ respectively, R is the Gas constant, T is the absolute temperature, Z is the valency (=1 for K^+), and F is Faraday's constant.

It is convenient to work out the constants at 37°C in millivolts (mV), and to convert from natural logarithms to $\log_{10}$, giving:

$$E_K = 61 \text{ mV} \log_{10} ([K^+]_o / [K^+]_i)$$

So, at 37°C with an extracellular $[K^+]$ of 5 mM and intracellular $[K^+]$ of 140 mM, the potassium equilibrium potential will be -88 mV.

For an ion channel with particular concentrations of ions on each side, there will be a membrane potential at which the current through the channel is zero. This is known as the reversal potential. For highly selective channels, much more permeable to their chosen ion than to others, this reversal potential will be very close to the equilibrium potential for the permeant ion. This is the case for K^+ channels. Typically they let K^+ through between 50 and 100 times more easily than Na^+. So the reversal potential for current through K^+ channels, although not exactly equal to E_K, will be very close to its value.

The K^+ equilibrium potential also enables us to calculate the **electrochemical driving force** moving K^+ ions through K^+ channels. At E_K, the force on K^+ ions is of course zero, so there is no net flow of ions even through an open K^+ channel. The driving force at any membrane potential V_m will simply be V_m-E_K (fig. 4). Since the membrane potential will normally lie positive to E_K, this driving force on K^+ will be in an outward direction, so that K^+ flows outward through open K^+ channels. Opening K^+ channels will therefore move the membrane potential towards E_K, and make it harder to shift the membrane potential away from E_K. The driving force on K^+ will be much greater when the cell is depolarized, as during an action potential, than it is at the resting potential.

In contrast to E_K, the equilibrium potentials for both Na^+ and Ca^{2+}, which are present at much higher concentrations outside the cell than inside, are positive (Table 1) so that opening of Na^+ or Ca^{2+} channels leads to entry of Na^+ or Ca^{2+} and so to depolarization.

Potassium channels have stabilizing, hyperpolarizing, or repolarizing functions

Because the value of E_K is very negative, K^+ channels when open will tend to hold the membrane potential at a negative value, the resting potential. Opening of K^+ channels can hyperpolarize cells, or lead to repolarization at the end of an action potential. So K^+ channels either keep the membrane potential negative, or move it in a negative direction. The channels that lead to electrical excitation, Na^+ or Ca^{2+} channels, are activated at depolarized membrane potentials, and so open K^+ channels keep the membrane potential in (or move it towards) a range where very few such channels will be open. For this reason, K^+ channels are often said to have a stabilizing effect on cellular electrical activity.

K^+ channels determine resting potentials

Biological membranes usually have a higher permeability to K^+ than other ions and K^+ channels provide the pathway for this permeability. Because the membrane is more permeable to K^+, the resting membrane potential becomes quite close to the equilibrium potential for K^+, E_K. Although K^+ channels dominate the resting membrane ionic permeability, channels permeable to other ions, for example Na^+ and Cl^- will also be open. (In skeletal muscle Cl^- channels actually contribute more than K^+ channels to resting permeability; in this tissue the equilibrium potential for Cl^- has a value close to the resting potential.) What happens to the membrane potential in a cell where more than one type of channel is open? We can often predict what the membrane potential will be by taking into account the relative ease with which different ions can cross the membrane using a constant field equation, the Goldman-Hodgkin-Katz equation:

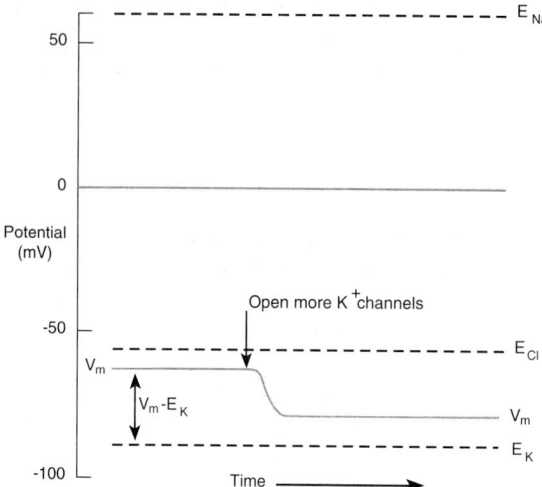

Fig. 4. Ionic equilibrium potentials and the membrane potential. Typical values for the equilibrium potentials for Na$^+$, K$^+$ and Cl$^-$ are illustrated (E_{Na}, E_K and E_{Cl} respectively). The membrane potential, V_m, is less negative than the equilibrium potential for E_K. The electrochemical driving force on K$^+$ is outward and is given by V_m-E_K. If more K$^+$ channels open (arrow), the membrane potential moves closer to E_K, in other words the membrane hyperpolarizes

$$V_m = 61 \text{ mV} \log_{10} \frac{P_K [K^+]_o + P_{Na} [Na^+]_o + P_{Cl} [Cl^-]_i}{P_K [K^+]_i + P_{Na}[Na^+]_i + P_{Cl} [Cl^-]_o}$$

where V_m is the membrane potential, P_K represents the permeability of the cell membrane for potassium, a function of the number of open K$^+$ channels as well as of how readily K$^+$ passes through those that are open, and similarly P_{Na} and P_{Cl} are permeabilities for Na$^+$ and Cl$^-$ respectively. It is not necessary to have absolute values for these permeabilities to use the equation, the ratio of P_K: P_{Na}: P_{Cl} is sufficient to predict the membrane potential. Although the Goldman-Hodgkin-Katz equation looks similar to the Nernst equation, it is derived under much more restricted assumptions [1], and does not represent a true equilibrium situation as V_m is not at the equilibrium potential for any of the ion species considered. Nevertheless, it often fits experimental membrane potentials well, and is very useful in understanding what will happen in a cell membrane where more than one type of channel is open. It is not surprising that if the relative permeability for a particular ion is high, corresponding to a relatively large number of open channels, the equilibrium potential for that ion will make a correspondingly large contribution to the membrane potential.

Opening K$^+$ channels can hyperpolarize cells

If, for simplicity, we consider a membrane with only K$^+$ and Na$^+$ channels open (in other words if $P_{Cl} = 0$), it is clear that the membrane potential will lie between the equilibrium potentials for K$^+$ (about -90 mV) and Na$^+$ (about +60 mV). If P_K and P_{Na} are equal, V_m will lie exactly half way between E_K and E_{Na}, while if P_K is much greater than P_{Na} the membrane potential will be much nearer E_K than E_{Na}, but will still be more positive than E_K. This approximates the situation found in most cells, where E_K is somewhat more negative than the resting membrane potential. For example, smooth muscle cells have resting potentials in the range -45 to -70 mV, while E_K will be around -85 to -90 mV. If more K⁺ channels open, for example in response to a neurotransmitter, so that P_K rises, the membrane potential will move closer to E_K, so becoming more negative (fig. 4). Thus activation of K⁺ channels can hyperpolarize cells. Conversely, closure of K⁺ channels will lead to depolarization of the cell membrane.

Some channels are less selective

A similar approach to that used for a membrane with channels for more than one type of ion can be used to understand the effect on membrane potential of an individual ion channel which is permeable to more than one type of ion. As for the entire cell membrane discussed above, the reversal potential for the channel will reflect the equilibrium potentials for the permeant ions, with ions that pass through the channel more easily contributing more to the reversal potential. It is often possible to use the Goldman-Hodgkin-Katz equation to fit the reversal potential for a single channel in terms of its relative permeability to different ions. Thus, a non-selective cation channel, for example, might have a reversal potential close to 0 mV. The effect of opening such a channel in a cell membrane will be, as always, to move the membrane potential towards the reversal potential for the channel, in this case 0 mV. In contrast to the situation for a highly selective channel, however, this reversal potential does not lie near the equilibrium potential for a particular ion.

K⁺ channels repolarize action potentials

In cells that fire action potentials, the rapid upstroke or depolarization, as we have seen, is caused by influx of either Na⁺ or Ca^{2+} ions through channels activated by membrane depolarization. The process is regenerative, so that an initial depolarization, caused for example by the action of a neurotransmitter or a pacemaker potential, leads to opening of voltage-dependent Na⁺ or Ca^{2+} channels, which in turn allow entry of positive ions causing further depolarization. Na⁺ and Ca^{2+} channel proteins also have intrinsic mechanisms which close them after a delay, a process called inactivation. Repolarization, that is the return of the membrane potential to its resting value, requires an outward current, and this is carried by K⁺ ions flowing through K⁺ channels. These channels are also voltage-activated, opening in response to the depolarization of the action potential, and are usually of the type called delayed rectifiers, discussed below. Their kinetic properties are such that the number of open K⁺ channels reaches a peak somewhat later than the number of open Na⁺ or Ca^{2+} channels, leading to the rapid downstroke that terminates the action potential. Since these K⁺ channels move the membrane potential to negative values where they themselves will close, their opening represents a self-inhibiting process.

Voltage-clamp and patch-clamp provide powerful methods for the study of channels

Much of our detailed knowledge of the properties and behaviour of ion channels comes from experiments using the methods of electrophysiology, and in particular the voltage-

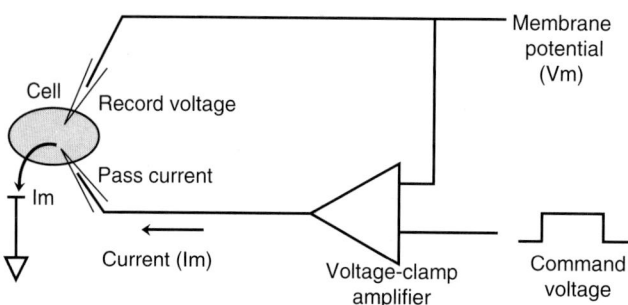

Fig. 5. Voltage clamp. The basic method of voltage clamp involves an electrode that measures membrane potential (V_m) and a negative feedback amplifier that controls the membrane potential by passing current through a second electrode. A command pulse supplied to the amplifier is used to change the membrane potential as desired, and the current flowing through the cell membrane (I_m) is measured either as it flows out of the amplifier or as it flows into the recording bath

clamp method and its modern variant, the patch-clamp. Descriptions of channels often use terms derived from this type of study, for example single channel conductance or inward rectification. The methods use the fact that it is possible to record the flow of ions through channels by measuring the electric current that they carry through the cell membrane, and that such measurements can be made with great sensitivity and good time resolution. One of the problems in studying the behaviour of channels during an action potential, for example, is that the membrane potential affects the opening and closing of many types of channel as well as the electrochemical driving force on permeant ions, while the current carried by ions flowing through these channels can change the membrane potential. The voltage-clamp overcomes this complex interdependence by using feedback from an amplifier to control, or clamp, the membrane potential. The behaviour of channels can then be studied using simple voltage changes under the control of the experimenter, for example square steps to study voltage-gated channels. For ligand-gated channels it is often useful to clamp the membrane potential at a fixed value while the relevant chemical ligand is applied. These techniques have also provided the means to study the action of drugs that affect ion channels, for example channel blockers or openers, in great detail, often at the single channel level.

Voltage-clamp methods require a means of measuring the membrane potential and a way of delivering current to the cell so as to control it (fig. 5). Methods for clamping relatively large or elongated cells, like giant axons, large neurones, or nerve fibres have been available for about forty years. Such methods often use wire electrodes which can be inserted along the axis of giant axons or muscle fibres, or glass microelectrodes which penetrate the cell membrane to measure voltage and deliver current. The patch-clamp, introduced by Neher and Sakmann in 1976, and for which they won the 1991 Nobel Prize in Medicine and Physiology, both enabled a much wider range of cells to be studied and allowed currents to be recorded from single ion channel proteins as well as from whole cells. In this technique a glass micropipette, with a tip diameter usually less than 1 μm, is pressed against the cell membrane. The tip of the pipette, which has been fire-polished to make it smooth, does not penetrate but its rim seals tightly to the cell membrane, electri-

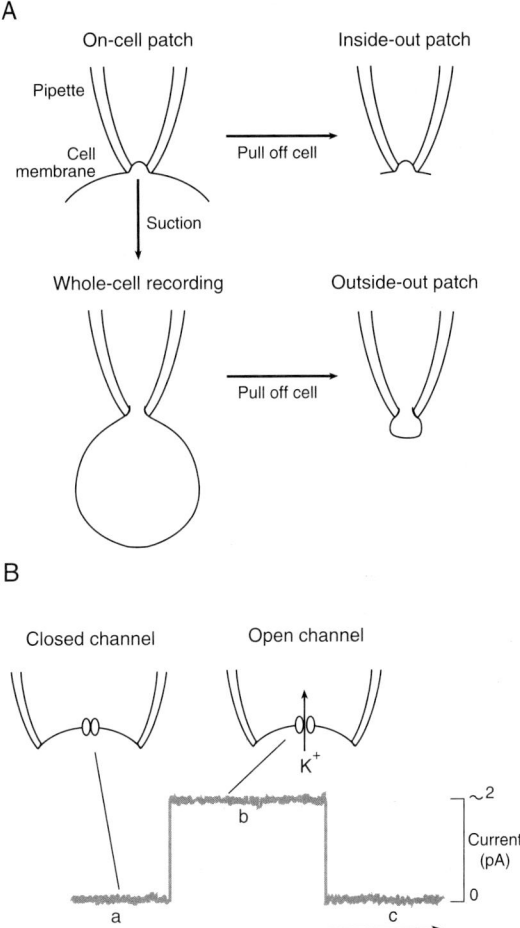

Fig. 6 A, B. The patch-clamp technique. **A** The different patch recording configurations. **B** Current recorded from a single channel. The current is zero when the channel is closed *(a)*. Channel opening leads to an abrupt change in the current as ions flow through the channel *(b)*. The single channel current is the size of this current, usually in the range 0-5 pA. Records of single channel current often look noisy because of the high amplification needed to record currents of this size. When the channel closes the current moves abruptly back to zero *(c)*. Channels stay open for varying periods of time, but open times are often in the order of msecs

cally isolating a small patch of membrane in the electrode tip (fig. 6). At this point several variants of the method are possible. The membrane patch normally contains just a few channels, sometimes only one and sometimes none, and it is possible to record currents from these channels with the pipette attached to the cell (**cell-attached or on-cell recording**). By applying suction to the pipette the membrane patch in the tip can be broken, so that the pipette interior makes contact with the inside of the cell. In this configuration currents are recorded from the whole of the remainder of the cell membrane (whole-cell recording). Finally, it is possible to pull off, or excise, patches of membrane attached to the pipette tip. These may be either **inside-out**, that is with the cytoplasmic face of the membrane facing the bathing solution, or **outside-out** (fig. 6A). These configurations give the opportunity of applying substances of interest to either side of an ion channel in its native membrane. A recent development is the permeabilized patch method, in which a pore-forming antibiotic such as nystatin is used to make the membrane patch within the pipette tip permeable to monovalent cations. This allows whole cell recording, but has the advantage that the biochemical machinery inside the cell is not washed out by the

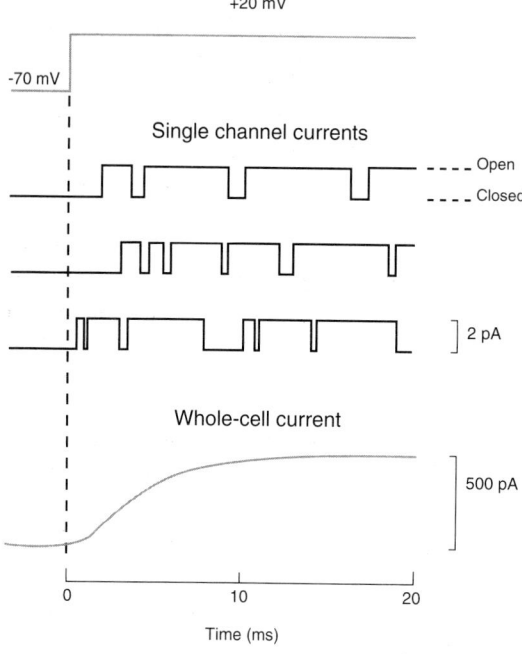

Fig. 7. Single channel and whole-cell currents. The figure shows three examples of single channel currents recorded from delayed K^+ channels in response to a step in membrane potential from -70 to +20 mV (top trace). Since the voltage during the step is +20 mV, and the K^+ equilibrium potential about -90 mV, the electrochemical driving force on K^+ is 110 mV and K^+ flows outward through the channel when it is open (by convention, outward currents are given a positive sign and plotted upward). The bottom trace shows a whole-cell recording of current through a large number of the same channels in response to the same voltage step. The current is the sum of the currents through all the individual channels, but the gain of the recording is now too low to see steps in current as individual channels open or close. Thus the whole-cell current shows how channel open probability changes with time (though it does not measure the absolute value of open probability). The size of the whole-cell current will depend on the number of channels in the cell being studied, and so can vary widely between different types of cell

pipette solution [2] and is often useful in studying effects dependent on second messengers. A good introduction to patch clamp can be found in reference [3] and more detailed information on the methods is given in references [4-6].

Both single channel and whole-cell currents are used to study channel behaviour

Although methods were available to record whole-cell currents flowing through many channels long before it became possible to record from single channels, it is simpler to consider first the current recorded through a single channel, and then the way in which many such channels will give rise to a whole-cell current. Fig. 6B shows the current recorded through a single channel as it opens and closes. The transition from closed to open or back again happens so rapidly that the current changes in a square step fashion between zero (when the channel is closed) and the single channel current (when the channel is open). The single channel current depends on the rate of flow of ions through the open channel, so that the constant single channel current seen in fig. 6B corresponds to a constant rate of ion flow. Single channel currents are usually in the order of picoamperes (1 pA = 10^{-12} A). 1 pA corresponds to a flow rate of 6.25 million monovalent ions per second.

At the level of a single channel the pattern of opening and closing is a random process, so that individual recordings of currents made under the same conditions will show different patterns of opening and closing. Fig. 7 shows examples of such records from a delayed rectifier K^+ channel, a voltage-activated channel that is involved in action potential repolarization in many tissues. The channel is activated by a depolarizing voltage-clamp step. The current recorded from a whole cell represents the sum of many such single channel recordings, and the whole-cell current rises smoothly with time, representing the average behaviour of the delayed rectifier channels (fig. 7). It is thus much easier to see overall channel kinetics from a whole-cell recording, since the current is already summed from many channels. Whole-cell currents are sometimes called macroscopic currents.

Single channel and whole-cell currents

The current carried by ion flow through a single ion channel is usually given the symbol i. The current through a large number of the same channels, called the macroscopic or whole-cell current is usually called I, and will be the product of i, the probability that a channel is open (P_o), and the total number of channels in the membrane (N), so that:

$$I = N \, i \, P_o$$

At a fixed voltage, for example in a voltage-clamp experiment, N and i will usually be constant, so that a change in whole-cell current represents a change in channel open probability.

Single channel conductance and open probability

In considering the functioning of a single ion channel we are interested in two things: how likely the channel is to be open (its open probability), and how much current will flow through when it is. These two factors will determine the average current flowing through the channel over a period of time and thus the contribution it will make to membrane current of a cell.

The open probability is the property that changes when a channel is activated, for example by a change in membrane voltage or binding of a ligand. It can be measured directly as the total time the channel spends open divided by the total duration of recording. The random nature of channel behaviour at the level of a single channel, however,

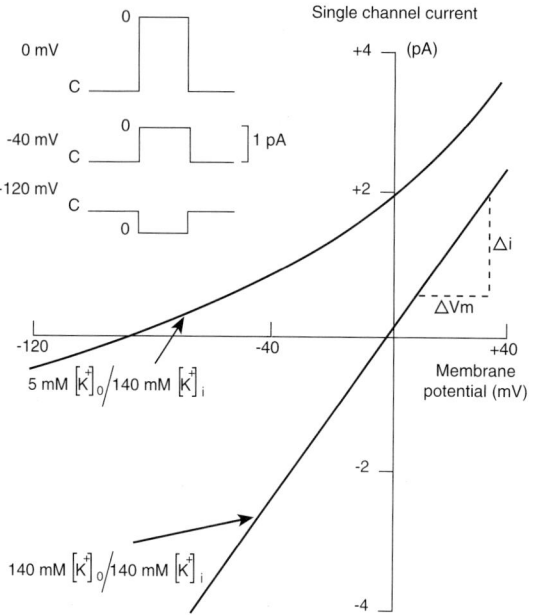

Fig. 8. Single channel current-voltage relations. The current flowing through a single open K$^+$ channel is plotted against the membrane potential. The inset shows examples of the single channel current recorded at different membrane potentials; such measurements are used to generate the current-voltage relation. The left hand I-V relation is measured with a physiological K$^+$ gradient, external 5 mM K$^+$ and internal 140 mM K$^+$; the direction of current flow reverses negative to -80 mV, close to E_K. The right hand relation is measured with 140 mM K$^+$ on both sides of the membrane, E_K is now 0 mV. The single channel conductance is given by the slope of the I-V relation, [$\Delta i / \Delta V_m$], 60 pS in the case of the high K$^+$ relation shown here

sometimes means that it is necessary to average over some time to get an accurate measure of open probability from single channel recordings.

The single channel current will also depend on voltage because this determines the driving force on K$^+$. For this reason **single channel conductance**, which takes account of this effect, is commonly used as a measure of how easily ions pass through the pore of an ion channel, and it is a property often used in describing a channel type. Single channel conductance is usually estimated from the **current-voltage relation** for a single open channel, obtained either by measuring single channel current at a series of different voltage-clamped membrane potentials or by changing the membrane potential in a ramp fashion. Fig. 8 shows current-voltage relations for a single K$^+$ channel, both with physiological K$^+$ concentrations on either side of the membrane and in symmetrical high K$^+$. Positive to E_K the current is outward, while at potentials negative to E_K the direction of the driving force on K$^+$ is reversed and the current is inward. The relation under physiological conditions shows the single channel currents that may be expected in vivo, but is usually curved. From the point of view of defining a value for the single channel conductance this creates a problem, since this conductance, which is equal to the slope of the current-voltage relation (see fig. 8), will change with voltage. For this reason, the conductance of single channels is often quoted under conditions of symmetrical high K$^+$, usually with a [K$^+$] of 140 mM (the normal intracellular value) on either side of the membrane. Under these conditions the current-voltage relation reverses at 0 mV and is usually close to a straight line, so that the single channel conductance is the same at different potentials and can be measured easily. Single channel conductances are expressed in

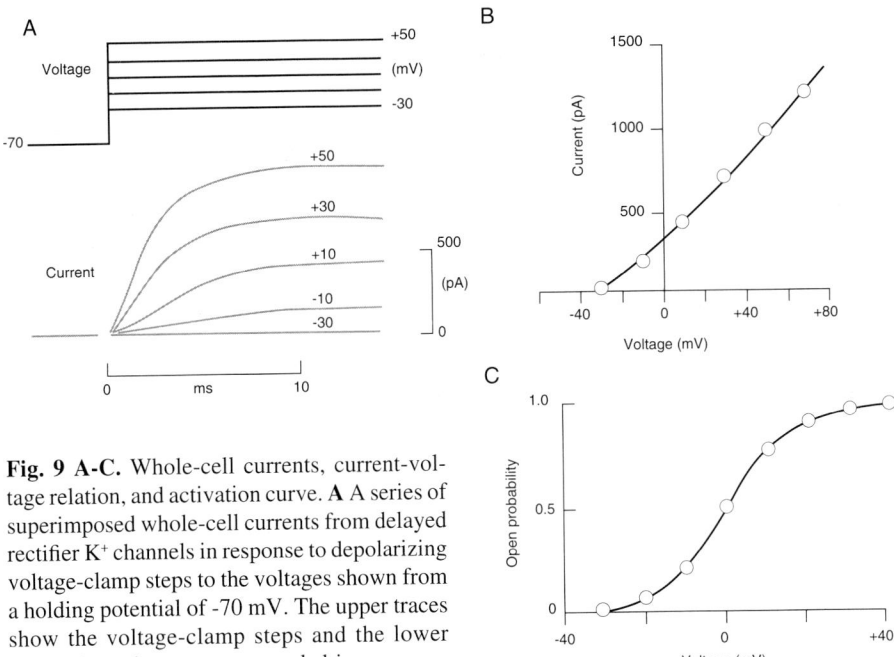

Fig. 9 A-C. Whole-cell currents, current-voltage relation, and activation curve. **A** A series of superimposed whole-cell currents from delayed rectifier K$^+$ channels in response to depolarizing voltage-clamp steps to the voltages shown from a holding potential of -70 mV. The upper traces show the voltage-clamp steps and the lower traces show the currents recorded in response. The number above each current trace shows the voltage during the corresponding voltage step. **B** The whole-cell current-voltage relation for the currents shown in A. The relation is obtained by plotting the final amplitude of the current, which is assumed to represent its steady-state value, against the voltage during the step. **C** An activation curve for delayed rectifier channels. The curve shows the way in which the relative number of open channels, or the relative channel open probability, changes with membrane potential

picosiemens (1 pS = 10^{-12} Siemens, Siemen = 1/ohm) and range from less than 10 pS up to 300 pS depending on the type of channel.

Whole-cell current-voltage relations and activation curves

Recording current from a whole cell has both advantages and problems compared to single channel recording. The main advantages are that the average behaviour of many channels is measured in one recording, and that the size of the whole-cell current through a particular channel type may give an indication of the abundance of the channel in the membrane. A problem is separation of currents flowing through different ion channels, since what is recorded is the net electrical current carried by all the ions moving across the membrane. By manipulating ionic conditions, choosing appropriate voltage protocols, and by using agents that specifically block certain types of channel it is often possible to arrange things so that one type of channel dominates the whole cell current [7].

Whereas in single channel recording the single channel current and open probability can be measured separately, both of these factors will affect the whole-cell current measured through a population of channels. Thus whole-cell current changes with voltage both as the current flowing through single open channels changes, and as the number of

Table 2. Cardiovascular K⁺ channels and currents

Channel	Current	Equivalent or similar channels and currents
Delayed rectifier K⁺ channel	i_K	$i_{K(V)}$; i_{DR}; i_x; RHK1
Transient outward K⁺ channel	i_{to}	i_A; A-current; RHK4
Inward rectifier	i_{K1}	i_{bg}; i_{IR}
Ca²⁺-activated K⁺ channel	i_{K-Ca}	maxi K; BK channels (large conductance)
ATP-sensitive K⁺ channel	i_{K-ATP}	
Muscarinic K⁺ channel	i_{K-ACh}	i_m; m channel
	$i_{K-adenosine}$	
Na⁺-activated K⁺ channel	i_{K-Na}	
Arachidonate-activated K⁺ channel	$i_{K-arachidonate}$	

open channels changes. Fig. 9 shows a series of whole-cell currents flowing through delayed rectifier K⁺ channels like those of fig. 7 measured when voltage clamp is used to step the cell membrane to a series of different voltages. Plotting the current after it has had time to reach a steady state against voltage gives the whole-cell current-voltage (or I-V) relation for the delayed rectifier current. The current increases as the membrane potential becomes more positive both as more channels open with depolarization and as the driving force on K⁺, and so the single channel current, increases. To obtain information on the dependence of the number of channels open (or the probability that a channel is open) on voltage it is usual to allow for the driving force on K⁺ ions, either by dividing the current at each voltage by $(V-E_K)$, or by rapidly shifting the voltage to a fixed potential and measuring the current at this voltage, in other words with a fixed driving force, before the K⁺ channels have time to close. This latter method is known as tail current measurement. These methods give a plot of the voltage dependence of channel open probability, which is usually known as an activation curve (fig. 9). Activation curves for voltage-dependent channels are usually sigmoid in shape. For channels that show voltage-dependent inactivation an equivalent inactivation curve can be generated which describes the way in which channel open probability declines with voltage.

Potassium channels come in many varieties

K⁺ channels occur in a particularly wide variety of types, reflecting the many functions that they perform. In the tissues of interest here, there are at least eight different types of K⁺ channel, and more probably await discovery. Certain properties of the channels are conserved between the types, especially the selectivity for K⁺ over other ions, and the permeability sequence for other ions that can pass through K⁺ channels such as Tl⁺ and Rb⁺. This implies a high degree of similarity in the part of the protein that forms the channel pore. The channels vary, however in such properties as single channel conductance, and especially in the factors that control their opening and closing. Even within the types given here, there are several isoforms of the channel, so that each type may be said to form a family of channels. The properties, roles, and pharmacology of the varieties of K$^+$ channel introduced here will be discussed much more fully in appropriate chapters of this book; the sections below are intended merely to provide an overview of the different K$^+$ channels of interest in the cardiovascular system.

A complication is that each type of K$^+$ channel often has several names. In this book, we have named the channels as set out in Table 2, and use abbreviated names in terms of the current that passes through the channel. Thus the transient outward K$^+$ channel, for example, is abbreviated to the i$_{to}$ channel, since the current that passes through it is i$_{to}$.

Voltage-activated K$^+$ channels: delayed rectifiers and transient outward channels

Voltage-activated K$^+$ channels occur in both cardiac and smooth muscle tissues. They comprise a large family of channels [8], each channel being formed as a tetramer of subunits. The channels open in response to a membrane depolarization; their probability of being open is a steep function of membrane potential (figs. 7, 9). Their opening is not instantaneous; the open probability rises with time, leading to a characteristic delayed rise of the whole-cell current with time and to the name used for one group of these channels, delayed rectifiers (for the meaning of rectification see below). Such channels repolarize action potentials in most excitable cells including those smooth muscles that fire action potentials. A channel of this type, but with particularly slow kinetics, also occurs in cardiac muscle, where the current flowing through it is sometimes called i$_K$. Many class III antiarrhythmic agents prolong cardiac action potentials primarily by blocking delayed rectifiers of heart muscle.

Rectification

Certain types of K$^+$ channel are referred to as rectifiers, for example the delayed rectifier and inward rectifier channels. In electronics the term rectifier is used for a device that allows current to flow more easily in one direction than another. Rectifier channels were so named because they give rise to whole cell currents that rectify, either because channel open probability depends on membrane voltage or because the current through a single open channel itself rectifies, or both. An example of a rectifying current (through inward rectifier channels) can be seen in fig. 10, where current flows more easily inward than outward.

Some voltage-activated K$^+$ channels open on depolarization, but then close again quite rapidly. These are variously called transient K$^+$ channels or A channels. In heart they are sometimes called i$_{to}$ channels because they underlie the transient outward current. i$_{to}$ gives rise to an early phase of rapid repolarization in the action potential of some tissues, for example atria and Purkinje fibres [9].

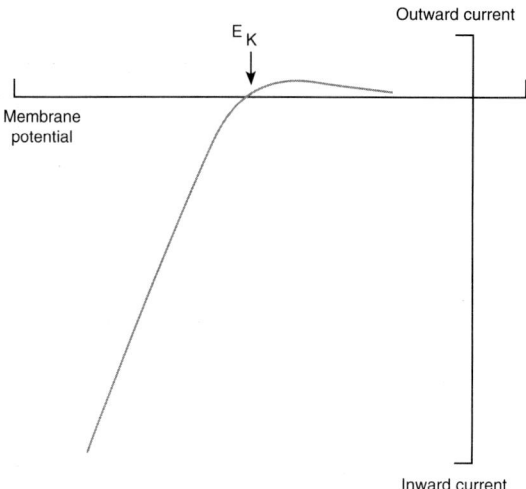

Fig. 10. Inward rectification. The figure shows an inwardly rectifying whole-cell current-voltage relation, such as may be found in the resting membrane of cardiac muscle and arterioles. Positive to E_K currents through K^+ channels are outward and small, while negative to E_K currents are inward and large

Inward rectifiers (i_{K1} channels)

Inward rectifier K^+ channels give rise to current-voltage relations like that of fig. 10, which shows inward rectification; inward currents are much larger than those in an outward direction. Inward rectifier channels (which underlie the current named i_{K1} in heart) [10] have the property of opening on hyperpolarization, but closing with depolarization. In addition, outward currents are small even through open channels because of blocking by physiological intracellular concentrations of magnesium ions. In fact, other types of K^+ channel in heart, for example ATP-sensitive K^+ channels show inward rectification because of such a Mg^{2+} block. Inward rectification is a very important property in tissues like cardiac muscle that spend long periods depolarized. The low K^+ conductance at the depolarized voltage of the action potential plateau reduces K^+ efflux and so K^+ current, so that only a small inward current, mainly Ca^{2+}, is needed to counteract this current and maintain the membrane potential at the plateau. This greatly reduces the metabolic cost of such long depolarizations, since energy must be used continuously to maintain the differences in ionic concentration between the inside and outside of the cell. When action potential repolarization begins, the K^+ conductance increases, so increasing K^+ efflux and contributing to rapid repolarization. At the resting potential the K^+ conductance due to inward rectifiers is also large, so that they contribute to the maintenance of a stable resting potential.

Ca^{2+}-activated K^+ channels

Ca^{2+}-activated K^+ channels are opened by binding of Ca^{2+} to the cytoplasmic face of the channel (for reviews see [11, 12]). Thus they are sensitive to the intracellular Ca^{2+} concentration, but also respond to membrane voltage, being activated by depolarization. There are several types of channel; channels of large single channel conductance, called BK (for big conductance) which are blocked by charybdotoxin, a peptide toxin from scorpion venom. They occur in a wide variety of smooth muscles, but probably not in some types of cardiac muscle. A small conductance, or SK, channel is blocked by another peptide, apamin. Channels of intermediate conductance have also been reported. Ca^{2+}-activated K^+ channels provide a link between membrane potassium permeability and intracellular Ca^{2+} concentration. They have functions in the hyperpolarizations that follow action potentials or trains of action potentials in nerve tissues, and may act to prevent intracellular Ca^{2+} rising too high under certain conditions by opening to oppose membrane depolarization.

ATP-sensitive K^+ channels

These channels are characterized by sensitivity to intracellular adenosine triphosphate, but may also be sensitive to other intracellular metabolites (for reviews see [13, 14]). They show little voltage sensitivity. ATP inhibits the channels, so that they are closed when ATP rises, with half closure in excised patches occurring at ATP concentrations of 10-200 µM, much lower than the intracellular [ATP] of around 5 mM. The inhibition is reduced in various tissues by ADP and by a fall in intracellular pH. ATP-sensitive K^+ channels again form a family with quite diverse properties. They seem to have a broad function of sensing the metabolic state of the cell in whose membrane they lie, but also form targets for some neurotransmitters. In cardiac muscle ATP-sensitive K^+ channels open to shorten the action potential in ischaemia or hypoxia, while in smooth muscle they may open to produce membrane hyperpolarization. ATP-sensitive K^+ channels are blocked by sulphonylurea drugs like tolbutamide and glibenclamide, but it is not yet clear whether the receptor for these blockers lies on the channel itself.

Muscarinic K^+ channels

These are so called because they are opened by muscarinic receptors [15, 16]. The current flowing through them in heart is also sometimes called i_{K-ACh}, as the channels are normally activated by acetylcholine binding to those receptors. The muscarinic receptor activates the channel by way of a GTP-binding protein, and the same channel appears to be activated by adenosine binding to a separate receptor. Current through these channels plays a major part in the negative chronotropic and inotropic response to parasympathetic nerve stimulation.

Na^+-activated K^+ channels

K^+ channels activated by intracellular Na^+ have been described in cardiac muscle and neurones [17]. The channels are of large conductance, and it has been suggested that they may open when $[Na^+]_i$ rises, as for example when the Na^+ pump is inhibited by cardiac glycosides, to shorten the action potential.

Arachidonate-activated K^+ channels

K^+ channels activated by intracellular arachidonic acid or phosphatidylcholine have been found in cardiac myocytes, where they may contribute to the increase in K^+ permeability seen in ischaemic cells and may respond to fatty acid derivatives acting as second messengers [18].

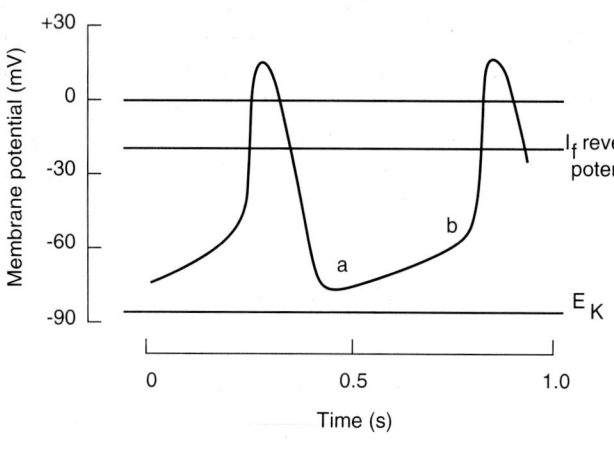

Fig. 11. Pacemaking in a sinoatrial node cell. The pacemaker potential is the progressive drift of the membrane potential in a positive direction that begins after action potential repolarization (between a and b in the diagram). During this time K$^+$ channels are closing and I$_f$ channels with a reversal potential around -20 mV are opening, so that the membrane potential moves further from E$_K$ and towards -20 mV. When it reaches the threshold for firing an action potential, about -55 mV, the next action potential is initiated by a regenerative opening of Ca^{2+} channels. Modified from a recording from a rabbit sinoatrial node cell by DiFrancesco et al. [19]

Potassium channels are modulated by neurotransmitters and drugs

In addition to its basic gating properties, the behaviour of almost every type of channel can be modulated by the action of transmitter substances released from nerves, endocrine or other cells. In a few cases, the receptor of a ligand-gated channel forms part of the channel protein itself. An example is the nicotinic acetylcholine receptor of the vertebrate neuromuscular endplate, and the transmitter in these cases is usually referred to as opening or gating the channel rather then modulating its behaviour. The distinction is rather artificial, but the term modulation is normally reserved for effects in addition to the basic mechanism of channel gating. Modulation usually involves a membrane receptor protein that communicates with the channel by way of a GTP-binding protein in the membrane. The diversity of possible modulatory pathways downstream from the G-protein is great [1]; in some cases the G-protein interacts directly with the channel after it has been activated by the receptor, while in others a biochemical cascade is initiated by the G-protein, leading eventually to an intracellular second messenger that affects the channel by way of a protein kinase. Many examples of K$^+$ channel modulation will be considered in this book.

The diversity of K$^+$ channels, and the involvement of the different types in different aspects of cardiovascular function, provides opportunities for increasingly selective and specific pharmacological intervention in these processes. Drugs can act on channels in many ways. The simplest actions are those which occur directly. Channel blockers may plug the pore of open channels, or may prevent the protein from reaching its open conformation. Such actions can often be measured directly by single channel recording. Many blockers of K$^+$ channels are known. The first was the quaternary ammonium ion TEA$^+$

which blocks all K⁺ channels, though with varying effectiveness. For several K⁺ channels much more selective blockers are now available, and more are continually being developed. In contrast to blockers, K⁺ channel openers drive the channel protein into its open configuration, and again the range of specific openers for different K⁺ channels can be expected to increase. In addition, the modulatory processes targeted on ion channels provide ways for drugs to affect channels more indirectly; a drug can affect a channel by way of its effect on some part of the modulatory pathway.

Pacemaking involves a complex interplay of potassium and other channels

Tissues that show pacemaking behaviour fire action potentials at regularly spaced intervals. Some neurones have this property, but we are concerned here with pacemaker activity in cardiac tissue, and especially in the sinoatrial node. Pacemaking involves a pacemaker potential; a steady drift of the membrane potential in the depolarizing direction from an initial negative value that it reaches just after an action potential (fig. 11). The pacemaker potential is terminated when it reaches a sufficiently depolarized level for voltage-activated Ca^{2+} or Na^+ channels to give rise to the next action potential. The detailed mechanisms underlying pacemaker potentials can differ in different tissues, but the underlying principles are the same (see also chapter 4). The pacemaker depolarization can arise in two ways. First, channels with a relatively positive reversal potential may open progressively with time, contributing more to the membrane permeability, so that the membrane potential moves steadily in the direction of their reversal potential. Secondly, channels with a very negative reversal potential, usually K⁺ channels, may close with time. Providing there are channels with more positive reversal potentials also open in the membrane, closure of K⁺ channels will lead to movement of the membrane potential away from E_K as the relatively permeability of the cell membrane to K⁺ falls. Often both types of mechanism occur simultaneously.

This appears to be the case for the pacemaker cells of the sinoatrial node. The K⁺ channels that close with time are slow delayed rectifiers that are activated at the depolarized potential of the action potential. On repolarization they begin to close, but do so quite slowly, leading to an outward K⁺ current that declines progressively. At the same time, channels with a more positive reversal potential are opening. These are i_f channels; cation-selective channels permeable to both Na^+ and K⁺ that have a reversal potential around -20 mV. These channels have a voltage dependence different from that of most voltage-gated channels; they are activated by hyperpolarization. Thus they open at the negative potential that the membrane reaches after action potential repolarization, but do so with quite slow kinetics, leading to a steadily increasing inward current, called i_f. A background level of opening of Na^+ channels may also contribute a steady depolarizing current. The pacemaker leads finally to an action potential upstroke caused by Ca^{2+} entry through voltage-activated Ca^{2+} channels, and opening of these channels also contributes to the final stages of the pacemaker depolarization itself.

Modulation of the channels involved in the pacemaker potential provides the means by which the heart rate is altered by the actions of sympathetic and vagus nerve stimulation. The modulatory actions are complex, and involve several channel types [1], but the basic effect of noradrenaline released from sympathetic nerves, and acting by way of ß-adrenoceptors, is to increase the open probability of channels (i_f and Ca^{2+} channels) which depolarize the cells of the sinoatrial node, so leading to a more rapid pacemaker depolarization. In contrast, acetylcholine released by vagal stimulation activates muscarinic channels, increasing the membrane K^+ permeability and slowing the pacemaker potential, and at the same time reduces the open probability of i_f channels. Both transmitters also modulate delayed rectifier K^+ channels so as to shorten the duration of the action potential when the heart rate increases and vice versa.

References

1. Hille B (1992) Ionic channels of excitable membranes. Sinauer, Sunderland
2. Horn R, Marty A (1988) Muscarinic activation of ionic currents measured by a new whole-cell recording method. J Gen Physiol 92: 145-159
3. Neher E, Sakmann B (1992) The patch clamp technique. Scientific American March 1992: 44-51
4. Hamill OP, Marty A, Neher E, Sakmann B, Sigworth F (1981) Improved patch-clamp techniques for high-resolution current recording from cells and cell-free membrane patches. Pflügers Arch 391: 85-100
5. Ogden DC, Stanfield PR (1987) Introduction to single channel recording. In: Standen NB, Gray PTA, Whitaker MJ (eds) Microelectrode Techniques. The Plymouth Workshop Handbook Company of Biologists, Cambridge, pp 63-81
6. Sakmann B, Neher E (1983) Single Channel Recording. Plenum, New York
7. Standen NB (1987) Separation and analysis of ionic currents. In: Standen NB, Gray PTA, Whitaker MJ (eds) Microelectrode Techniques. The Plymouth Workshop Handbook Company of Biologists, Cambridge, pp 29-40
8. Jan LY, Jan YN (1990) How might the diversity of potassium channels be generated? Trends Neurosci 13: 415-419
9. Escande D, Coulombe A, Faivre J-F, Deroubaix E, Coraboeuf E (1987) Two types of transient outward currents in adult human atrial cells. Am J Physiol 252: H142-H148
10. Trautwein W (1973) Membrane currents in cardiac muscle fibres. Physiol Rev 53: 793-835
11. Blatz AL, Magleby KL (1987) Calcium-activated potassium channels. Trends Neurosci 10: 463-467
12. Latorre R, Oberhauser O, Labarca P, Alvarez O (1989) Varieties of calcium-activated potassium channels. Ann Rev Physiol 51: 385-399
13. Ashcroft FM (1988) Adenosine 5'-triphosphate-sensitive potassium channels. Ann Rev Neurosci 11: 97-118
14. Davies NW, Standen NB, Stanfield PR (1991) ATP-dependent potassium channels of muscle cells: their properties, regulation, and possible functions. J Bioenergetics Biomembranes 23: 509-535
15. Sakmann B, Noma A, Trautwein W (1983) Acetylcholine activation of single muscarinic K^+ channels in isolated pacemaker cells of the mammalian heart. Nature 303: 250-253
16. Soejima M, Noma, A (1984) Mode of regulation of the ACh-sensitive K-channel by the muscarinic receptor in rabbit atrial cells. Pflügers Arch 400: 424-431
17. Luk H-N, Carmeliet E (1990) Na^+-activated K^+ current in cardiac cells: rectification, open probability, block and role in digitalis toxicity. Pflügers Arch 416: 766-768.
18. Kim D, Clapham DE (1989) Potassium channels in cardiac cell activated by arachidonic acid and phospholipids. Science 244: 1174-1176
19. DiFrancesco D, Ferroni A, Mazzanti M, Tromba C (1986) Properties of the hyperpolarizing-activated current (if) in cells isolated from the rabbit sino-atrial node. J Physiol (Lond) 377: 61-88

CHAPTER 2
The molecular structure of potassium channels

O Pongs and W Stühmer

Molecular biological techniques provide information about K⁺ channel structure in relation to function

Application of the techniques of molecular biology is currently providing a wealth of information on the structure of K⁺ channels, on the structural basis of their diversity, and is beginning to identify structural features of the channel proteins that are involved in the different aspects of their functional behaviour. A picture is emerging both of the regions of the protein that form the pore through which ions flow, and so determine ionic selectivity, and of the regions involved in channel gating.

The initial step was the molecular cloning of a gene encoding for a K⁺ channel in the fruit fly *Drosophila melanogaster*. A classical behavioural mutant, called Shaker because the mutant flies shook when anaesthetized with ether, was found to involve an alteration in the properties of a K⁺ channel. Cloning of the Shaker gene was followed rapidly by the cloning of a variety of K⁺ channels from this and other sources. Copy DNA (cDNA) may be cloned either from genomic DNA or from messenger RNA (mRNA), and so the primary amino-acid sequence of the channel protein may be obtained. Functional K⁺ channels can be expressed either from mRNA, primarily in oocytes of the frog *Xenopus*, or from cDNA in one of several in vitro expression systems. These methods provide several ways to answer questions about structure-function relationships in the channel protein. First, hydrophobicity analysis of the sequence gives an indication of those regions likely to lie in the membrane (the most hydrophobic ones) and so also about regions that may be intra- or extracellular. Comparison of the protein sequence with other known sequences may give indications, for example, of the regions that might be binding sites for channel ligands. Two broad approaches have been used to modify channel structure and look for resulting effects on function; domain swapping in which relatively large regions of the channel protein are substituted (or even removed), and site-directed mutagenesis aimed at changing a specific amino-acid in the channel protein. For voltage-gated K⁺ channels, these approaches have already given much information not only about the channel pore and gating, but also about structures important for the interaction of drugs with the channel protein.

Protein sequences of voltage-gated K⁺ channels indicate that they belong to a superfamily

The properties of K⁺ channels are remarkably diverse [1]. Their distribution varies among different cellular regions and cell types reflecting a spectrum of signalling capabilities [2]. K⁺ channels can be broadly classified according to their kinetics and pharmacological properties [1-3] as non-inactivating delayed rectifiers, rapidly inactivating transient A-channels, inward rectifiers, or as channels modulated by intracellular agents such as Ca^{2+}, ATP and G-proteins (see also chapter 1). A great number of K⁺ channels have now been cloned from various sources, for example *Drosophila*, mouse, rat and human (for review see [4]). Alignment of the primary K⁺ channel protein sequences derived in this way indicates that voltage-gated K⁺ channels belong to a superfamily of ionic channels. Despite the bewildering diversity of K⁺ channels, the results obtained by cloning voltage-gated and Ca^{2+}-activated K⁺ channels indicate they share basic structural features [5]. Over recent years the expression of channel cDNAs in in vitro systems has dramatically advanced our knowledge of the molecular structure of K⁺ channels and our understanding of the molecular basis of K⁺ channel diversity.

K⁺ channel protein sequences share a common basic design

K⁺ channel protein sequences have common features. Most notably, K⁺ channel forming proteins possess a basic design that comprises a core domain of six potentially membrane-spanning segments (S1-S6), together with a seventh segment (H5) that is probably tucked into the membrane between segments S5 and S6 (fig. 1) [6]. This core domain is flanked by hydrophilic amino- and carboxy-termini facing the intracellular side of the membrane. From structure-function analyses it can be inferred that variations in the amino-acid sequences of K⁺ channel proteins are sufficient to generate the diverse range of K⁺ channels with their distinct properties. This proposition will be explored in the following sections with reference to the major K⁺ channel superfamily outlined above. Another family of slowly activating voltage-gated K⁺ channels that are formed by small polypeptides containing only one or two potentially membrane-spanning segments [7] will not be discussed in this chapter.

Structure-function relations are best understood for the Shaker channel

The Shaker K⁺ channel was the first K⁺ channel to be cloned and characterized [8-12], and since then Shaker K⁺ channels have remained in the forefront of research on structure-function relations of K⁺ channels. Therefore, we shall discuss the structure and function of Shaker K⁺ channels in some detail in order to provide examples of our present knowledge about the functional K⁺ channel domains that are implicated in activation, inactivation and selectivity. Subsequently, we shall discuss other cloned K⁺ channels which have been found to be expressed in heart tissue.

Extensive molecular analysis of the Shaker locus [8-14] has shown that it encodes a large transcription unit which expresses at least ten different A-type K⁺ channel subunits. These subunits are translated from alternate mRNAs which are probably generated both by alternate transcription from different start sites within the K⁺ channel gene and by alternate splicing of primary transcript(s). Subunits assemble into a tetrameric K⁺ channel in the *Xenopus* oocyte expression system, giving rise to K⁺ channels with distinct pro-

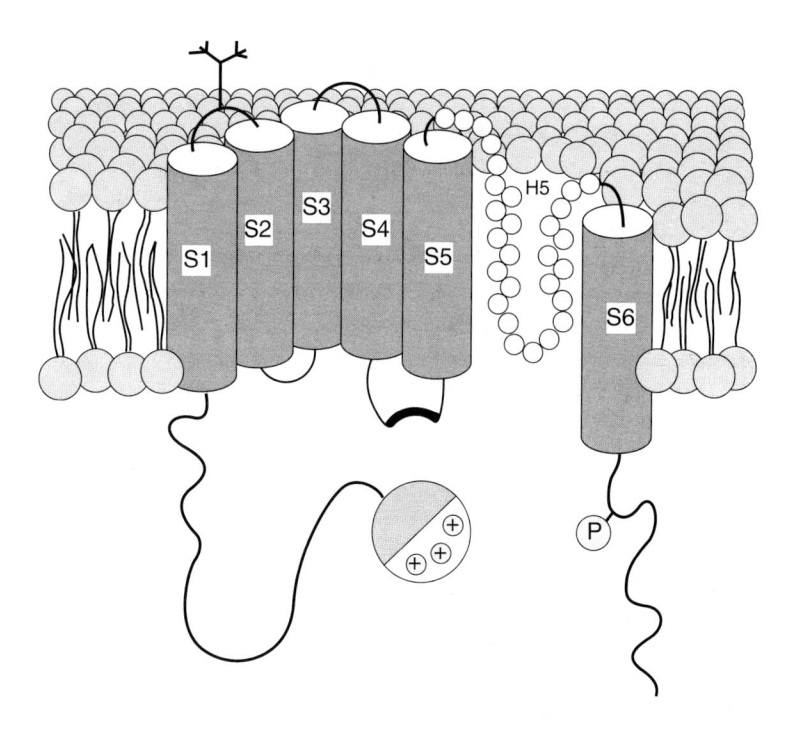

Fig. 1. The proposed membrane-spanning orientation of one protein subunit of a voltage-gated K$^+$ channel. Putative membrane-spanning segments are designated S1-S6. H5 designates a hydrophobic region that is highly conserved among K$^+$ channels. It is proposed that the H5 sequence enters and exits the lipid bilayer from the extracellular side. It is likely that four such subunits assemble to make a functional K$^+$ channel. The N-terminus of some K$^+$ channels contains an amphipathic structure (illustrated by a shaded ball with positive charges), which may occlude the pore of the channel. C-termini often contain phosphorylation sites, indicated by P. The sequence between segments S1 and S2 is frequently N-glycosylated, indicated in the diagram by small branches

perties. The subunit structures have a common core domain which contains part of the cytoplasmic amino-terminal domain as well as segments S1 to S5 and H5 (fig. 1). This core domain is flanked by variant amino-termini and/or by variant S6 segments and carboxy-termini. It is the particular combination of these amino- and carboxy-terminal variants which gives the K$^+$ channel subunit its distinct properties. The functional diversity of Shaker channels probably represents a mechanism for generating physiological diversity within excitable cells. Consistent with this idea is the differential distribution of splicing variants from the Shaker locus in the *Drosophila* nervous system as well as in muscle [15, 16]. An example of the biological implementation of the extensive diversity of Shaker transcripts has recently been provided by a characterization of Shaker K$^+$ channels in photoreceptor cells of *Drosophila* [16]. Here, single channel properties are similar to those reported in muscle [17] with respect to channel mean open time and single channel slope conductance. However, the photoreceptor and muscle A currents differ very significantly in the voltage range over which the channels operate. Whereas the voltage dependence of Shaker A-currents both in muscle and in oocyte expression studies is very similar, the voltage range over which A currents in photoreceptors activate and inactivate is shifted by approximately 40-50 mV in the negative direction. How these differences in A-current are related to the presence of different splice variants of Shaker transcripts in photoreceptor and muscle cells [16] is not yet clear and requires further study.

The properties of Shaker K$^+$ channels in the *Xenopus* oocyte expression system vary most notably in their inactivation behaviour, variations that are also reflected in differences in mean channel open time [18, 19]. These observations led to the hypothesis that at least two types of inactivation [18-20], termed C-type and N-type inactivation, operate in Shaker K$^+$ channels. After inactivation, K$^+$ channels are refractory for a period until recovery takes place. During recovery the K$^+$ channel protein presumably undergoes a conformational change that moves the channel back into the resting state. The route of recovery taken by an inactivated channel during reactivation passes via the open-channel state [21].

N-type inactivation occurs by a ball-and-chain mechanism

N-type inactivation in Shaker K$^+$ channels, as well as in other K$^+$ channels, is governed by the amino-terminal sequences which are encoded in different exons of the Shaker transcription unit. Variant N-type inactivation behaviour can be correlated with the variant amino-termini of Shaker K$^+$ channel subunits [18-22]. C-type inactivation is governed by the carboxy-terminal S6 segment [23, 24]. In vitro mutagenesis studies have shown that the substitution of one particular amino-acid within S6 has a profound influence on C-type inactivation. N-type inactivation appears to be correlated with slow recovery from inactivation, while C-type inactivation appears to be correlated with rapid recovery. N-type inactivation is thought to involve an occlusion of the ion-conducting pore of the channel from the intracellular side. Probably the amino-terminal cytoplasmic inactivation domain of the channel swings into and out of the pore in a similar way to a tethered ball. Apparently this ball-and-chain mechanism for N-type inactivation does not require a specific primary sequence [21], since several N-terminal inactivation balls of rapidly inactivating K$^+$ channels are able to function interchangeably to close K$^+$ channels from the inside. The binding site for the inactivation ball is located within the domain that connects segments S4 and S5 of the channel protein [25]. In vitro mutagenesis studies on Shaker K$^+$ channels have shown that substitution of amino-acids in this region has a profound influence on channel inactivation behaviour.

The S4 segment is involved in voltage gating

Within the core domain, segment S4 appears to be a hallmark of voltage-dependent ion channels. This membrane-spanning segment carries a number of regularly positioned positively charged amino-acid side chains (lysine or arginine) exhibiting the general sequence pattern [Arg/Lys-X-X]n, where n may vary between 4 and 7 [4]. Mutations in the S4 segment of the Shaker channel cause a shift of the midpoints of the conductance-voltage curve [26]. From the slope of these curves it was estimated that the charge carried by the channel-gating particles across the membrane was reduced by neutralizing some of the charged amino-acid chains within the S4 sequence. However, the presence or absence of positively charged amino-acid chains might not be correlated in a simple way

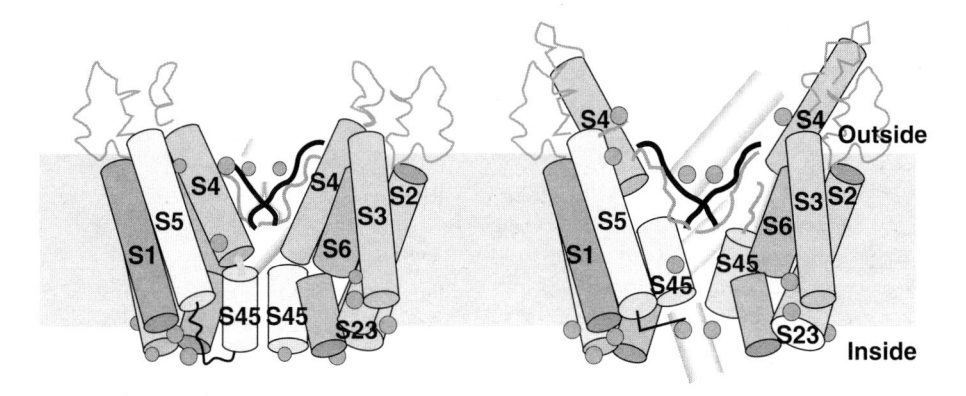

Fig. 2. Detailed topological model of the transmembrane region of voltage-gated K^+ channels. The model is adopted from Durrell and Guy [6]. Transmembrane segments S1-S6 are presented as α-helical cylinders. Proposed helical domains connecting segment S2 and S3 (S23) and segments S4 and S5 (S45) are also indicated. The N-terminal half of the hydrophobic extracellular pore region, H5, is in light grey, the C-terminal half in black. Dots represent positively or negatively charged amino acid sidechains commonly found in voltage-gated K^+ channels. Two possible channel structures are illustrated: left, closed resting state; and right, open state. Two subunits are shown for each channel, with the other two (of the probable total per channel of four) shaded lightly

with gating charges, since substitutions of positively charged amino-acid side chains do not cause equal shifts in voltage dependence and the slope of channel activation. Also, the nature of the charged amino-acid, arginine or lysine, may influence the gating characteristics of the channel. More recently, similar effects on the voltage dependence of activation were found when hydrophobic amino-acids, rather than charged ones, were mutated in the S4 sequence [27]. Conservative substitutions of hydrophobic residues in the S4 sequence of Shaker channels caused comparable or even larger shifts of the conductance curves along the voltage axis than did the neutralization of basic amino-acid residues. These results suggest that the entire S4 sequence plays an important role in the conformational change that accompanies gating. This conformational change probably involves different parts of the channel that cooperatively alter their relative distribution within the electric field. Perhaps the S4 sequence provides both the interactions between charged residues and the membrane electric field that underlie the gating current, and the interactions between other parts of the channel that are involved in the voltage-dependent conformational transition of the closed channel into an activated state. Such a key role for the S4 segment of ionic channels has been proposed in structural models of ionic channels (see fig. 2).

The H5 sequence forms a major part of the K^+ channel pore

The combination of mutagenesis with in vitro expression studies has also pinpointed the primary sequences of the channel protein that are implicated in forming the K^+ channel pore. Characterization of the binding sites for the open K^+ channel blockers tetraethylammonium (TEA) and charybdotoxin (CTX) has been instrumental in these studies.

The block of K$^+$ channels by TEA has been studied in great detail (see [28] for review). Electrophysiological studies on native K$^+$ channels showed that TEA can block the K$^+$ channel pore from the outside and from the inside, acting at distinct binding sites [2]. Binding of external TEA is not usually affected by a change in membrane potential, whereas binding of internal TEA is sensitive to membrane potential. Extensive mutational analysis of Shaker K$^+$ channels has provided important clues about the structure of the K$^+$ channel pore [29-32]. Characterization of the Shaker related RCK family of voltage-gated K$^+$ channels expressed in the rat nervous system indicated that different members of this family had distinct affinities for external TEA [33]. The protein sequences differed substantially at a particular residue in the conserved H5 region of the core domain (fig. 3). Substitution of this amino-acid residue in Shaker K$^+$ channels by site-directed mutagenesis showed a good correspondence between the presence of a particular RCK-like amino-acid residue and the TEA sensitivity of the mutant Shaker channel [30]; for example, replacement of Thr by Lys (which occurs in the H5 region of RCK4 protein) rendered the mutant Shaker channel insensitive to external TEA, as is the RCK4 channel. Replacement of the same Thr by Tyr (which is found in the H5 sequence of RCK1 protein) rendered the mutant channel more sensitive to TEA, in accordance with the behaviour of the RCK1 channel [33-34]. Furthermore, a reciprocal mutation of RCK4 protein gave similar results; for example replacement of the Lys in the H5 region of RCK4 by Tyr yielded an RCK4 channel which was very sensitive to external TEA [35].

Other mutations of the H5 sequence either abolished K$^+$ channel activity or, excitingly, specifically altered the sensitivity of the internal TEA binding site of Shaker K$^+$ channels [31, 32]. These mutations affected amino-acids which were six to eight residues away from the ones affecting the external TEA binding site. The studies of TEA block described above suggested that the H5 sequence might be an important structural component of the K$^+$ channel pore. Indeed, the same mutations that specifically alter the external or internal TEA binding sites also affect such important properties of the K$^+$ channel pore as conductance and ion selectivity. These results lend strong support to the view that the H5 sequence is the major determinant in K$^+$ channel proteins for forming the pore [36].

A model for K$^+$ channel structure

A simple model of K$^+$ channel subunits (fig. 1) has been derived from the studies described above. It proposes six membrane-spanning segments, S1 to S6, and a seventh pore-forming segment (H5) which enters and exits the extracellular side of the membrane. The amino- and carboxy-terminal hydrophilic sequences are cytoplasmic and the amino-terminal inactivation ball can bind to a structure close to the cytoplasmic entrance of the pore. This structure is quite assymmetrical because most of the amino-acid residues of K$^+$ channels are located on the intracellular side of the membrane. Only a few residues are extracellular, namely those between segments S1 and S2, S3 and S4, and S5 and S6.

A detailed structural model of the transmembrane topology of the core domain of Shaker related K$^+$ channels has been developed by Guy and co-workers (fig. 2). This model is based on extensive sequence similarities detected by the alignment of the derived sequences of transmembrane regions of different cloned K$^+$ channel proteins. In

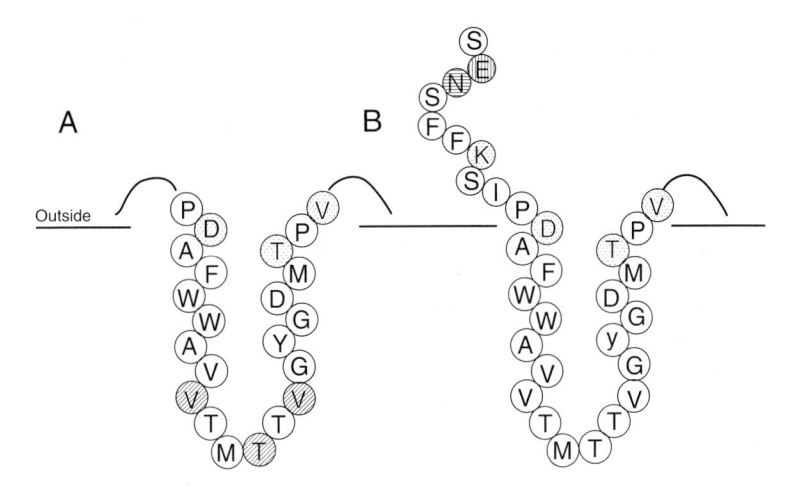

The molecular structure of potassium channels

Fig. 3 A, B. Diagram in single-letter code of amino acid sequence in the H5 region of voltage-gated Shaker K⁺ channels. The H5 region is inserted into the membrane from the outside. **A** Dotted circles mark amino acids that apparently form contact sites for externally bound tetraethylammonium, hatched circles amino acids for the contact sites for internally bound tetraethylammonium. **B** Dotted circles mark amino acids that apparently form contact sites for both charybdotoxin and dendrotoxin; vertically striped circles those for charybdotoxin alone, and horizontally striped circles those for dendrotoxin alone. Data have been compiled from [29-32, 34-36, 49]

addition, the model has been influenced by results of in vitro site-directed mutagenesis studies (described above), which have provided some insight into the K⁺ channel domains that are involved in gating, selectivity and inactivation. According to this model, Shaker-like K⁺ channels are composed of four subunits. The transmembrane portion of the channel protein comprises an outer cylinder of 16 α-helices, which surrounds a middle cylinder of 8 α-helices, which in turn surrounds an 8-stranded antiparallel ß-barrel. Helical segments S1, S2, S3 and S5 form the outer cylinder. These segments traverse the entire membrane.

Segments S6 and S4 form part of the middle cylinder. S6, which contains prolines that might disrupt the helix, is partitioned into two helices, S6n and S6c. Segment S4 is most likely the voltage sensor, moving by a helical screw mechanism toward the extracellular surface during the transition of the closed channel into the activated state. S45, which connects segments S4 and S5, probably forms an amphipathic helix, which is also part of the middle cylinder. S45 has been oriented in the model so that positively charged residues point towards the centre of the pore. This orientation may form an electrostatic barrier to the passage of cations. Accordingly, the movement of S45 to and from the inner pore of a K⁺ channel would regulate closing and opening of the activated channel. Like the movement of helical segment S4, the S45 movement implies a rotation of this amphipathic helix. This rotation moves the positive charges of S45 away from the entrance of the pore and brings in a negative charge, which could take part in the formation of a negatively charged ring at the intracellular entrance of the pore. The pore itself is formed from four ß-hairpin structures of H5, one from each subunit of the channel. The first eight and the last three amino acids of H5 are positioned near the extracellular entrance to the pore.

Table 1. Pharmacological profiles of cloned rat voltage-gated K+ channels

K+ channel	TEA (mM)	4-AP (mM)	DTX (nM)	MCDP (nM)	CTX (nM)	Activation	Inactivation
RCK1	0.6	1.0	12	45	22	fast	very slow
RCK2	7	1.5	20	10	1	—	slow
RCK3	50	1.5	> 600	> 1000	> 1000	—	intermediate
RCK4	> 150	12.5	> 200	> 2000	> 40	—	rapid
RCK5	129	0.8	4	175	6	—	slow
RCK7	> 150	0.5	> 200	> 200	> 200	—	very slow
Raw1	0.1	0.9	> 100	> 200	> 200	—	very slow
Raw2	0.1	0.25	> 100	> 200	> 200	—	slow
Raw3	0.3	0.5	> 100	> 200	n.d.	—	rapid
DRK1	6	0.5	n.d.	n.d.	> 200	slow	very slow

Data are IC_{50} values, i.e. they indicate the drug concentration at which the peak amplitude at +20 mV is halved. Data for RCK1, RCK3, RCK4, RCK5 are from [33], for RCK2 from [50], for RCK7(KV1) from [51], for Raw1,2,3 from [52] and for DRK1 from [53]. RCK K+ channels are Shaker-related, Raw K+ channels Shaw-related and DRK1 K+ channels Shab-related. Activation was scored as fast if the rise time to peak amplitude at +20 mV is less than 10 msec, and as slow if the rise time is more than 20 msec. Inactivation was scored as rapid if K+ currents inactivated at a +20 mV test potential within 100 ms, as intermediate if they inactivated within 1 s, as slow if they inactivated within 5 s, and as very slow if they did not inactivate within 5 s. TEA, tetraethylammonium; 4-AP, 4-aminopyridine; DTX, dendrotoxin; MCDP, mast cell degranulating peptide; CTX, charybdotoxin

These amino acids are not highly conserved among voltage-gated K+ channels. However, the model suggests that these amino acids may be involved in binding K+ channel blockers that block the entrance to the channel from the outside.

K+ channels of Shaker relatives

Shaker related K+ channels have been characterized by isolating and expressing homologous cDNAs from *Xenopus,* mouse, rat, bovine and human cDNA libraries [4]. Alignment of the derived primary protein sequences reveals a high degree of primary sequence conservation. Sequence identities are 70% and higher, indicating a substantial evolutionary pressure to preserve Shaker related K+ channel structures. K+ channel diversity may result not only from sequence variations within the Shaker K+ channel protein family, but also from the presence of an extended gene family within the genome. Three Shaker relatives have been identified in *Drosophila* encoding the Shal, Shaw and Shab proteins [37, 38]. Like the products of the Shaker gene, the Shal, Shaw, and Shab gene products express voltage-gated K+ channels in the oocyte expression system. As for Shaker channels, within these subfamilies members differ markedly in their kinetics and voltage-sensitivity. However, very similar structural principles to those for Shaker K+ channels underlie the properties of the other channels [4, 5].

The pharmacology of voltage-gated K$^+$ channels

Although voltage-gated K$^+$ channels show a striking diversity in their electrophysiological properties, they do not vary much in their pharmacological properties (Table 1) [1]. Thus many more distinct K$^+$ channels can be measured in excitable cells than can be discerned by their pharmacological profile. To put it simply, not enough different K$^+$ channel blockers are currently available to match the diversity of K$^+$ channel proteins. This situation makes it very difficult at present to compare the properties of cloned K$^+$ channels with those of native K$^+$ channels. Moreover, it is not even clear whether the pharmacology of a cloned K$^+$ channel expressed in the *Xenopus* oocyte system is in fact like that of the corresponding native K$^+$ channel in its natural environment [40]. As discussed above (see also fig. 3), the study of the effects of blockers and toxins that bind to K$^+$ channels has proved to be very informative in establishing structure-function relationships for cloned K$^+$ channels. But it is not feasible to use these K$^+$ channel blockers for the identification of native K$^+$ channels because they do not show sufficient selectivity between channels. In view of these limitations in K$^+$ channel pharmacology it is, however, interesting to note that dendrotoxin and mast cell degranulating peptide might be employed in the functional characterization of some members of the Shaker related K$^+$ channel family. So far, cloned K$^+$ channels of the other K$^+$ channel subfamilies have not been found to be blocked by dendrotoxin or by mast cell degranulating peptide (Table 1). In combination with molecular biological analyses this observation may be of potential use in physiological studies of K$^+$ channel function.

Both differences in protein sequence and different combinations of subunits may contribute to the diversity of K$^+$ channels

The diversity of K$^+$ channels does not only arise from the expression of distinct but related genes encoding distinct members of the K$^+$ channel superfamily. K$^+$ channel diversity may also arise from the formation of heteromultimeric K$^+$ channels [41, 42]. Each of the known K$^+$ channel proteins resembles one of the four internally homologous repeats of Na$^+$ or Ca^{2+} channels [5, 11, 43]. By analogy, K$^+$ channels are multimeric membrane proteins formed by the aggregation of several independent subunits [44]. By studying the interaction of charybdotoxin with co-expressed wild-type and toxin-insensitive mutant subunits, Shaker K$^+$ channels have been found to have a tetrameric structure [44], and all cloned K$^+$ channel subunit proteins probably assemble into multimeric structures with a similar stoichiometry.

Functional K$^+$ channels may also be formed by the assembly of different subunits. The assembly of heteromultimers may yield K$^+$ channels with properties different from those of the corresponding homomultimeric channels. Two examples of the formation of heteromultimeric K$^+$ channels have been studied in the *Xenopus* oocyte expression system [41, 42]. Co-expression of different Shaker K$^+$ channels gave a kinetic behaviour distinct from that expressed by the homomultimeric K$^+$ channels [41]. Similarly, two vertebrate K$^+$ channel subunits have been co-expressed either in oocytes or in HeLa cells [42]. Here, homomultimeric K$^+$ channels (RCK1 and RCK4) had distinct single-channel conductances, kinetic behaviour and pharmacology, while heteromultimers resembled

Table 2. Voltage-gated K$^+$ channels in rat heart

K$^+$ channel	Ventricle	Atrium	Aorta	Skeletal muscle
RCK1	-	+	-	+
RCK4	-	+	-	-
RCK5	-	+	-	-
RCK7	+	+	+	+
RShal	-	-	-	-
DRK1	-	+	+	-

Data are taken from [46]. A "+" sign indicates that the mRNA is relatively abundant in the corresponding tissue. A "-" sign indicates that the mRNA is of low abundance or is absent

RCK1 in conductance but RCK 4 in gating properties [42]. These experiments suggest that co-assembly of different K$^+$ channel subunits is another source of K$^+$ channel diversity. It is possible that alternative splicing of K$^+$ channel primary RNA transcripts, as observed for Shaker RNA [8-12], further increases the diversity. Whether interactions with other proteins, or post-translational modifications, such as glycosylation and phosphorylation, also contribute to K$^+$ channel diversification is not known.

K$^+$ channels have also been cloned from heart tissue

Voltage-gated K$^+$ channels may be responsible for limiting cardiac action potential duration, as discussed in chapters 4 and 5. Consequently, agents that affect K$^+$ currents modulate cardiac rhythm. The characterization of cardiac voltage-gated K$^+$ channels is therefore of great interest, since the development of drugs that target cardiac K$^+$ channels may provide an effective therapy for heart diseases such as arrhythmias. Altogether seven different cDNAs encoding voltage-gated K$^+$ channels have so far been detected in heart tissue as follows (see also Table 2): RK1 (RCK1), RK2 (RCK5), RK3 (RCK4), RK4 (RCK7), DRK1, RShal, and mini K1 [45-48]. With the exception of RK4 and mini K, all other K$^+$ channel mRNAs are more abundantly expressed in brain than in heart or skeletal muscle tissue [46,48]. mRNA from different regions of the heart has also yielded different combinations of these K$^+$ channels. Thus in total RNA isolated from rat ventricle, RK4 and DRK1 were detected, in RNA from rat atrium RK1, RK2, RK3, RK4, and DRK1, and in RNA from rat aorta RK4 and DRK1. Two full-length cDNA clones have been isolated from human ventricular cDNA libraries [47]. The clones encoded the human RCK4 analogue HK1 and the human RCK7 analogue HK2, respectively. Northern blot analysis of HK1 and HK2 mRNA expression indicated that the HK1 and HK2 channel mRNAs were expressed equally in ventricle, whereas the HK2 channel mRNA was much more abundant in atrium mRNA than in ventricle. These results may indicate that RCK4- and RCK7-like mRNAs have different expression patterns in rat as compared with human heart tissue. Unfortunately, a good correlation of these K$^+$ channels with native K$^+$ channels in heart tissue is not yet possible and remains to be developed. Thus the physiological significance of cloned voltage-dependent K$^+$ channels with respect to the fast repolarizing phase of cardiac action potentials is still unclear.

Curiously, though class III antiarrhythmic agents have been intensely studied for their potential to block cardiac K^+ channels, these agents have not yet been used to characterize cloned K^+ channels. Hopefully, experiments in the near future will provide information on the pharmacology of cloned K^+ channels with respect to class III antiarrythmic agents.

The work in our laboratories on the molecular biology of voltage-gated K^+ channels was supported by grants of Deutsche Forschungsgemeinschaft, European Economic Community, and the Fond der Chemischen Industrie.

References

1. Rudy B (1988) Diversity and ubiquity of K channels. Neuroscience 25: 729-749
2. Hille B (1992) Ionic channels of excitable membranes. Sinauer, Sunderland
3. Moczydlowski E, Lucchesi K, Ravindran A (1988) An emerging pharmacology of peptide toxins targeted against potassium channels. J Membr Biol 105: 95-111
4. Pongs O (1992) Molecular biology of voltage-dependent potassium channels. Physiol Rev 72: S69-S88
5. Jan LY, Jan YN (1992) Tracing the roots of ion channels. Cell 69: 715-718
6. Durrell SR, Guy R (1992) Atomic scale structure and functional models voltage-gated potassium channels. Biophys J 62: 238-250
7. Philipson LH, Miller JR (1992) A small K^+ channel locus looms large. Trends Pharmacol Sci 13: 8-11
8. Papazian DM, Schwarz TL, Tempel BL, Jan YN, Jan LY (1987) Cloning of genomic and complementary DNA from Shaker, a putative potassium channel gene from Drosophila. Science 237: 749-753
9. Baumann A, Krah-Jentgens I, Müller R, Müller-Holtkamp F, Seidel R, Kecskemethy N, Canal J, Ferrus A, Pongs O (1987) Molecular organization of the maternal effect region of the Shaker complex of Drosophila: characterization of an IA channel transcript with homology to vertebrate Na^+ channel. EMBO J 6: 3419-3429
10. Schwarz TL, Tempel BL, Papazian DM, Jan YN, Jan LY (1988) Multiple potassium-channel components are produced by alternative splicing at the Shaker locus in Drosophila. Nature 331: 137-142
11. Pongs O, Kecskemethy N, Müller R, Krah-Jentgens I, Baumann A, Kiltz HH, Canal I, Llamazares S, Ferrus A (1988) Shaker encodes a family of putative potassium channel proteins in the nervous system of Drosophila. EMBO J 7: 1087-1096
12. Kamb A, Tseng-Crank J, Tanouye MA (1988) Multiple products of the Drosophila Shaker gene may contribute to potassium channel diversity. Neuron 1: 421-430
13. Timpe LC, Jan YN, Jan LY (1988) Four cDNA clones from the Shaker locus of Drosophila induce kinetically distinct A-type potassium currents in Xenopus oocytes. Neuron 1: 659-667
14. Iverson LF, Tanouye MA, Lester HA, Davidson N, Rudy B (1988) A-type potassium channels expressed from Shaker locus cDNA. Proc Natl Acad Sci USA 85: 5723-5727
15. Schwarz TL, Papazian DM, Carretto RC, Jan YN, Jan LY (1990) Immunological characterization of K^+ channel components from the Shaker locus and differential distribution of splicing variants in Drosophila. Neuron 2: 119-127
16. Hardie RC, Voss D, Pongs O, Laughlin SB (1991) Novel potassium channels encoded by the Shaker locus in Drosophila photoreceptors. Neuron 6: 477-486

17. Zagotta WN, Aldrich RW (1990) Voltage-dependent gating of Shaker A-type potassium channels in Drosophila muscle. J Gen Physiol 95: 29-60
18. Hoshi T, Zagotta WN, Aldrich RW (1990) Biophysical and molecular mechanisms of Shaker potassium channel inactivation. Science 250: 533-538
19. Stocker M, Stühmer W, Wittka R, Wang X, Müller R, Ferrus A, Pongs O (1990) Alternative Shaker transcripts express either rapidly inactivating or non-inactivating K^+ channels. Proc Natl Acad Sci USA 87: 9411-9415
20. Choi KL, Aldrich RW, Yellen G (1991) Tetraethylammonium blockade distinguishes two inactivation mechanisms in voltage-activated K^+ channels. Proc Natl Acad Sci USA 88: 5092-5095
21. Ruppersberg JP, Frank R, Pongs O, Stocker M (1991) Cloned neuronal IK(A) channels reopen during recovery from inactivation. Nature 353: 657-660
22. Zagotta WN, Hoshi T, Aldrich RW (1990) Restoration of inactivation in mutants of Shaker potassium channels by a peptide derived from Shaker B. Science 250: 568-571
23. Hoshi T, Zagotta WN, Aldrich RW (1991) Two types of inactivation in Shaker K^+ channels: effects of alterations in the carboxy-terminal region. Neuron 7: 547-556
24. Wittka R, Stocker M, Boheim G, Pongs O (1991) Molecular basis for different rates of inactivation in the Shaker potassium channel family. FEBS Lett 286: 193-200
25. Isacoff EY, Jan YN, Jan LY (1991) Putative receptor for the cytoplasmic inactivation gate in the Shaker K^+ channel. Nature 353: 86-90
26. Papazian DM, Timpe LX, Jan YN, Jan LY (1991) Alteration of voltage-dependence of Shaker potassium channel by mutations in the S4 sequence. Nature 349: 305-310
27. Lopez GA, Jan YN, Jan LY (1991) Hydrophobic substitution mutations in the S4 sequence alter voltage-dependent gating in Shaker K^+ channels. Neuron 7: 327-336
28. Pongs O (1992) Structural basis of voltage-gated K^+ channel pharmacology. Trends Pharmacol Sci 13: 359-365
29. MacKinnon R, Miller C (1989) Mutant potassium channels with altered binding of charybdotoxin. Science 245: 1382-1385
30. MacKinnon R, Yellen G (1990) Mutations affecting TEA blockade and ion conductance in voltage-activated K^+ channels. Science 250: 276-279
31. Yool HJ, Schwarz TL (1991) Alteration of ionic selectivity of a K^+ channel by mutation of the H5 region. Nature 349: 700-704
32. Yellen G, Jurman ME, Abramson T, MacKinnon R (1991) Mutations affecting internal TEA blockade identify the probable pore-forming region of a K^+ channel. Science 251: 939-942
33. Stühmer W, Ruppersberg JP, Schröter KH, Sakmann B, Stocker M, Giese KP, Perschke A, Baumann A, Pongs O (1989) Molecular basis of functional diversity of voltage-gated potassium channels in mammalian brain. EMBO J 8: 3235-3244
34. Heginbotham L, MacKinnon R (1992) The aromatic binding site for tetraethylammonium ion on potassium channels. Neuron 8: 483-491
35. Stocker M, Pongs, O, Hoth M, Heinemann S, Stühmer W, Schröter K-H, Ruppersberg JP (1991) Swapping of functional domains in voltage-gated K^+ channels. Proc R Soc Lond B 245: 101-107
36. Kirsch GE, Drewe JA, Hartmann HA, Taglialatela M, De Biasi M, Brown AM, Joho RH (1992) Differences between the deep pores of K^+ channels determined by an interacting pair of nonpolar amino acids. Neuron 8: 499-505
37. Butler A, Wei A, Baker K, Salkoff L (1989) A family of putative potassium channel genes in Drosophila. Science 243: 943-947
38. Wei A, Covarrubias M, Butler A, Baker K, Pak M, Salkoff L (1990) K^+ channel diversity is produced by an extended gene family conserved in Drosphila and mouse. Science 248: 599-603
39. Salkoff L, Baker K, Butler A, Covarrubias M, Pak MD, Wei A (1992) An essential set of K^+ channels conserved in flies, mice and humans. Trends Neurosci 15: 161-166

40. Zagotta WN, Germeraad S, Garber SS, Hoshi T, Aldrich RW (1989) Properties of Shaker B A-type potassium channels expressed in shaker mutant Drosophila by germline transformation. Neuron 3: 773-782
41. Isacoff EY, Jan YN, Jan LY (1990) Evidence for the formation of heteromultimeric potassium channels in Xenopus oocytes. Nature 345: 530-534
42. Ruppersberg JP, Schröter KH, Sakmann B, Stocker M, Sewing S, Pongs O (1990) Heteromultimeric channels formed by rat brain potassium-channel proteins. Nature 345: 535-537
43. Catterall W (1988) Structure and function of voltage-sensitive ion channels. Science 242: 50-61
44. MacKinnon R (1991) Determination of the subunit stoichiometry of a voltage-activated potassium channel. Nature 350: 232-235
45. Tseng-Crank JC, Tseng GN, Schwartz A, Tanouye MA (1990) Molecular cloning and functional expression of a potassium channel cDNA isolated from a rat cardiac library. FEBS Lett 268: 63-68
46. Roberds SL, Tamkun M (1991) Cloning and tissue-specific expression of five voltage-gated potassium channel cDNAs expressed in rat heart. Proc Natl Acad Sci USA 88: 1798-1802
47. Tamkun MM, Knoth KM, Walbridge JA, Kroemer H, Roden DM, Glover DM (1991) Molecular cloning and characterization of two voltage-gated K^+ channel cDNAs from human ventricle. FASEB J 5: 331-337
48. Honore E, Attali B, Romey G, Heurteaux C, Ricard P, Lesage F, Lazdunski M, Barhanin J (1991) Cloning, expression, pharmacology and regulation of a delayed rectfier K^+ channel in mouse heart. EMBO J 10: 2805-2811
49. Hurst RS, Busch AE, Kavanaugh MP, Osborne PB, North RA, Adelman JP (1991) Identification of amino-acid residues involved in dendrotoxin block of rat voltage-dependent potassium channels. Mol Pharmacol 40: 572-576
50. Grupe A, Schröter KH, Ruppersberg JP, Stocker M, Drewes T, Beckh S, Pongs O (1990) Cloning and expression of a human voltage-gated potassium channel. A novel member of the RCK potassium channel family. EMBO J 9: 1749-1756
51. Swanson R, Marshall J, Smith JS, Williams JB, Boyle JB, Folander K, Luneau CL, Antanavage J, Oliva C, Buhrow SA, Bennett C, Stein RB, Kaczmarek LK (1990) Cloning and expression of cDNA and genomic clones encoding three delayed rectifier potassium channels in rat brain. Neuron 4: 929-939
52. Rettig J, Wunder F, Stocker M, Lichtinghagen R, Mastiaux F, Beckh S, Kues W, Pedarzani P, Schröter KH, Ruppersberg JP, Veh R, Pongs O (1992) Characterization of Shaw-related potassium channel family in rat brain. EMBO J 11: 2473-2486
53. Frech GC, van Dongen AM, Schuster G, Brown AM, Joho RH (1989) A novel potassium channel with delayed rectifier properties isolated from rat brain by expression cloning. Nature 340: 642-645

Physiology
The heart muscle

CHAPTER 3
The role of potassium channels in maintaining resting potential in normal and anoxic cardiac muscle

A Noma and H Matsuda

Under physiological conditions, differences in K$^+$ permeability between the various cardiac tissues account for most of the differences in the shape of the cardiac action potential. For example, ventricular myocytes show a very high K$^+$ permeability at the resting potential (mainly through inward rectifier i_{K1} channels) that is shut off immediately on depolarization as a consequence of rectification (see chapter 1) to support the long-lasting plateau potential. In contrast, sinoatrial node pacemaker cells have very few i_{K1} channels but possess in their membrane numerous delayed rectifier K$^+$ channels that activate on depolarization. In addition, muscarinic K$^+$ channels that are specific to atrial tissues may come into play under neuronal regulation and so modify the resting K$^+$ permeability.

Under anoxic pathological conditions, major changes in the K$^+$ permeability occur and so modify the membrane excitability in injured tissue. Experimentally, intracellular ATP depletion profoundly affects the K$^+$ channels responsible for the resting K$^+$ conductance: the inward rectifier K$^+$ channels inactivate and the ATP-sensitive K$^+$ channels activate.

In this chapter, the respective role of these two K$^+$ channels in generating the resting K$^+$ permeability under both normoxic and hypoxic conditions is discussed.

The background K$^+$ current in cardiac cells as described before the introduction of the patch-clamp technique

The measurement of membrane currents using classical voltage-clamp techniques, such as the sucrose gap and the double microelectrode voltage-clamp methods in multicellular preparations, already revealed the inward-going rectification of the cardiac background current, i_{K1} [1]. The K$^+$-sensitive nature of i_{K1} was well demonstrated by K$^+$ flux measurements [2], blockade of the current by Cs$^+$, Rb$^+$ and Ba^{2+}, and by the shift of the current-voltage relation with changes in the K$^+$ equilibrium potential (E_K). The conductance of i_{K1} is voltage-dependent inasmuch as it is maximum at potentials slightly negative to the resting potential, so stabilizing the resting membrane potential near the K$^+$ equilibrium potential, and decreases at potentials more positive than the reversal potential as an energy saving mechanism during the plateau of the action potential (see chapter 1).

When energy metabolism is disrupted, the membrane K$^+$ permeability increases dramatically, especially at potentials positive to the resting potential, and the plateau of the ventricular action potential shortens markedly [3]. The interaction between the membrane K$^+$ permeability and intracellular energy metabolism was suggested by the recovery of the action potential produced by simply increasing the extracellular glucose concentration in the poisoned papillary muscle [4]. The exact nature of this anoxia-induced K⁺ permeability remained unresolved until the development of the patch-clamp technique.

The patch-clamp technique identified several types of K⁺ channels in single cardiac cells

The combination of the newly developed techniques of single cardiac cell preparation [5] and patch-clamp led to revolutionary progress in the field of cardiac electrophysiology. The most significant advances were made either by the intracellular perfusion of single cells or by recording of single channel currents, leading to the identification of several types of K⁺ channel. The modulation of channel activity by intracellular factors was also clarified using these techniques.

At least 8 classes of K⁺ channels have been identified so far in mammalian cardiac muscles by their conductance and kinetic properties (see chapter 1): the inward rectifier K⁺ channel, the delayed rectifier K⁺ channel, the transient outward channel, the muscarinic K⁺ channel, the ATP-sensitive K⁺ channel, the Na⁺-activated K⁺ channel, the Ca²⁺-activated K⁺ channel, and the arachidonate-activated K⁺ channel [6]. The inward rectifier K⁺ channel accounts for most of the background K⁺ current responsible for the high level of the resting potential of cardiac muscle [7-9]. In addition, every type of K⁺ channel may also participate in the resting K⁺ permeability to some extent. For example, in atrial cells the muscarinic K⁺ channel shows spontaneous openings even in the absence of any muscarinic stimulation. The delayed rectifier K⁺ channel activates significantly at potentials positive to -60 mV and can thereby contribute to the steady membrane potential of cells such as nodal cells that show low resting potentials. The increase in the intracellular Na⁺ concentration caused by the application of ouabain opens Na⁺-activated K⁺ channels.

Compared to these K⁺ channels, however, the inward rectifier K⁺ channel provides the major part of the resting K⁺ permeability of ventricular cells under normoxic conditions, whereas the ATP-sensitive K⁺ channel is mainly responsible for the membrane K⁺ permeability in situations of impaired cellular metabolism.

The inward rectifier K⁺ channel is the main determinant of the resting potential under normoxic conditions

The single channel conductance of the inward rectifier K⁺ channel, i_{K1}, depends on extracellular [K⁺]

The inward rectifier single channel current is usually recorded from cell-attached membrane patches of isolated ventricular cells using a pipette solution containing a high extracellular K⁺ concentration (150 mM KCl). Inward-going currents through these channels are observed at transmembrane potentials more negative than the reversal potential. However, at positive potentials clear transitions from the open to the closed

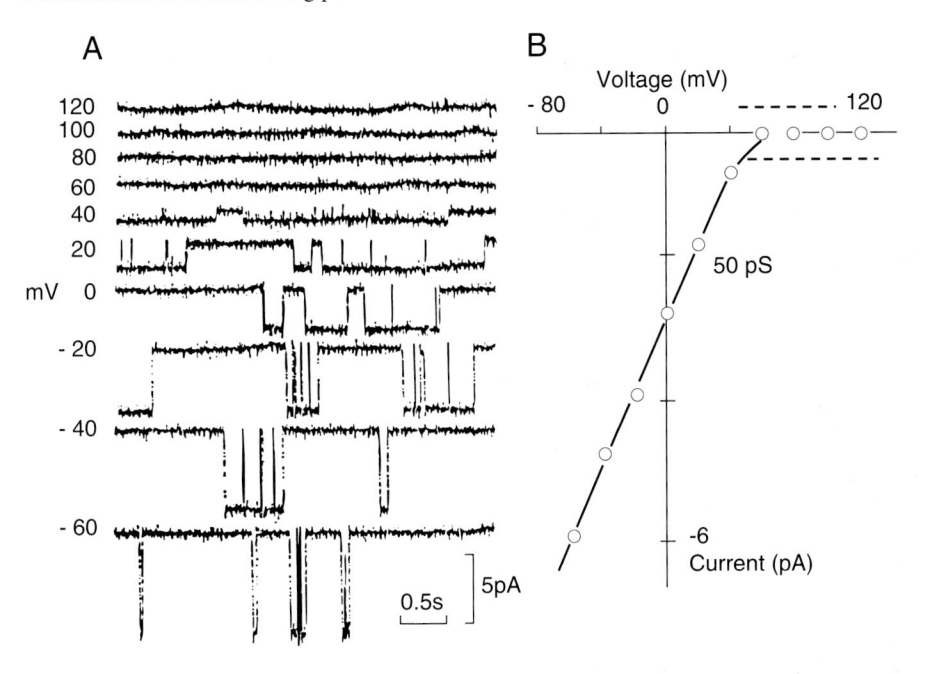

Fig. 1. **A** Single channel currents of the inward rectifier K$^+$ channel recorded in a cell-attached patch on a guinea-pig ventricular myocyte. The voltage deviation from the resting potential (-82 mV) is indicated at the left of each trace. **B** The single channel current-voltage relationship. Broken lines indicate the noise level. Modified from [9]

state are seldom observed. A typical set of records at different potentials is illustrated in fig. 1. At negative potentials, the single channel current-voltage relation is linear with a slope conductance of about 50 pS at 35° C (150 mM KCl external solution). The unitary conductance decreases to about 10 pS as the extracellular K$^+$ concentration ([K$^+$]$_o$) is decreased to physiological levels. The direct consequence of the strong dependence of i_{K1} channel conductance on [K$^+$]$_o$ is that during hypokalaemia the i_{K1} current is decreased and thus less outward current is available to repolarize the cell, leading to a slowed repolarization of the cardiac action potential and to increased excitability. The i_{K1} channel is thought to be highly selective for K$^+$, and is blocked by external Cs$^+$, Rb$^+$, Sr^{2+} and Ba^{2+}. Similar arrhythmias to those observed during hypokalaemia can be induced experimentally in dogs by intravenous injections of Cs$^+$ (see chapter 4).

Rectification of the i_{K1} channel is principally caused by intracellular Mg^{2+} block

If channel gating is defined as the opening and closing of the conducting pore by conformational changes of the channel protein, the block of the K$^+$ efflux through the channel by intracellular Mg^{2+} is of a different nature. However, Mg^{2+} block may be included in the gating mechanisms of the channel under physiological conditions since Mg^{2+} is present intracellularly at 1-3 mM. Intracellular Na$^+$ has essentially the same blocking action on K$^+$ channels as Mg^{2+} though at much higher concentrations. In addition, i_{K1} channels are slowly closed on strong hyperpolarization beyond the reversal potential for K⁺ by a mechanism that is usually called inactivation. However, it should be kept in mind that, physiologically, no cellular mechanism exists to drive the membrane potential more negative than E_K.

K⁺ channel block by intracellular Mg^{2+} has now been demonstrated for most categories of cardiac K⁺ channels including inward rectifier K⁺ channels [10], ATP-sensitive K⁺ channels [11], muscarinic K⁺ channels, and Na⁺-activated K⁺ channels. This behaviour was first reported for the ATP-sensitive K⁺ channel [11] and was shown to be consistent with the model proposed for the ionic block of K⁺ channels by cations such as Na^+, Li^+, Rb^+, tetraethylammonium, Ba^{2+} and Sr^{2+}. It is usually considered that the blocking ion enters the mouth of the channel pore, crossing an energy barrier, but that it is unable to move on through the aqueous pore to the opposite side of the membrane because of the presence of an additional barrier ; in such a way the blocking ion interferes with the passage of the permeant K⁺. The block by internal cations is voltage-dependent inasmuch as it increases when the membrane potential is made more positive.

Essentially the same mechanism of open-channel block by Mg^{2+} has been demonstrated for the inward rectifier K⁺ channel (fig. 2) [10]. However, in the inward rectifier K⁺ channel Mg^{2+} is effective at much lower concentrations and the blocking kinetics are much slower so that the transitions of channel blocking and unblocking can be directly measured from individual events.

The inward rectifier K⁺ channel is a time- and voltage-dependent channel

In the absence of Mg^{2+} block, single channel recording shows that the inward rectifier K⁺ channel closes rapidly on stepping the membrane to potentials more positive than E_K. Decay of the average current during depolarization is exponential with a time course that depends on voltage. Therefore, the channel has an intrinsic gating property in addition to Mg^{2+} block and is thus a time- and voltage-dependent channel. This contrasts with what was previously thought on the basis of results obtained with classical voltage-clamp techniques. Indeed, the above gating behavior of the inward rectifier K⁺ channel is far too fast to be resolved even in the whole cell patch-clamp configuration. This is the reason why the current i_{K1} was initially defined as a time-independent current. Its time course has now been accurately analysed using an oil-gap method to obtain a fast voltage-clamp of single ventricular cells [12]. Furthermore, analysis of the inward-going current over the potential range negative to E_K revealed the presence of an inactivation gate which closes the channel on strong hyperpolarization.

The role of the inward rectifier K⁺ current during the plateau and repolarization phases of the action potential has recently been investigated by Shimoni et al. [13]. The membrane potential of single ventricular cells was clamped using the action potential as a command signal and the i_{K1} current was determined as the component suppressed by superfusing the cell with a K⁺-free solution. The results are shown in fig. 3. At the level of the resting membrane potential, the inward rectifier K⁺ channels contribute an outward current of approximately 16 pA that quickly shuts off upon depolarization. The outward current rapidly increases during repolarization and subsequently declines during early diastole.

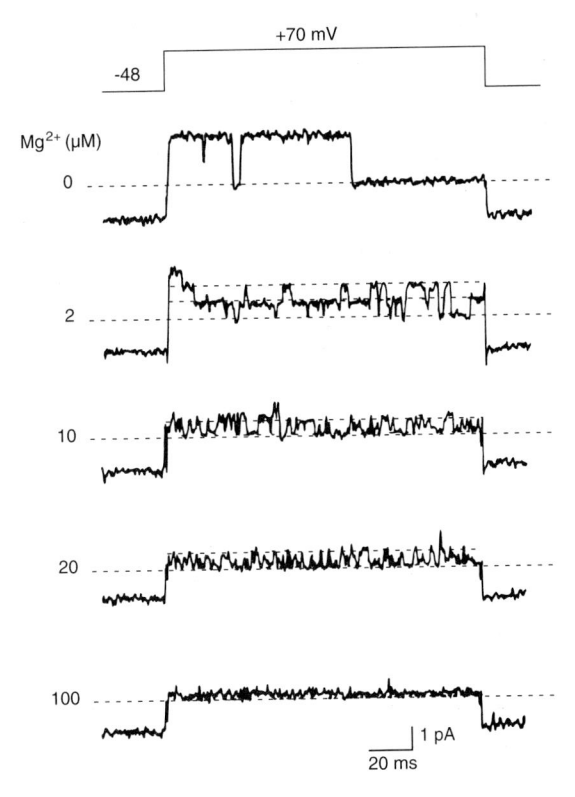

Fig. 2. Blockade by internal Mg^{2+} of outward single channel currents through inward rectifier channels. Concentrations of Mg^{2+} in the internal solution for the open-cell attached patch membrane are indicated at the left of each trace. The voltage protocol is indicated at the top. The capacitive and linear leakage current are subtracted, and the horizontal lines indicate zero current level, or are separated with an interval of one third of the full amplitude of the single channel current. Cited from [10]

The ATP-sensitive K⁺ channel contributes to the resting potential under anoxic conditions

The ATP-sensitive K⁺ channel also shows inward rectification due to intracellular Mg^{2+} block

Cardiac ATP-sensitive K⁺ channels show high K⁺ selectivity and a single channel conductance of 80 pS in 150 mM K⁺ solution at 35°C, decreasing as $[K^+]_o$ is lowered. The single channel current-voltage relation shows inward-going rectification due to partial block of the outward current by internal Mg^{2+} and Na⁺ ions.

The block of the channel by Mg^{2+} can be well explained by saturation kinetics. For the ATP-sensitive K⁺ channel, the Hill coefficient for the block by Mg^{2+} is 1 and the $[Mg^{2+}]$ for half saturation is 0.5 and 2.0 mM Mg^{2+} at +40 and 0 mV, respectively. These concentrations are much higher than the values of 0.6 and 8.8 μM Mg^{2+} (at +30 and +90 mV respectively) for the inward rectifier K⁺ channel, and of 0.29 and 0.11 mM (at +40 and +60 mV respectively) for the muscarinic K⁺ channel. The block of the Na⁺-activated K⁺ channel occurs over the range of 0-10 mM Mg^{2+}, similar to that for the ATP-sensitive K⁺ channel.

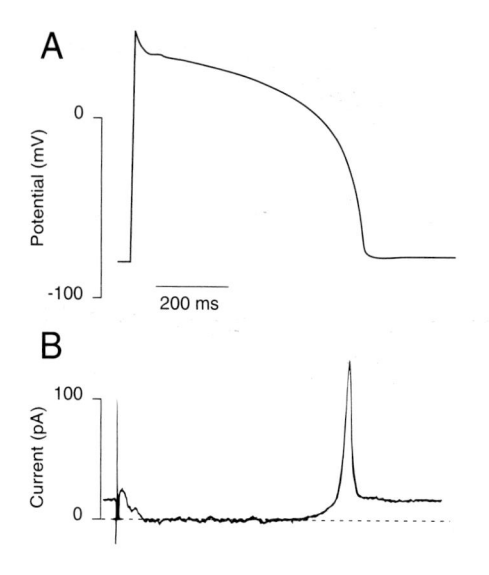

Fig. 3 A, B. Contribution of the inward rectifier K⁺ current to the ventricular action potential. **A** The action potential recorded from a rabbit ventricular myocyte, which was used for the voltage-clamp command signal. **B** Subtraction of the currents before and after removal of external K⁺ gives the net K⁺-sensitive i_{K1} current during the action potential. Note that the current is outward at the resting potential and decreases on depolarization. Modified from [13]

The ATP-sensitive K⁺ current is time-independent

In the absence of blocking cations, i.e. Mg^{2+} and Na^+, in the intracellular medium, the current-voltage relationship of the ATP-sensitive K⁺ current is linear. The kinetics of ATP-sensitive channels depends on the intracellular concentration of ATP and related nucleotides. The channel scarcely discriminates between free ATP and MgATP. The ATP dose-response curve for channel activity shows a Hill coefficient of more than 2. However, the channel is closed when one molecule of ATP is bound to a binding site and additional binding of ATP to closed conformations of the channel protein may result in the larger Hill coefficient seen in the steady state dose-response relation [15].

In the absence of both ATP and blocking ions on the cytoplasmic side of the membrane, the ATP-sensitive K⁺ channel still shows open-closed kinetics leading to long lasting bursting behaviour. The intrinsic gating behaviour during bursts of ATP-sensitive K⁺ channel openings depends on the direction of unitary current. Outward currents have relatively longer open times than inward currents, which show more evident bursting kinetics. However, this inherent gating has little influence on the mean amplitude of the channel current so that the ensemble average of the single channel current does not show obvious time- and voltage-dependent changes in response to voltage jumps.

ATP both blocks the channel and maintains its function

Two different mechanisms are present that underlie the channel regulation by ATP. These are clearly shown in experiments using the oil-gate concentration jump technique (fig. 4) [16]. Quick removal of ATP from the intracellular side of the inside-out patch opens the channel within a few seconds, but the channel activity gradually runs-down after peak activation. The responsiveness of the channel to the ATP-free solution can be restored gradually by prolonging the subsequent application of MgATP. The latter potentiating action of ATP requires Mg^{2+}, and cannot be substituted by non-hydrolysable analogues of ATP, in contrast to the blocking effect of ATP which does not discriminate

Fig. 4 A, B. Effects of removing ATP on the ATP-sensitive K$^+$ channel (**A**) and the inward rectifier K$^+$ channel (**B**). The variation of the ATP concentration in the internal solution for the inside-out patch is indicated above each current recording. Note the rapid activation of the ATP-sensitive K$^+$ channel current on depleting ATP and the subsequent run-down of the channel activity. The perfusion of 1 mM MgATP restored the channel activity visible during the next ATP depletion. The activity of the inward rectifier K$^+$ channel simply runs-down in the ATP-free solution and is restored by the reapplication of MgATP. Modified from [16]

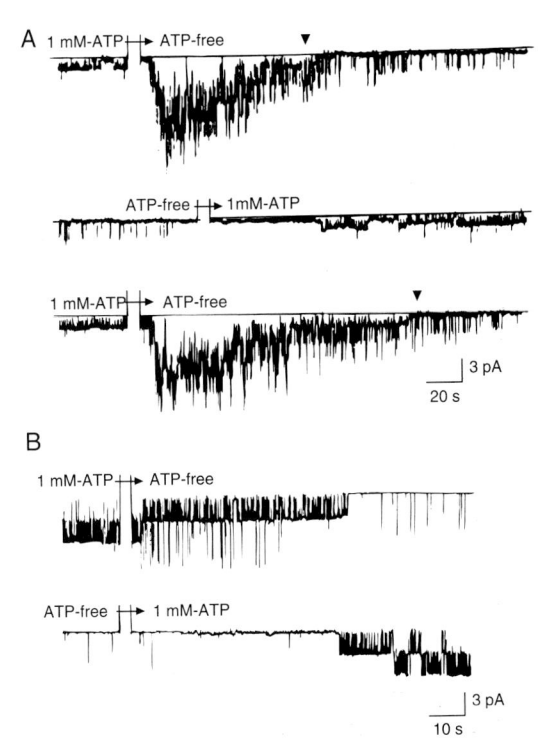

between MgATP, free ATP and non-hydrolysable analogues of ATP. Binding of nucleotide triphosphates such as ATP, GTP, CTP and UTP most probably determines the open-closed kinetics of the channel, while phosphorylation of the channel protein or associated structures may determine the responsiveness of the channel. Although there is still no concrete evidence for channel phosphorylation, the relation between phosphorylation and activity may be essentially the same as for other types of ion channels such as the L-type Ca^{2+} channel and the delayed rectifier K$^+$ channel.

It is still not known whether the binding site for nucleotide triphosphates is located on the channel protein itself or on an associated protein. Concentration jump experiments show that the time course of closure of the ATP-sensitive K$^+$ channel is exponential and has no obvious delay. This finding supports a binding site on the channel protein. Nucleotide diphosphates have at least two different effects on the ATP-sensitive K$^+$ channel. Firstly, nucleotide diphosphates can substitute for ATP in gating the K$^+$ channel, and they decrease the ATP sensitivity when the channel is partially closed by ATP. This effect may be explained by assuming that nucleotide diphosphates bind competitively and do not close the channel as effectively as ATP. The other effect of nucleotide diphosphates is typically observed in the absence of ATP. The activity of the ATP-sensitive K$^+$ channel decays in ATP-free solutions, but the subsequent and immediate application of Mg-nucleotide diphosphates re-opens the K$^+$ channel. The relative potency of various nucleotide diphosphates to cause this latter effect at 10 mM is UTP>IDP>CDP.

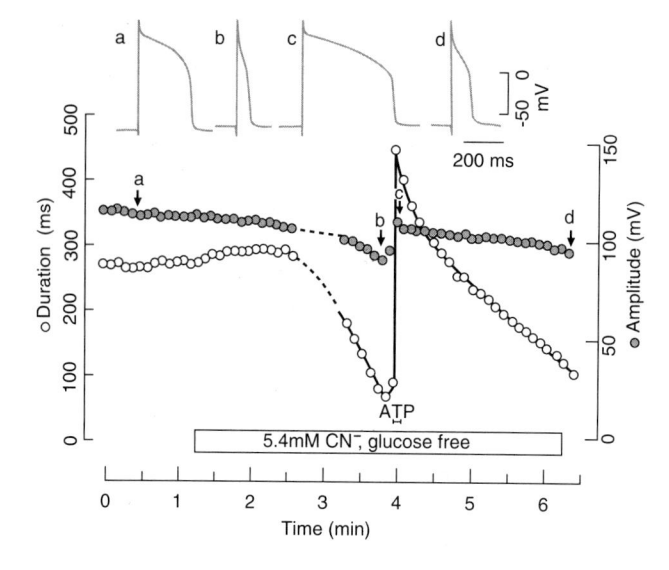

Fig. 5. Depression of the duration and the amplitude of the action potential during superfusion with 5 mM CN⁻ containing Tyrode solution and recovery in response to microinjection of ATP. ATP was injected intracellularly by applying positive pressure to the interior of a glass microelectrode containing 0.5 mM ATP. Modified from [17]

During intracellular ATP depletion, the resting K^+ permeability switches from inward rectifier to ATP-sensitive K^+ channels

Single channel recordings in the cell-attached situation demonstrate that ATP-sensitive K⁺ channels open during metabolic impairment caused by either anoxia or metabolic inhibitors, such as CN⁻ or dinitrophenol. However, to evaluate the contribution of the ATP-sensitive K⁺ channel to the resting membrane conductance under situations of impaired metabolism, findings from whole cell experiments should also be considered. When ventricular cells are exposed to a glucose-free external solution containing CN⁻, the action potential shortens markedly (fig. 5) [17], an effect which is readily reversed by intracellular microinjection of ATP. This strongly suggests the involvement of ATP-sensitive K⁺ channels. Fig. 6 shows current-voltage relations for the initial current (left panel) and the late current (right) elicited by voltage steps applied from - 35 mV before and during metabolic poisoning of the cell with CN⁻. The zero current potential (corresponding to the resting membrane potential) is only slightly modified during the course of CN⁻-treatment (fig. 6). However, close inspection of the inward current at voltages more negative than the reversal potential reveals that the membrane conductance is actually decreased in spite of the additional opening of ATP-sensitive K⁺ channels. This may be explained by a concomitant depression of the inward rectifier K⁺ channel under conditions of metabolic impairment as has been revealed at the single channel level. In open cell-attached patch experiments, complete intracellular ATP depletion decreased the open state probability of the inward rectifier K⁺ channel from 0.46 to 0.09 while the ATP-sensitive K⁺ channel was massively activated [16].

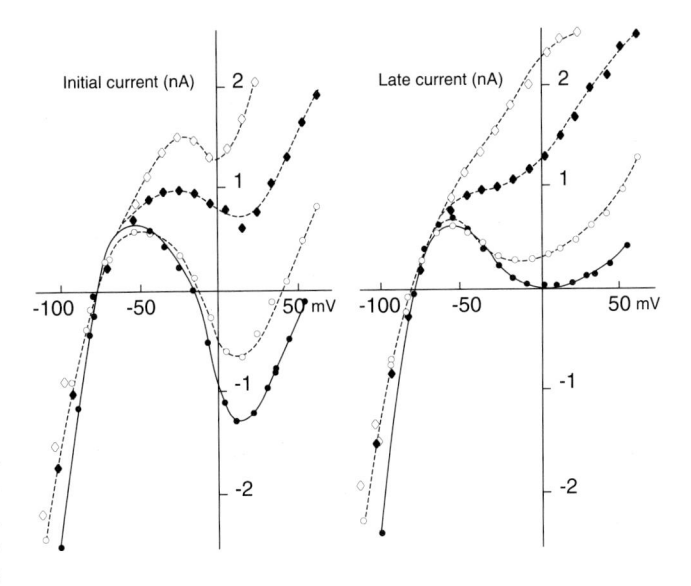

Fig. 6. Current voltage relationships for the initial and late currents of a single guinea-pig ventricular cell. During the whole cell voltage clamp, a 5 mM CN⁻ glucose-free solution was applied and currents in control *(filled circle)*, 6-8 min *(open circle)*, 10-12 min *(filled diamond)*, and 15-17 min *(open diamond)* were recorded. Note that the amplitude of the Ca^{2+} current is markedly depressed in the initial current. Cited from [14]

Why do ATP-sensitive K^+ channels open during ischaemia?

During hypoxia, the level of creatine phosphate significantly decreases, but the intracellular ATP concentration remains close to its normal level, and is much higher than the concentration for half maximal block of ATP-sensitive K^+ channels reported in excised membrane patches. Thus, whether the opening of the ATP-sensitive K^+ channel is responsible for the shortening of the action potential is questionable. However, experiments performed with the ATP-sensitive K^+ channel blocker, glibenclamide, strongly suggest that this channel opens during hypoxia (see chapter 13). Furthermore, the shortening of the cardiac action potential can easily be explained if the amplitude of the net current is calculated during the plateau phase of the action potential. If dV/dt is 10 mV/100 ms during the plateau phase in a 140 pF single guinea-pig ventricular myocyte, the amplitude of the net outward current needed for repolarization is only 14 pA. If we consider a K^+ channel current amplitude of 1.5 pA at the plateau potential and a channel density of 5000 per cell, an increase in open probability of only 0.002 can double the rate of repolarization [18]. The effects of modulation of the ATP-sensitivity of the channel on open probability should also be considered. Because the ADP level increases during hypoxia, the antagonistic relationship between ADP and ATP can partly explain the opening of the channel at relatively high $[ATP]_i$. Furthermore, intracellular acidification during the impairment of cellular energy metabolism may also decrease the sensitivity of the ATP-sensitive K^+ channel. It should also be noted that the ATP-sensitivity shows considerable variation from patch to patch, the value of $[ATP]_i$ for half inhibition ranging from 9 to 580 µM. It is also possible that the metabolic impairment is not homogeneous so that activation of the ATP-sensitive K⁺ channels in a localized region generates a large current that shortens the plateau of the action potential in surrounding tissues by way of gap junctions.

ATP is a compound of central importance in view of its participation in oxidative phosphorylation and its role as the source of high-energy phosphate for nearly all energy-requiring reactions in the cell. Therefore, it seems likely that transducing the ATP concentration within the cell into electrical signals is vital for most living cells. The opening of the ATP-sensitive K⁺ channel shortens the cardiac action potential and thereby decreases the magnitude of the twitch contraction. This may prevent further consumption of ATP in the injured cells, which might cause irreversible damage of the metabolic pathway. In this way, opening of ATP-sensitive K⁺ channels may protect the cell [19]. It should also be noted that the hyperpolarization or the maintenance of negative potentials caused by the opening of ATP-sensitive K⁺ channels should aid extrusion of Ca^{2+} through the Na/Ca exchanger under pathological conditions. Because of its electrogenic nature, Ca^{2+} extrusion is accelerated exponentially as the membrane potential becomes more negative.

References

1. Trautwein W (1973) Membrane currents in cardiac muscle fibers. Physiol Rev 53: 793-835
2. Vereecke J, Isenberg G, Carmeliet E (1980) K⁺ efflux through inward rectifying K channels in voltage clamped Purkinje fibres. Pflügers Archiv 384: 207-217
3. Carmeliet E (1978) Cardiac transmembrane potentials and metabolism. Circ Res 42: 577-587
4. McDonald TF, MacLeod DP (1973) Metabolism and the electrical activity of anoxic ventricular muscle. J Physiol (Lond) 229: 559-582
5. Powell T, Twist VW (1976) A rapid technique for the isolation and purification of adult cardiac muscle cells having respiratory control and a tolerance to calcium. Biochem Biophys Res Commun 72: 327-333
6. Noma A. (1987) Chemical-receptor-dependent potassium channels in cardiac muscles. In: Noble D, Powell T (eds) Electrophysiology of single cardiac cells. Academic Press, London, p 223
7. Sakmann B, Trube G (1984a) Conductance properties of single inwardly rectifying potassium channels in ventricular cells from guinea-pig heart. J Physiol (Lond) 347: 641-657
8. Sakmann B, Trube G (1984b) Voltage-dependent inactivation of inward-rectifying single-channel currents in the guinea-pig heart cell membrane. J Physiol (Lond) 347: 659-683
9. Kameyama M, Kiyosue T, Soejima M (1983) Single channel analysis of the inward rectifier K current in the rabbit ventricular cells. Jpn J Physiol 33: 1039-1056
10. Matsuda H (1988) Open-state substructure of inwardly rectifying potassium channels revealed by magnesium block in guinea-pig heart cells. J Physiol (Lond) 397: 237-258
11. Horie M, Irisawa H, Noma A (1987) Voltage-dependent magnesium block of adenosine-triphosphate-sensitive potassium channel in guinea-pig ventricular cells. J Physiol (Lond) 387: 251-272
12. Ishihara K, Mitsuiye T, Noma A, Takano M (1989) The Mg^{2+} block and intrinsic gating underlying inward rectification of the K⁺ current in guinea-pig cardiac myocytes. J Physiol (Lond) 419: 297-320
13. Shimoni, RB, Clark PB, Giles WR (1992) Role of an inwardly rectifying potassium current in rabbit ventricular action potential. J Physiol (Lond) 448: 709-727
14. Noma A, Shibasaki T (1985) Membrane current through adenosine-triphosphate-regulated potassium channels in guinea-pig ventricular cells. J Physiol (Lond) 363: 463-480

15. Qin D, Takano M, Noma A (1989) Kinetics of the ATP-sensitive K$^+$ channel revealed with oil-gate concentration jump method. Am J Physiol 257: H1624-H1633
16. Takano M, Qin D, Noma A (1990) ATP-dependent decay and recovery of K$^+$ channels in guinea-pig cardiac myocytes. Am J Physiol 258: H45-H50
17. Taniguchi J, Noma A, Irisawa H (1983) Modification of the cardiac action potential by intracellular injection of adenosine triphosphate and related substances in guinea pig single ventricular cells. Circ Res 53: 131-139
18. Lederer WJ, Nichols CG (1989) Nucleotide modulation of the activity of the rat heart ATP-sensitive K$^+$ channels in isolated membrane patches. J Physiol (Lond) 419: 193-211
19. Cole WC, McPherson CD, Sontag D (1991) ATP-regulated K$^+$ channels protect the myocardium against ischaemia-reperfusion damage. Circ Res 69: 571-581

CHAPTER 4
Potassium channels and the regulation of refractoriness and heart rate

E Carmeliet

Action potential duration is the major determinant of the refractory period

Under normal conditions a close relation exists in cardiac cells between the action potential duration and recovery of excitability [1, 2]. The threshold is determined by the current required to displace the membrane potential to a level where sufficient inward current is activated such that the membrane current becomes net inward. This starts the positive feedback cycle (inward current leads to cell depolarization which leads to more inward current) generating a full-sized upstroke of the action potential. Following the upstroke of the action potential the Na^+ channels remain inactivated for most of the plateau duration. It is only during the final phase that Na^+ channels recover from inactivation and may again become activated. Recovery from inactivation is a potential-dependent phenomenon that is very fast at negative membrane potentials but is much slower at positive potentials; i.e. during the plateau. Under those conditions action potential duration and refractory period are closely related (fig. 1A). Any change or difference in action potential duration is accompanied by a change in refractoriness.

This close relationship however may be broken in conditions such as ischaemia [3], high extracellular K^+ concentration [4] or presence of drugs acting on the Na^+ channel [5]. With a high extracellular K^+ concentration the cell membrane becomes partially depolarized and some of the Na^+ channels enter the state of inactivation at rest. The remaining channels that are still available for activation and which become inactivated during the action potential will recover more slowly because of the smaller (less negative) resting potential. Refractoriness then lags behind full repolarization. A similar situation is generated by the use of antiarrhythmic agents acting on the Na^+ channels, especially those that block the channel in the inactivated state. A decrease in Na^+ current especially at depolarized levels will reduce excitability, prolong the time course of recovery and cause a dissociation between action potential duration and refractoriness (fig. 1B). Antiarrhythmic drugs acting on the i_K current however do not exert this dissociating effect and any prolongation of the action potential will be accompanied by an equally important change in refractoriness, without modification of the recovery process during and following the final repolarization.

Repolarization has three phases

Repolarization in cardiac cells is usually described as consisting of 3 phases, phase 1 or initial rapid repolarization, phase 2 or plateau and phase 3 or final repolarization [1]. The relative importance of these phases is different according to the cell type. A rather positive plateau is seen in the sino-atrial and atrio-ventricular node cells and in the ventricle.

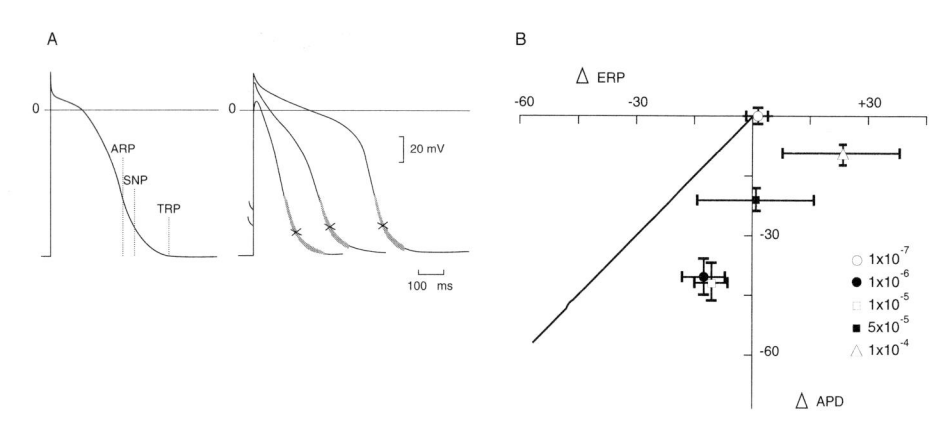

Fig. 1 A, B. Relationship between action potential duration and refractoriness. **A** Left: action potential time course in a canine Purkinje fibre with indication of excitability parameters. *TRP:* total refractory period; *ARP:* absolute refractory period; *SNP:* indicates the time at which a minimum current is required to excite during the supernormal period. Right: parallel changes in action potential duration and excitability parameters during changes in rhythm. The longest duration of action potential was evoked at a basic cycle length of 800 ms. The shorter action potentials were successively evoked as premature beats at cycle lengths of 460 ms and 251 ms. The shaded areas delineate the boundaries of the period of supernormal excitability. The x's within the shaded area indicate the point of minimum current requirements. For excitation from [2]. **B** Change in action potential duration *(ΔAPD)* relative to change in effective refractory period *(ΔERP)* in seven ventricular muscle fibres of the dog is shown as a function of lidocaine concentration (10^{-7} - 10^{-4} M). The diagonal line is the line of identical changes for APD and ERP. At all concentrations, except 10^{-7}M lidocaine, the shortening of ERP was less than the shortening in APD. Units are ms. From [5]

In atrial cells the plateau is less pronounced and repolarization in general is more rapid. In Purkinje fibres a long, rather negative plateau follows the pronounced initial repolarization or spike potential. This superficial description is far from complete and heterogeneity is much more pronounced than it suggests. In atrial tissue for instance marked differences have been described between cells in the roof, the pectinate muscle, the crista terminals and strands between trabecular muscles. More recently the existence of a marked heterogeneity in the ventricle has been emphasized [6]. Apart from the Purkinje system, different types of action potential have been distinguished in the endocardial, the epicardial and the mid-region layers of the ventricular myocardium (fig. 2). Epicardial myocytes are characterized by an action potential with a prominent initial rapid repolarization followed by a secondary depolarization, resulting in a characteristic spike-dome appearance. In the mid-region these two phases are still prominent but the secondary depolarization runs out in a much longer plateau; the rate of depolarization during phase 0 or upstroke is also much greater. The action potential in the mid-region resembles to a certain extent the action potential in Purkinje fibres.

Several K⁺ currents are involved in repolarization

In their original analysis Hodgkin and Huxley (see [1]) described the nerve action potential in terms of two successive currents; an inward Na⁺ current followed by an outward

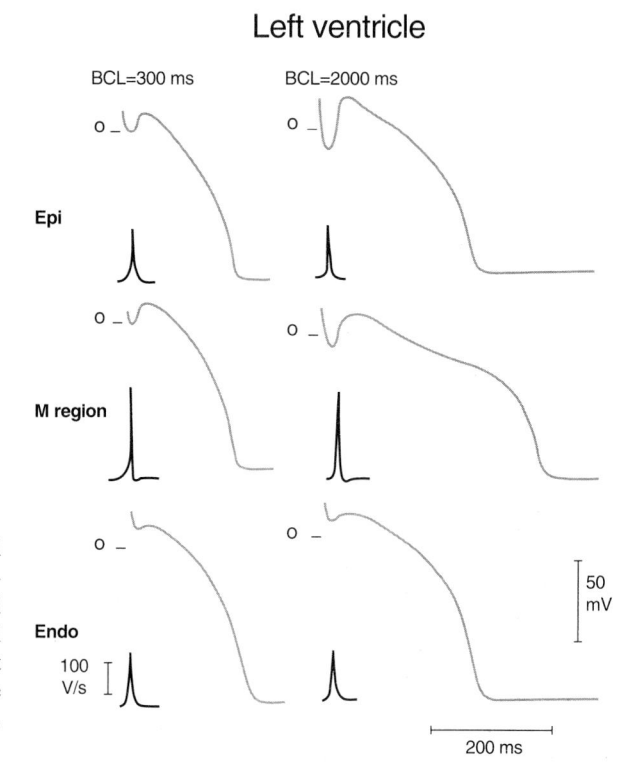

Fig. 2. Action potentials, and their first derivative of epicardial *(Epi)*, deep subepicardial *(M region)*, and endocardial *(Endo)* cells in the canine left ventricle at two basic cycle lengths (BCL) of 0.3 and 2.0 s. From [6]

K$^+$ current. Since then and especially following the development of the patch clamp, electrophysiology has become more complex. Not only has the number of currents increased dramatically but new mechanisms of activation have been described (see [7, 8]). Actually a distinction can be made between voltage-operated, ligand-operated and stretch-operated channels; important currents are also carried through exchanger and cotransporter molecules. In all these current systems the interaction between ions and transport molecules can be compared to enzymatic reactions, the difference being in the activation energy and the final outcome of the reaction. In contrast to classic enzymatic reactions, a channel or a transporter does not transform a molecule but translocates the ion across the membrane. For activation, energy is provided by a change in electrical gradient across the membrane (voltage-operated channels), by binding of a ligand (e.g. acetylcholine, Ca^{2+}, ATP) or by a mechanical deformation. In the case of exchangers and cotransporters the comparison with enzymatic reactions is more direct, the ion movement being determined by the concentration gradient across the membrane.

The present analysis will be restricted to K$^+$ currents during the plateau phase of the action potential under physiological conditions. K$^+$ currents activated during ischaemia are described in chapter 3 of this book. Under physiological conditions three to four main K$^+$ currents participate in the cardiac repolarization process (fig. 3): (i) the transient outward current, i_{to}, responsible for the early rapid repolarization from the crest of the action potential to the plateau level, (ii) the delayed rectifier K$^+$ current, i_K, involved in the overall repolarization process during the plateau, and (iii) the inward rectifier i_{K1}

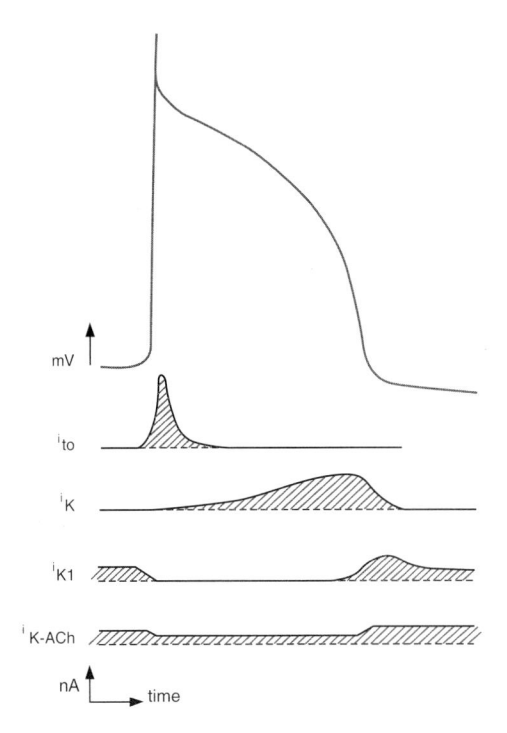

Fig. 3. Schematic representation of the four K$^+$ currents which participate in the repolarization of the cardiac action potential under physiological conditions

current mainly responsible for the final rapid repolarization and for the maintenance of the resting potential. In the sino-atrial node, the atrio-ventricular node, the atria and the Purkinje system, the muscarinic K$^+$ current, i_{K-ACh}, activated on vagal stimulation, participates in the repolarization process and the resting potential; a channel with the same characteristics is activated by adenosine and somatostatin (see chapter 1). These channels may be called physiological channels. In pathophysiological conditions, mainly ischaemia, ligand-operated channels such as the ATP-sensitive K$^+$ channel, the Na$^+$-activated K$^+$ channel and a K$^+$ channel activated by fatty acids or phospholipids may become active. They will not be discussed in this review.

The transient outward current contributes to the initial repolarization and to total action potential duration

The transient outward current shows rapid activation, followed by inactivation. The inactivation process is responsible for the transient nature of this current. On the basis of its kinetics and pharmacological characteristics the current has been described as composed of a purely voltage-activated and a smaller Ca^{2+} activated component [9]. Recently the latter component has been claimed to be a Cl$^-$ and not a K$^+$ current [10]. Activation of the voltage-dependent component is very rapid, which explains why the current plays an important role during the early phase of the repolarization, often called the spike potential. This spike potential will either lead smoothly to a plateau phase at a rather negative potential (e.g. in Purkinje fibres) or it may be followed by a secondary depolarization due to the inflow of Ca^{2+} ions. This secondary depolarization is especially expressed in subepicardial cells [6]. It is also prominent in Purkinje fibres treated with adrenalin [11]. The

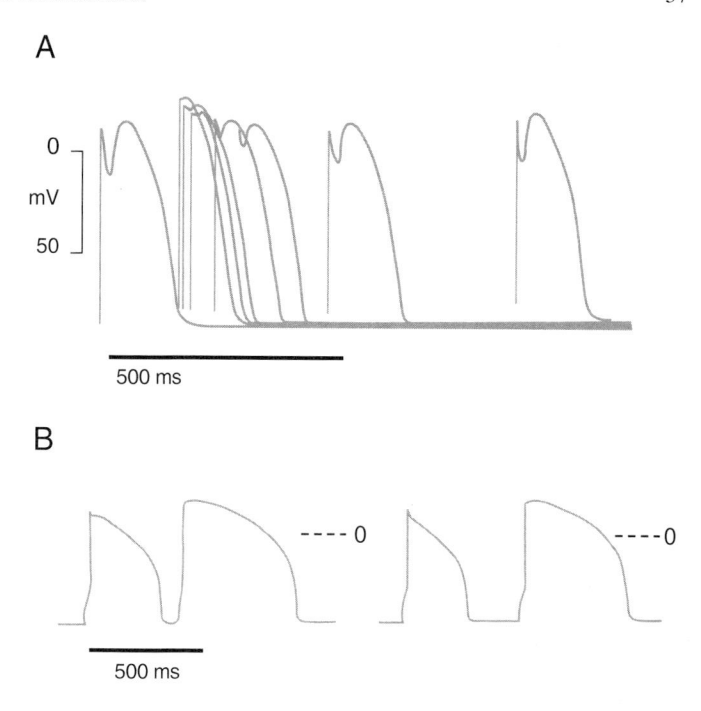

Fig. 4 A, B. Incomplete recovery of the i_{to} from inactivation changes the configuration of the extra-systolic action potential. In the two examples shown the plateau starts at a more positive level but total action potential duration is shortened in **A** and prolonged in **B**. **A** Restitution of the action potential parameters in subepicardial fibers of the canine heart. From [6]. **B** Similar experiment in rabbit ventricular cells. From [12]

existence of a secondary depolarization which gives the action potential its characteristic spike-dome appearance thus depends on the presence of functional Ca^{2+} channels in sufficient number. It is clear that any change in the balance between activation of the i_{to} and of i_{Ca} affects the expression of the spike and dome and secondarily also the total action potential duration. Two examples will serve to illustrate this statement.

The first example is provided by the changes in action potential configuration when an extra stimulus is applied during diastole (fig. 4). Since both current systems, i_{to} and i_{Ca}, undergo inactivation during systole an extra stimulus in early diastole may arrive at a time when both currents are incompletely recovered from inactivation. The kinetics of recovery however are different and since i_{to} recovery is slower, the fall in outward current will be of more importance. As a consequence the spike of the extrasystolic action potential will be less prominent and the plateau will start at a more positive level. At this level inactivation of i_{Ca} may be faster and this phenomenon together with summation of the delayed rectifier K^+ current may result in a shorter action potential (fig. 4A). However this is not the general rule. If the fall in i_{to} is more pronounced and inactivation of i_{Ca} and/or activation of i_K are less effective, prolongation of the extrasystolic action potential ensues (fig. 4B) [12]. In all cases the plateau is shifted to more positive membrane voltages.

A second example can be found in the peculiar effects of acetylcholine (ACh) on the mammalian ventricle (fig. 5) [6]. For a long time it has been an unsolved problem why ACh exogeneously applied to a mammalian ventricular preparation exerts significant effects on contractility and refractory period in vivo but causes little or no change in action potential characteristics in vitro. The reason for this paradox has to be found in the fact that in vitro experiments were done on endocardial myocardial preparations. Recently it has been discovered that the response of the canine ventricular subendocardium and subepicardium to ACh and isoprenaline are totally different. While the subendocardial cells are resistant to ACh concentrations up to 10^{-5}M, subepicardial cells show a non-negligible lengthening at 10^{-7} and 10^{-6}M, and a substantial shortening at 10^{-5}M. The greater sensitivity of the subendocardial cells (and probably also of the mid-region cells) to ACh has to be correlated with the presence of a spike-dome sequence, and thus of a pronounced i_{to} and i_{Ca}. The ionic mechanism by which ACh affects the ventricular action potential configuration is by decreasing i_{Ca}. ACh stimulates cardiac m_2 receptors, that are negatively linked to the generation of cAMP. Cyclic AMP is normally responsible for activation of protein kinase A and phosphorylation of the Ca^{2+} channel. With a reduction in i_{Ca} in the presence of ACh the spike repolarization becomes more prominent, while the secondary depolarization or dome is delayed and lower in amplitude. The total outcome is eventually a slight prolongation of the action potential. When i_{Ca} is further suppressed at high ACh concentration repolarization during the spike occurs to a membrane potential level outside the activation range of i_{Ca}. The secondary depolarization is absent and repolarization continues rapidly to the resting potential.

From the two examples discussed above it is clear that the result of pharmacological manipulation of the i_{to} on the action potential duration is difficult to predict; a block of i_{to} is not unequivocally linked to a prolongation of the ventricular action potential. A selective blocker of i_{to} may thus not be the ideal antiarrhythmic to use at the level of ventricular cells. On the other hand such a substance will be very active in atrial cells where i_{to} is very pronounced, while i_{Ca} and i_K are less important in determining the total action potential duration. Tedisamil, which is an efficient blocker of i_{to} and i_K, causes the atrial as well as the ventricular action potential to be prolonged.

The delayed rectifier K^+ current (i_K) affects plateau duration

The plateau of the cardiac action potential is characterized by a fine balance between inward and outward currents [8]. The delayed rectifier K^+ current is the main outward current in a number of species. Ca^{2+} current and Na^+-Ca^{2+} exchange currents are the main inward currents in nodal, atrial and ventricular myocytes, while a slowly inactivating Na^+ current is prominent in the Purkinje tissue [13]. Ca^{2+} and Na^+ currents are slowly inactivated while K^+ current slowly activates: the outcome is a small net outward current which increases with time. The kinetics of the delayed rectifier K^+ current are not identical in all species. In the guinea-pig the nature of the delayed rectifier K^+ current is complex and consists of two components [14]: one termed i_{Kr} is activated between -50 mV and 0 mV with time constants in the order of hundreds of ms, and a second termed i_{Ks} is only activated above 0 mV with time constants in the order of s. The latter component will only be of importance at elevated frequencies. The faster component is the only one present in the rabbit and cat ventricular myocytes. It is also of interest to mention that the two compo-

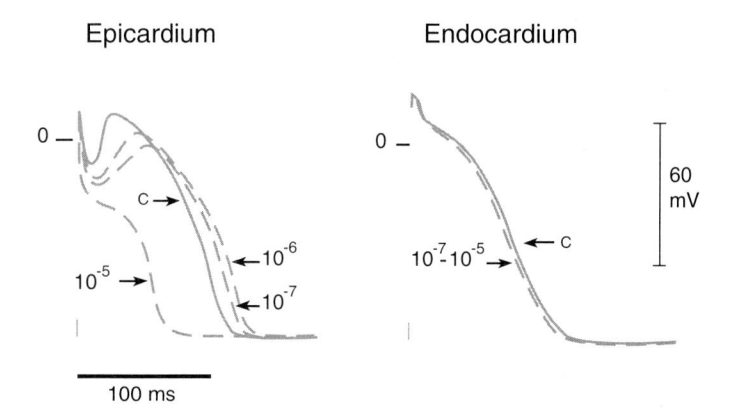

Fig. 5. Subepicardial cells of the canine heart are sensitive to acetylcholine (ACh), subendocardial cells are not. Transmembrane recordings were obtained before (solid lines) and after (dotted lines) exposure to 10^{-7} - 10^{-5} M ACh. In epicardium, ACh produced an accentuation of phase 1 magnitude and slowing of the second action potential upstroke, resulting in prolongation of APD at ACh concentrations of 10^{-7} and 10^{-6} M. ACh in a concentration of 10^{-5} M suppressed the plateau, thus causing a marked abbreviation of the action potential. Endocardium was largely unresponsive to ACh. Basic cycle length = 300 ms *C:* control. From [6]

nents are differently affected by drugs. In general the rapidly activated component is the most sensitive to class III drug inhibition (see chapter 11). The result will be a prolongation of the action potential, which is considered an important antiarrhythmic intervention. Indeed, in re-entry type arrhythmias with a short excitable gap only a small increase in refractoriness is needed to block the arrhythmia. Block of i_K is considered the best target mechanism for prolonging the action potential, and more appropriate than enhancing Na$^+$ and/or Ca^{2+} current, which may have the complication of Ca^{2+} overload.

A number of experimental drugs are currently in the process of evaluation (see chapter 11). Unfortunately none of these drugs seem to fulfill the requirements of the "ideal" drug (fig. 6A) [15]. Instead of acting preferentially at elevated frequencies they prolong the action potential at low frequencies to such an extent that very often early after-depolarizations are generated. In vivo this situation could result in *torsades de pointe* arrhythmia [16]. The ideal drug should block the i_K channel during activation with a time constant greater than 1 s, and show a recovery from block with a time constant in the order of 100 ms. Some of the drugs under investigation, such as amiodarone, almokalant and to a certain extent also dofetilide and E4031 show open channel and use-dependent block (fig. 6B) [17]. Recovery from block, however, is so slow that steady state frequency-dependent block is small or absent.

The inward rectifier current, i_{K1}, is important in final repolarization

The i_{K1} current is responsible for the resting potential and the final repolarization in most myocardial cells (the subscript 1 is used because it was the first K$^+$ current to be analysed in cardiac tissue). The current was characterized as a time-independent background current showing a pronounced inward rectification [8]. More recent analysis has shown that the i_{K1} current is not a time-independent background current but shows time-dependent

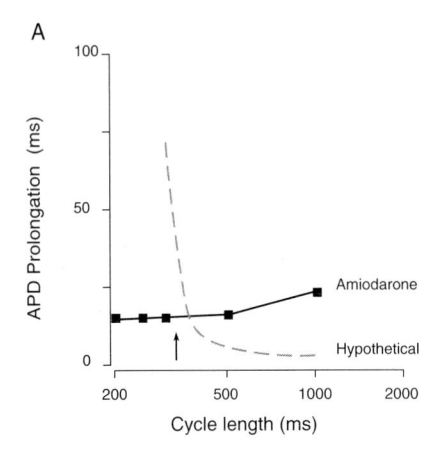

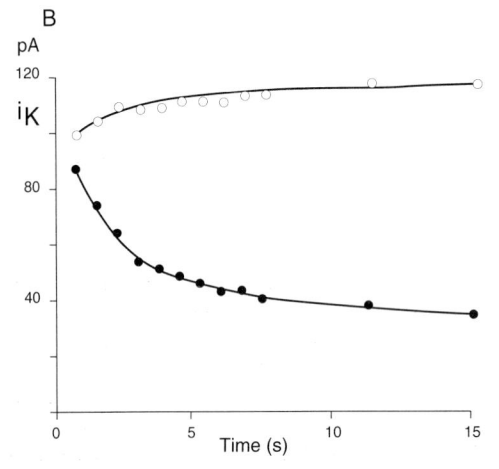

Fig. 6. A Action potential prolongation as a function of cycle length for amiodarone *(full line)* and an hypothetical "ideal" agent *(dashed line)*. From [15]. **B** Use-dependent block of i_K by 10^{-6} M amiodarone. Currents were obtained for 0.2 s depolarizations to +10 mV, applied at a frequency of 1.33 Hz following a 2 min rest at the holding potential of -50 mV. Under control conditions *(open circles)* tail currents show summation with repetition of the clamp depolarization. In the presence of the drug *(closed circles)* summation is not only absent but the reverse sequence is seen with tails declining on repetition. Rabbit ventricular myocyte; 37° C. Carmeliet, unpublished

activation on hyperpolarization [18]. In practical terms inward rectification means that the channel only passes current at membrane potentials negative to -20 mV, while large currents are carried at potentials close to the K⁺ equilibrium potential. During the plateau of the cardiac action potential the channel passes either no current or very little. When the cell repolarizes and especially during the final repolarization the i_{K1} current becomes important and is responsible for the regenerative increase in repolarization rate at about 40 mV positive to the maximum diastolic membrane potential. The regenerative increase in outward current through the i_{K1} channel is the consequence of more and more activation, and less and less blockade of the channel by intracellular Mg^{2+} and Ca^{2+} ions, a mechanism responsible for the inward rectification. Recent experiments by Shimoni et al. [19] have provided a direct evaluation of the contribution of i_{K1} to repolarization process in the ventricular myocyte; an example is given in fig. 7A and B.

The density of i_{K1} channels differs throughout the heart. In the sino-atrial and atrio-ventricular nodes the current is practically absent or is very small. The density is greater in the atrium, and it is the highest in the ventricle and the Purkinje system. The conductance of the i_{K1} channel is very sensitive to extracellular K⁺ concentration, a rise in $[K^+]_o$ elicits an increase in conductance and the reverse occurs as $[K^+]_o$ is lowered [20].

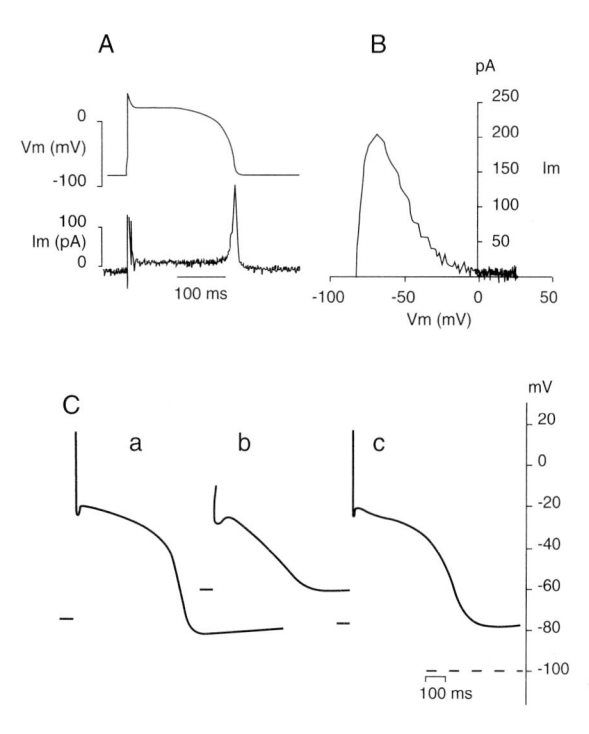

Fig. 7 A-C. Contribution of i_{K1} during the ventricular action potential. **A** Time course of the action potential in a rabbit ventricular myocyte at 32°C (upper record) and corresponding i_{K1} (lower record). The i_{K1} record was obtained using the action potential itself as the voltage-clamp command signal. The i_{K1} was estimated as the current sensitive to withdrawal of extracellular K+ (with ouabain present to block the Na/K pump current). **B** Current-voltage relation for i_{K1} during the action potential. Net current (i_m) from A is plotted against membrane potential (V_m) during the plateau and repolarization phase. Current through the i_{K1} channel is very small during the plateau but becomes pronounced during the final repolarization (phase 3), from [19]. **C** Cs+ ions block the inward rectifier K+ current i_{K1} in a sheep Purkinje fibre. *a* is control; *b* is 2 min after addition of 20 mM Cs+. Block of i_{K1} results in depolarization, reduction of upstroke amplitude, shorter plateau and slower final repolarization (phase 3). Hyperpolarization by a constant current *(c)* restores the resting potential, the action potential amplitude and the plateau but the final repolarization remains slow, from [21]

These changes explain to a certain extent the changes in action potential duration and in the T-wave of the ECG under those conditions. At the cellular level the rate of the final repolarization is increased at elevated external K+, while the reverse occurs at low K+ concentrations. In the ECG this is translated into a taller and narrower T-wave at high K+, while a flat and small amplitude T-wave is characteristic for hypokalaemia. Similar changes to those seen in low K+ occur when the i_{K1} channel is blocked by drugs or by Cs+ ions [21] (fig. 7C).

The role of the muscarinic K⁺ channel in the repolarization process

The muscarinic K⁺ channel (i_{K-ACh} channels) seems to be the only ligand-operated channel which plays a physiological role in determining resting and action potentials [22]. In mammals its role is restricted to the sino-atrial, atrio-ventricular node, the atrium and the Purkinje system. Ventricular myocytes with the exception of those in the ferret [23], do not show an increase of K⁺ current upon addition of acetylcholine. Parasympathetic stimulation, however, affects ventricular contractile activity and the subepicardial (and probably also the mid-region) electrical activity, via a negative effect on the Ca^{2+} current. This aspect of the effects of ACh has been discussed in the context of the spike-dome configuration of the ventricular action potential and the interaction of i_{to} and i_{Ca} [6].

The muscarinic K⁺ channel is activated by acetylcholine which binds to a muscarinic receptor, m_2, and secondarily activates a G-protein [24]. The G-protein is directly coupled to the K⁺ channel. As discussed above, coupling to the Ca^{2+} channel is more complex and occurs via a cascade of reactions and intracellular messengers. Other receptors, such as A1 receptors stimulated by adenosine or receptors stimulated by somatostatin are also coupled to a similar K⁺ channel. It is interesting to note that stimulation of the adenosine receptor and its resulting negative dromotropic effect is used as a diagnostic and therapeutic intervention for supraventricular tachycardias.

K⁺ currents and pacemaker activity

The delayed rectifier K+ current participates in determining firing frequency in the sino-atrial node

In contrast to the stable membrane potential during diastole in atrial and ventricular myocytes, a spontaneous diastolic depolarization occurs in pacemaker cells of the sino-atrial node which is the primary pacemaker of the heart. A similar spontaneous decrease in membrane potential is also seen in secondary pacemakers, localised in the atrium, the atrio-ventricular node and the Purkinje system. The firing rate or beat frequency of the sino-atrial node is determined: (i) by the time needed for the membrane potential to change from the maximum level to the threshold, which thus depends on this maximum level, the rate of diastolic depolarization and the threshold but also; (ii) by the time spent in the depolarized state, in other words by the duration of the action potential.

Diastolic depolarization involves three currents

Three currents participate in the generation of the pacemaker depolarization in the node [25]: the delayed rectifier K⁺-current which undergoes de-activation, the funny non-selective cation current, i_f, which is activated on hyperpolarization and the Ca^{2+} current (T- and L-type) which becomes activated during the second part of the diastolic phase when the membrane is sufficiently depolarized to reach the activation level of i_{Ca}. De-activation of a K⁺ current requires the simultaneous presence of a background inward current to generate a time-dependent depolarization. The existence of such an inward background current has been predicted on the basis of a low maximum diastolic potential, which is appreciably less negative than the theoretical equilibrium potential for K$^+$ ions. The presence of such an inward background current as well as poor expression of the inward rectifier K$^+$ current have been experimentally verified [26].

The rate of i_K de-activation changes with membrane potential. It is fast at negative potentials and slower at depolarized levels. At the level of maximum diastolic potential of the node the rate is appropriate to be responsible for the early phase of diastolic depolarization. Since activation of i_f is rather slow, inward current through this channel becomes more important with time. The relative importance of both phenomena, de-activation of i_K and activation of i_f, has been the object of controversial statements and discussions in the literature (see [8, 25]). It is safe to say that the activation of i_f will contribute more the more negative the maximum diastolic potential, while de-activation of i_K becomes more important at less negative potentials.

The duration of the action potential in pacemaker cells

The time spent in the depolarized state is equally important as modulator of the firing frequency as is the duration of the diastolic period. Prolongation of the action potential is thus an efficient way to reduce beating frequency; conversely any shortening of the plateau phase will enhance firing rate.

Sympathetic stimulation results in an increase of firing rate by modulation of a number of currents: the activation curve of the i_f current is shifted in the depolarizing direction; i_{Ca} amplitude is increased and the threshold is moved in the hyperpolarized direction; also the i_K current is enhanced with the consequence that the plateau is shortened. Vagal stimulation on the other hand reduces the firing rate: all three currents, i_f, i_{Ca} and i_K are modulated in the opposite way by low concentrations of acetylcholine. At high concentrations acetylcholine furthermore opens specific muscarinic K$^+$ channel, resulting in hyperpolarization and stabilization of the membrane potential at more negative levels. The physiological importance of K$^+$ channel activation has been questioned by DiFrancesco et al. [27], who explain the negative chronotropic effects by a selective reduction in i_f. Hirst et al. [28] on the other hand have proposed that vagal stimulation inhibits a background inward current.

Class III agents or drugs that block i_K and prolong the ventricular action potential may be expected to exert a negative chronotropic effect by causing lengthening of the systolic period. The effect on the diastolic period, however, should not be neglected. Due to the fall in i_K the maximum diastolic potential will be decreased and this change in potential may secondarily affect i_f and i_{Ca}. Recent estimations for a sinus venosus model [25] have shown that an inhibition of i_K to half its normal value prolongs the action potential duration, decreases the maximum diastolic potential and reduces the rate of diastolic depolarization (fig. 8A). The fall in frequency is primarily due to the prolongation of the repolarization process, and less to the reduction in the rate of diastolic depolarization. These effects predicted on theoretical grounds have been found experimentally. Unpublished records by Oexle et al. [29] of the effect of tedisamil on the rabbit sinus node (fig. 8B) show that some predictions based on the sinus venosus model (no i_f included) are valid for a mammalian preparation: prolongation of the action potential was the main mechanism of the bradycardic effect. The importance of a change in the duration of the

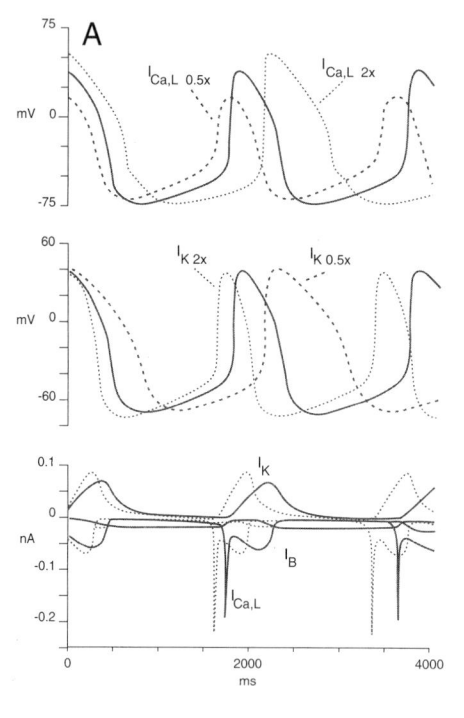

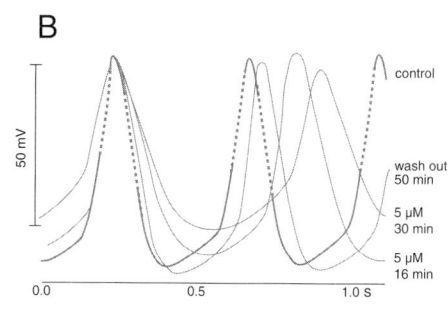

Fig. 8. A Predicted changes in membrane potential and current based on the sinus venosus model, when i_{Ca} or i_K are selectively changed. Effects of increasing 200% and decreasing 50% $i_{Ca,L}$ (upper panel) and i_K (middle panel). Lower panel: alterations in other currents as a result of selectively increasing i_K 200%. *Solid lines:* control; *dashed lines:* i_K increased. The model simulations indicate that a selective increase of $i_{Ca,L}$ produces a decrease in pacemaker frequency, while an increase of i_K produces an increase in pacemaker frequency. From [25]. **B** Effect of tedisamil 5×10^{-6} M, a blocker of i_K, on the action potential in the sino-atrial node of the rabbit. The main reason for the bradycardic effect is prolongation of the plateau duration. Antoni, unpublished (see also [29])

action potential as a chronotropic effect is further emphasized by calculations based on the same model but taking changes in i_{Ca} into account. A selective increase in i_{Ca} does not enhance the firing rate, as might be intuitively expected, but reduces the beating frequency by lengthening the action potential duration (fig. 8 A).

Block of i_K and/or i_{K1} channels may generate pacemaker activity

It is common clinical experience that antiarrhythmic drugs which excessively prolong the action potential duration may become arrhythmogenic. The ventricular tachyarrhythmia which appears following an excessive QT prolongation has been called *torsades de pointe*, because of its undulating peaks of QRS complexes. It is a classic complication of quinidine treatment, especially in the presence of hypokalaemia or hypomagnesaemia [16]. At the cellular level the occurrence of early afterdepolarizations (EADs) or/and the dispersion of repolarization have been put forward as possible mechanisms. Recent experiments with an animal model have favored the mechanism of EAD

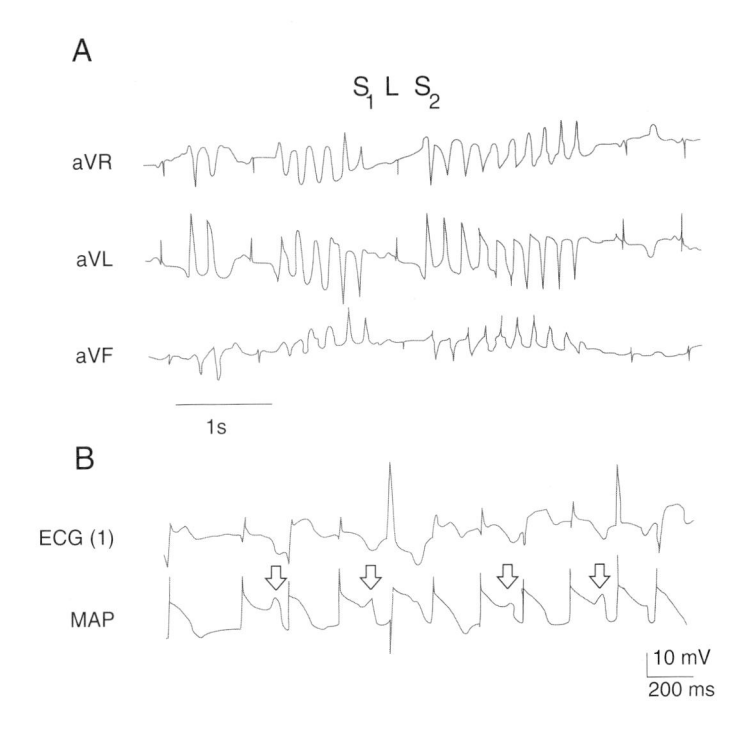

Fig. 9. A Representative electrocardiograms demonstrating initiation of *torsades de pointe* in an anaesthetized rabbit given clofilium (63 nmol/kg/min i.v.). Tachyarrhythmia was induced after administration of a cumulative dose of 0.43 µmol/kg clofilium. Note characteristic pause-dependent mode of initiation ("short-long-short" sequence, S_1 L S_2) and undulating morphology of sequential QRS complexes. **B** Electrocardiograms (ECG, lead I) and endocardial monophasic action potentials (MAP) recorded during runs of ventricular extrasystoles in anaesthetized rabbit given infusion of a cumulative dose of 0.85 µmol/kg clofilium. Note existence of deflections (arrows) consistent with early afterdepolarizations in repolarization phase of monophasic action potential. From [30]

generated in the Purkinje system and conducted to the contractile myocytes [30] (fig. 9). From in vitro experiments it was already known that excessive prolongation of the Purkinje action potential by substances which slow Na^+ current inactivation, or by drugs which block i_K, resulted in EADs that were eventually conducted to associated ventricular muscles. The ionic mechanism of EADs has been analysed in detail by January and Riddle [31]. These investigators have shown that when the rate of repolarization is critically reduced, Ca^{2+} channels of the L-type can recover from the inactivated state during the slow repolarization and return back to the activated state. These reactions occur in a range of membrane potentials where the activation and inactivation steady-state curves overlap. This range of membrane potentials is called the "window". It also exists for the Na^+ current, especially in the Purkinje system. In Purkinje fibres there exists furthermore an important slowly inactivating Na^+ current [13], which may be responsible for slowing down the repolarization rate and thus creating the conditions to re-activate L-type Ca^{2+} channels. To induce tachyarrhythmia in the ventricular mass there must be enough activation of Ca^{2+} current so that oscillatory depolarizations of sufficient amplitude are gene-

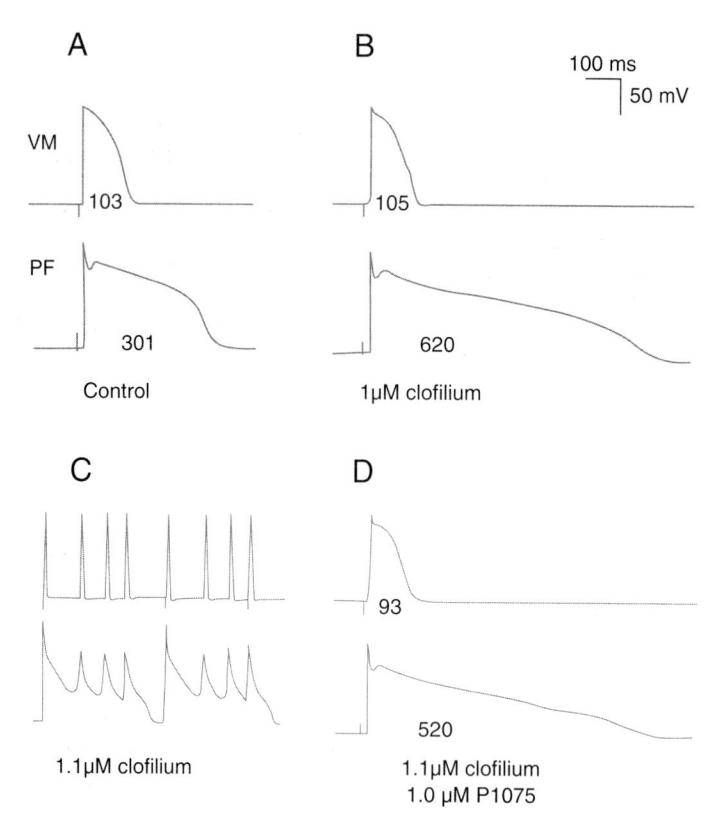

Fig. 10 A-D. Original tracings of action potentials simultaneously recorded from ventricular muscle *(VM)* and Purkinje fibre *(PF)* of rabbit show effects of clofilium and combination of clofilium and P1075, a K$^+$ channel opener. Panel **A** Control. Clofilium caused a concentration-dependent prolongation of the action potential duration in the PFs (panel **B**), and eventually, early afterdepolarizations appeared (panel **C**). Addition of P1075 abolished EADs, whereas APD was still extensively prolonged (panel **D**). Horizontal bar denotes 100 msec for panels **A**, **B** and **D** and 1 second for panel **C**. From [30]

rated superimposed on the prolonged action potential. Abolition of the Ca^{2+}-dependent oscillations in the Purkinje system by application of a critical concentration of K$^+$ channel openers stopped the dysrhythmia, without necessarily reversing the delayed repolarization in the Purkinje fibres [31, 32] (fig. 10). These experiments thus demonstrate that the simple difference in action potential length between Purkinje fibres and ventricular myocytes does not itself generate enough current to reach threshold but that oscillatory depolarizations grafted on the slow repolarization are necessary to start and support the dysrhythmia.

References

1. Hoffman BF, Cranefield PF (1960) Electrophysiology of the Heart. McGraw-Hill, New York
2. Spear JF, Moore EN (1974) Supernormal excitability and conduction in the His-Purkinje system of the dog. Circ Res 35: 782-792
3. Janse MJ, Wit AL (1989) Electrophysiological mechanisms of ventricular arrhythmias resulting from myocardial ischaemia and infarction. Physiol Rev 69: 1049-1169
4. Gettes LS, Reuter H (1974) Slow recovery from inactivation of inward currents in mammalian myocardial fibres. J Physiol (Lond) 240: 703-724
5. Bigger JT, Mandel WJ (1970) Effect of lidocaine on the electrophysiological properties of ventricular muscle and Purkinje fibres. J Clin Invest 49: 63-77
6. Antzelevitch C, Litovsky SH, Lukas A (1990) Epicardium versus endocardium: electrophysiology and pharmacology. In : Zipes DP, Jalife J (eds) Cardiac Electrophysiology. From cell to bedside. Saunders, Philadelphia, London, Toronto, pp 386-395
7. Hille B (1992) Ionic channels of excitable membranes, 2nd ed. Sinauer, Sunderland, Mass
8. Noble D (1984) The surprising heart: a review of recent progress in cardiac electrophysiology. J Physiol (Lond) 353: 1-50
9. Corabœuf E, Carmeliet E (1982) Existence of two transient outward currents in sheep cardiac Purkinje fibres. Pflügers Arch 392: 352-359
10. Zygmunt AC, Gibbons WR (1991) Calcium-activated chloride current in rabbit ventricular myocytes. Circ Res 68: 424-437
11. Carmeliet E, Vereecke J (1967) Adrenalin and the plateau phase of the cardiac action potential. Importance of Ca^{2+}, Na^+ and K^+ conductance. Pflügers Arch 313: 300-315
12. Hiraoka M, Kawano S (1987) Mechanism of increased amplitude and duration of the plateau with sudden shortening of diastolic intervals in rabbit ventricular cells. Circ Res 60: 14-26
13. Carmeliet E (1987) Slow inactivation of the sodium current in rabbit cardiac Purkinje fibres. Pflügers Arch 408: 18-26
14. Sanguinetti MC, Jurkiewicz NK (1990) Two components of cardiac delayed rectifier K^+ current. Differential sensitivity to block by class III antiarrhythmic agents. J Gen Physiol 96: 195-215
15. Hondeghem LM, Snyders DJ (1990) Class III antiarrhythmic agents have a lot of potential but a long way to go. Reduced effectiveness and dangers of reverse use-dependence. Circulation 81: 686-690
16. Surawicz B (1989) Electrophysiologic substrate of torsades de pointe: Dispersion of repolarization or early afterdepolarizations? J Am Coll Cardiol 14: 172-184
17. Carmeliet E (1992) Use-dependent block of the delayed K^+ current in cardiac cells. A comparison of different drugs with Class III activity. J Mol Cell Cardiol 24 (Suppl I): S108
18. Kurachi Y (1985) Voltage-dependent activation of the inward-rectifier potassium channel in the ventricular membrane of guinea-pig heart. J Physiol (Lond) 366: 365-385
19. Shimoni Y, Clark RB, Giles WR (1992) Role of inwardly rectifying potassium current in rabbit ventricular action potential. J Physiol (Lond) 448: 709-728
20. Carmeliet E (1989) K^+ channels in cardiac cells: mechanisms of activation, inactivation, rectification and K^+e sensitivity. Pflügers Arch 414 (Suppl I): S88-S92
21. Isenberg G (1976) Cardiac Purkinje fibres: Cesium as a tool to block rectifying potassium currents. Pflügers Arch 365: 99-106
22. Heidbüchel H, Vereecke J, Carmeliet E (1990) Three different potassium channels in human atrium. Contribution to the basal potassium conductance. Circ Res 66: 1277-1286
23. Boyett MR, Kirby MS, Orchard CH (1988) The negative inotropic effect of acetylcholine on ferret ventricular myocardium. J Physiol (Lond) 404: 613-635
24. Kurachi Y, Tung RT, Ito H, Nakajima T (1992) G protein activation of cardiac muscarinic K^+ channels. Prog Neurobiol 39: 229-246

25. Campbell DL, Rasmusson RL, Strauss HC (1992) Ionic current mechanisms generating vertebrate primary cardiac pacemaker activity at the single cell level: an integrative view. Ann Rev Physiol 54: 279-302
26. Hagiwara N, Irisawa H, Kasanuki H, Hosoda S (1992) Background current in sino-atrial node cells of the rabbit heart. J Physiol (Lond) 448: 53-72
27. DiFrancesco D, Ducouret P, Robinson RB (1989) Muscarinic modulation of cardiac rate at low acetylcholine concentrations. Science 243: 669-671
28. Hirst GDS, Bramich NJ, Edwards FR, Klemm M (1992) Transmission at autonomic neuroeffector junctions. TINS 15: 40-46
29. Oexle B, Weirich J, Antoni H (1987) Electrophysiological profile of KC 8857, a new bradycardic agent. J Mol Cell Cardiol 19 (Suppl III) : S65
30. Carlsson L, Abrahamsson C, Drews L, Duker G (1992) Antiarrhythmic effects of potassium channel openers in rhythm abnormalities related to delayed repolarization. Circ Res 85 : 1491-1500
31. January CT, Riddle JM (1989) Early after-depolarizations: mechanisms of induction and block. A role for L-type Ca^{2+} current. Circ Res 64: 977-990
32. Fish FA, Prakash C, Roden DM (1990) Suppression of repolarization-related arrhythmias in vitro and in vivo by low-dose potassium channel activators. Circulation 82: 1362-1369

CHAPTER 5
Potassium channels and cardiac contraction[1]

MR Boyett, SM Harrison and E White

The contractility of atrial and ventricular muscle of the heart is variable. Normally, it is under the control of the autonomic nervous system and circulating humoral factors, but, in addition, it is altered in disease states and can be controlled therapeutically by the administration of drugs. Inotropic agents alter contraction in various ways, such as by changing the Ca^{2+} sensitivity of the myofilaments, Ca^{2+} influx via Ca^{2+} channels or Ca^{2+} flux via Na/Ca exchange. Although both Ca^{2+} influx via Ca^{2+} channels and Ca^{2+} flux via Na/Ca exchange can be changed directly, they can also be changed indirectly by altering action potential duration. This chapter is concerned with inotropic agents which affect contraction by altering K^+ channels and thus action potential duration. Inotropic agents, which regulate contraction in this way under normal physiological conditions (ACh, adenosine and α-agonists), under pathological conditions (ischaemia and intracellular Na^+) and which are administered therapeutically (class III antiarrhythmic agents), are considered. First, the importance of action potential shape for contraction will be considered.

Action potential shape and contraction

The shape of the action potential influences contraction

The shape of the action potential is determined by the size and timecourse of the inward (depolarizing) currents, carried by Na^+ and Ca^{2+}, and the outward (repolarizing) currents, carried by K^+. An increase in inward current or a decrease in outward current will elevate the action potential plateau and prolong its duration. The importance of the shape of the action potential to the strength of the contraction was established in the late 1960's [1]. The effects of increasing and decreasing repolarizing currents on the action potential can be mimicked by applying appropriate current to a preparation. Fig. 1A shows the injection of repolarizing current during the action potential (simulating an increase in K^+ current). The plateau is depressed and action potential duration reduced. A decrease in contraction occurs immediately, but further reductions follow even though action potential shape does not change further. A new steady-state contraction is eventually reached (in this example after 7 stimulations). When the current is withdrawn (fig. 1B) there is a recovery which follows a similar timecourse. Injection of depolarizing current (simulating decreased K^+ current) elevates the plateau, prolongs the action potential and

---

[1] In this chapter contraction of cardiac muscle is shown in various ways. In experiments performed with multicellular cardiac preparations tension is measured and the increase in tension during contraction is shown as an upward going deflection. In experiments with single cardiac cells, cell shortening during contraction is measured and this is shown as either an upward or downward going deflection. Irrespective of the direction of the deflection, an increase in the amplitude of the deflection represents an increase in contraction.

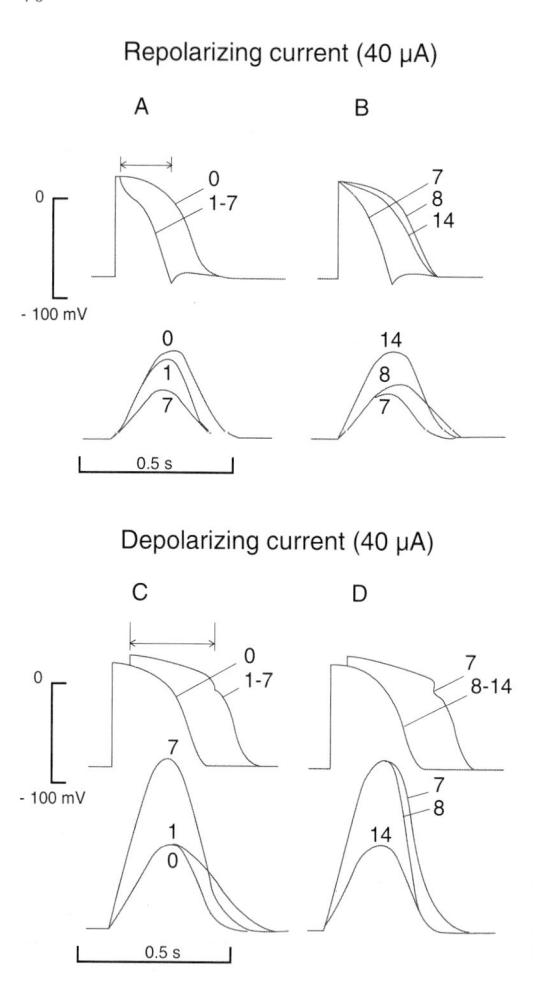

Fig. 1. A Application of repolarizing current to the action potential of a cat papillary muscle. Action potential "0" was obtained under control conditions. Repolarizing current was injected for the time shown by the arrow above the record during action potentials 1 to 7. Current injection resulted in an immediate reduction in the amplitude and duration of the action potential: action potentials 1 and 7 are superimposable and both are smaller and shorter than the control. In contrast, current injection resulted in a decrease in the contraction over several beats: contraction 1 is smaller than the control (0) but contraction 7 is further reduced. **B** Removal of repolarizing current. Following action potential 7, the subsequent action potentials were not modified by current injection. This resulted in an immediate increase in the amplitude and duration of the action potential and an increase in the contraction over several beats. **C, D** Application (**C**) and removal (**D**) of depolarizing current. The format of panels **C** and **D** is similar to that of **A** and **B**. Injection of depolarizing current resulted in an immediate increase in the amplitude and duration of the action potential and an increase in the contraction over several beats. On removal of the current the changes occurred in reverse. Taken from [45] with permission

increases the contraction with a similar timecourse (fig. 1C, D). These slow changes in contraction are a result of changes in the loading of intracellular Ca^{2+} stores [2].

Both the amplitude and duration of the action potential are important in controlling contraction

Fig. 1 shows that the injection of repolarizing current decreases both the amplitude and the duration of the action potential, whereas depolarizing current does the reverse. Fig. 2 (taken from voltage-clamped single ventricular cells) shows that both these effects are important. Fig. 2A shows that if the duration of the voltage-clamp pulse is increased (corresponding to an increase in action potential duration), the size of the contraction is also increased. Fig. 2B shows the importance of the action potential amplitude. Here, the duration of the depolarization is the same in each case but as the membrane potential is decreased during the pulse (simulating a depression of the action potential plateau) the contraction becomes smaller.

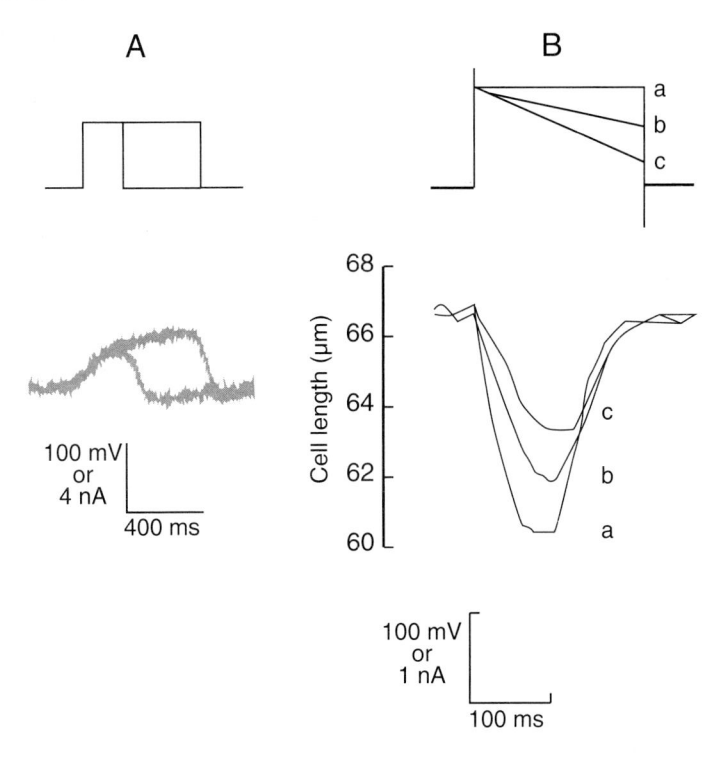

Fig. 2. A The effect of increasing the duration of a voltage clamp pulse (upper trace) on the contraction (lower trace, upward going deflection) of a guinea-pig ventricular cell. The cell was held at -40 mV and then clamped to +60 mV for either 200 or 600 ms. The contraction elicited during the 200 ms pulse is smaller than that during the 600 ms pulse. **B** The effect of reducing the membrane potential during a voltage clamp pulse of constant duration (upper trace) on the contraction (lower trace, downward going deflection) of another guinea-pig cell. The cell was held at -73 mV and then clamped to +40 mV for 200 ms. During the pulse the membrane potential was ramped to a more negative value. In different pulses the membrane potential was ramped at different speeds to simulate changes in the amplitude of the action potential plateau. The greater the rate of repolarization during the pulse, the smaller the accompanying contraction. **A** taken from [46], **B** taken from [47], with permission

Excitation-contraction coupling in the heart

A brief summary of the current concept of excitation-contraction coupling in the heart is given here to introduce the mechanisms which are modulated by the action potential (for a full review of cardiac excitation-contraction coupling, see [3]). Fig. 3 shows a schematic diagram of the process. During the plateau of the action potential there is an influx of Ca^{2+} ions into the cell via voltage-dependent L-type Ca^{2+} channels. This Ca^{2+} influx can lead to the development of contraction directly but in mammalian heart it is probably more important as a trigger for the release of greater quantities of Ca^{2+} ions into the cytoplasm from the sarcoplasmic reticulum (the intracellular store of Ca^{2+}). The intracellular calcium concentration $[Ca^{2+}]_i$, is elevated and interaction of Ca^{2+} ions with the myofilaments results in contraction. As Ca^{2+} ions leave the myofilaments the muscle relaxes. Ca^{2+} ions

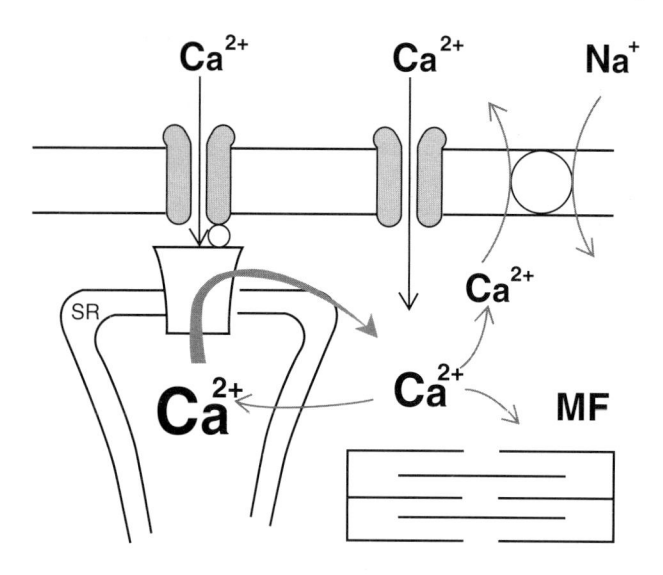

Fig. 3. Schematic diagram of excitation-contraction coupling in the heart. Influx of Ca^{2+} into the cell via Ca^{2+} channels triggers release of Ca^{2+} from sarcoplasmic reticulum stores *(SR)*. The Ca^{2+} released into the cytosol activates the myofilaments *(MF)*. When Ca^{2+} is released back into the cytosol it is either taken back into the SR or pumped out of the cell by the Na/Ca exchanger. Taken from [3] with permission

are removed from the cytoplasm either by being taken back up into the sarcoplasmic reticulum by an ATP dependent Ca^{2+} pump or are removed from the cell by a membrane Na/Ca exchanger. The strength of contraction of mammalian cardiac muscle is largely determined by the Ca^{2+} content of the sarcoplasmic reticulum, the greater the Ca^{2+} content, the greater the amount of Ca^{2+} that can be released to activate the myofilaments. The Ca^{2+} content of the sarcoplasmic reticulum depends both on Ca^{2+} influx via Ca^{2+} channels and on Ca^{2+} flux via Na/Ca exchange. For example, a decrease in Ca^{2+} influx or an increase in Ca^{2+} efflux results in less Ca^{2+} being available for uptake into the sarcoplasmic reticulum. Loading of the stores decreases over several beats and a decrease in the contraction results (fig. 1A).

Action potential shape influences L-type Ca^{2+} channels

A change in K^+ current can indirectly influence Ca^{2+} influx via Ca^{2+} channels. A reduction in the amplitude and duration of the action potential like that seen in fig. 1A attenuates the influx of Ca^{2+} into the cell via Ca^{2+} channels by speeding the return of the membrane potential to voltages at which the Ca^{2+} channels close. Conversely, the elevation and prolongation of the action potential plateau as shown in fig. 1C enhances Ca^{2+} influx via Ca^{2+} channels by prolonging the time available for channel opening.

Action potential shape influences the Na/Ca exchanger

The Na/Ca exchanger is a membrane protein which exchanges 3 Na^+ for 1 Ca^{2+}. Because the numbers of charges carried are not equal, the exchanger is electrogenic and an electric current flows in the same direction as the movement of Na^+ ions. The exchanger can operate in both directions. In the forward mode Ca^{2+} is extruded from the cell and Na^+ enters. In the reverse mode Ca^{2+} enters the cell in exchange for Na^+, which is extruded.

The operating mode of the exchanger is dependent on the intracellular and extracellular concentrations of Na^+ and Ca^{2+} and (more importantly for our discussion on the

Fig. 4. Schematic diagram showing the relationship between the reversal potential of the Na/Ca exchanger $(E_{Na/Ca})$ and the membrane potential (V_m) during the rabbit ventricular action potential. When $E_{Na/Ca}$ is negative to V_m *(shaded area)*, Na/Ca exchange causes Ca^{2+} entry. When $E_{Na/Ca}$ is positive to V_m, Na/Ca exchange causes Ca^{2+} efflux. The magnitude of the difference between $E_{Na/Ca}$ and V_m determines the rate of Ca^{2+} influx or efflux via Na/Ca exchange. The dotted line shows the internal calcium concentration, scale at the right in μM Ca^{2+}. Taken from [3] with permission

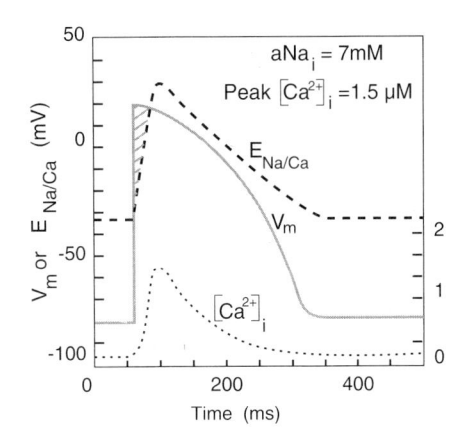

influence of K^+ channels on contraction) on the membrane potential. When the membrane potential of the cell is negative to the reversal potential of the exchanger ($E_{Na/Ca}$, the potential at which the activity of the exchanger in the forward and reverse modes is equal) the exchanger operates in the forward mode, Ca^{2+} is extruded from the cell and Na^+ enters. When the membrane potential is positive to $E_{Na/Ca}$ the exchanger operates in the reverse mode. Fig. 4 shows the relationship between the membrane potential and $E_{Na/Ca}$ during the action potential. The transient rise and fall of $[Ca^{2+}]_i$ during the action potential causes similar changes in $E_{Na/Ca}$. The membrane potential becomes positive to $E_{Na/Ca}$ for a short period of time at the start of the action potential because of the rapid change in membrane potential during the upstroke of the action potential. During this time, Ca^{2+} may enter the cell via the Na/Ca exchanger. Apart from this brief period, the membrane potential is more negative than $E_{Na/Ca}$ and the exchanger acts in the forward mode resulting in Ca^{2+} extrusion.

If repolarizing K^+ currents are reduced, the action potential plateau will be elevated and prolonged. This will reduce or even reverse the difference between the membrane potential and $E_{Na/Ca}$. As a result Ca^{2+} efflux via the exchanger will be decreased or even reversed (i.e. Ca^{2+} will be brought into the cell) during the plateau of the action potential. If K^+ currents are increased and the action potential amplitude and duration are decreased, the difference between the membrane potential and $E_{Na/Ca}$ will be increased and consequently Ca^{2+} efflux via the exchanger during the plateau of the action potential will be increased.

Summary of the effects of action potential shape on contraction

Increasing or decreasing the repolarizing K^+ currents that flow during the action potential will change the timecourse of the action potential. The Ca^{2+} content of the sarcoplasmic reticulum depends on both the amount of Ca^{2+} entering the cell (via Ca^{2+} channels) and that leaving the cell (via the Na/Ca exchanger). Because both these mechanisms are voltage-dependent, alterations in the shape of the action potential will alter the Ca^{2+} content of the sarcoplasmic reticulum and thus the size of the contraction.

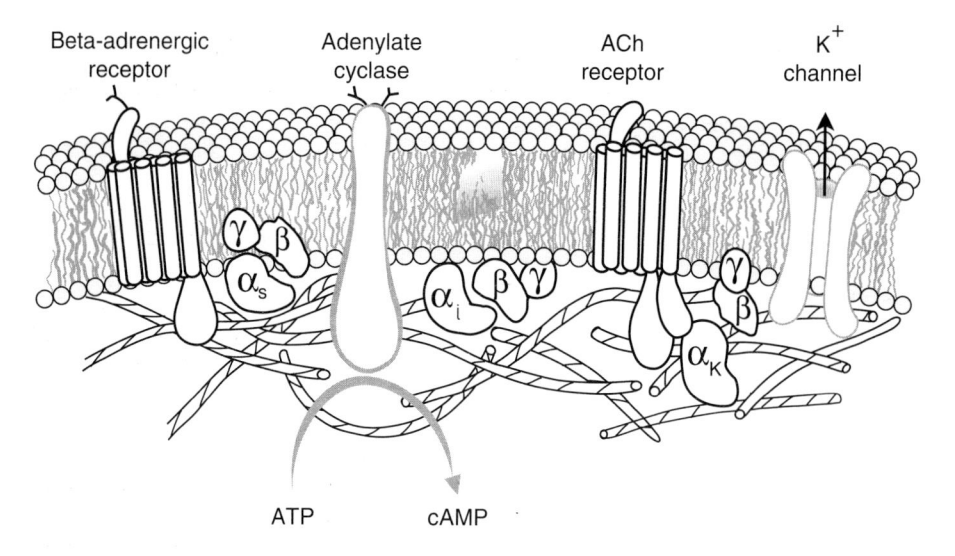

Fig. 5. Schematic diagram of the membrane of a cardiac cell showing the interaction of the muscarinic ACh receptor with adenylate cyclase and the muscarinic K⁺ channel via G proteins (made up of α, ß and γ subunits). Also shown is the ß-adrenergic receptor, which stimulates adenylate cyclase via another G protein. When adenylate cyclase is active, it produces cAMP from ATP. Modified from [48] with permission

The role of muscarinic K⁺ channels in modulating contraction

ACh decreases the contraction of cardiac muscle

In the heart acetylcholine (ACh) is released from parasympathetic nerves and it binds to muscarinic (M2) receptors on the heart cells (see fig. 5). This results in various responses including a decrease in the contractility of both the atrial and ventricular muscle (a negative inotropic effect). When ACh binds to the muscarinic receptor, two G proteins (G_i and G_K) are activated (fig. 5). G_i inhibits the activity of adenylate cyclase and thereby lowers the level of cAMP in the cell. The reduction of cAMP has many consequences including a dephosphorylation of L-type Ca^{2+} channels and therefore a reduction in the Ca^{2+} current, i_{Ca}, (Ca^{2+} influx) through them. In atrial muscle and ferret ventricular muscle there is some evidence that ACh produces a small decrease in i_{Ca} under "basal" conditions (i.e. in the absence of a ß-adrenergic agonist). However, in ventricular muscle of other species ACh does not inhibit i_{Ca} under basal conditions but only inhibits i_{Ca} once adenylate cyclase (and therefore i_{Ca}) has been stimulated by a ß-agonist (the ß-agonist binds to the ß-receptor which then stimulates adenylate cyclase via the G protein, G_s; fig. 5). The negative inotropic effect of ACh could be the result of this decrease in Ca^{2+} influx through the Ca^{2+} channels. The second of the two G proteins activated by the muscarinic receptor, G_K, activates the muscarinic K⁺ channel. The muscarinic K⁺ channel is not expressed in all parts of the heart: it is found in atrial muscle and the sino-atrial and atrio-ventricular nodes, but not in ventricular myocytes (ferret ventricular muscle being an exception). The muscarinic K⁺ channel is half-maximally activated by about 0.3 µM ACh, is a K⁺ selective channel and exhibits inward rectification. One characteristic feature of i_{K-ACh}

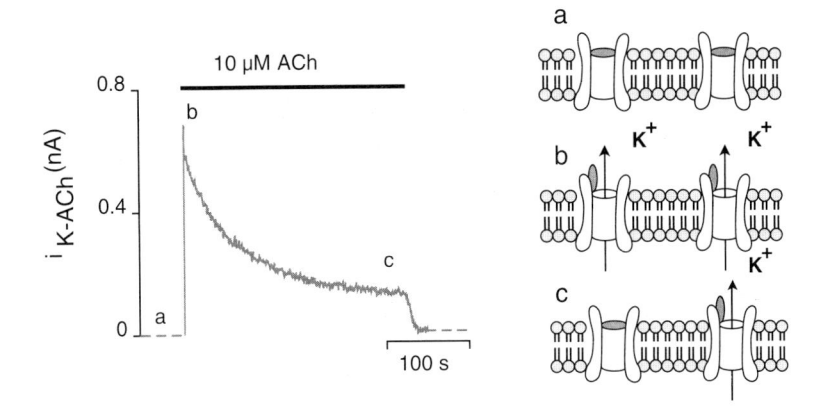

Fig. 6. Muscarinic K$^+$ current in a guinea-pig atrial cell. The left-hand panel shows the muscarinic K$^+$ current (an outward flow of K$^+$ through muscarinic K$^+$ channels) on application of ACh (holding potential, -53 mV). The right-hand panel shows a schematic diagram of two muscarinic K$^+$ channels at the times a, b, and c in the left-hand panel. In the absence of ACh, the muscarinic K$^+$ channels are closed (a). On application of ACh, the channels rapidly open, resulting in a large K$^+$ efflux (b). In the continuous presence of ACh, the muscarinic K$^+$ current gradually decays, because the muscarinic K$^+$ channels open less frequently and perhaps for less time (c). This phenomenon is known as desensitization. Modified from [4] with permission.

(the K$^+$ current through the muscarinic K$^+$ channel) is that in the continuous presence of ACh, it decays as a result of a marked time-dependent desensitization to ACh (fig. 6) [4]. The negative inotropic effect of ACh could also be the result of the activation of the muscarinic K$^+$ channel.

In atrial muscle and ferret ventricular muscle the negative inotropic effect of ACh is the result of the activation of the muscarinic K$^+$ channel

In atrial muscle and ferret ventricular muscle, exposure to ACh results in a shortening of the action potential. An example from dog atrial muscle is shown in fig. 7. The shortening is the result of the activation of K$^+$ current through muscarinic K$^+$ channels, although any decrease in basal i_{Ca} will also contribute. Another example (fig. 8A), from a ferret ventricular cell, shows the time course of the change in action potential duration. On addition of ACh there is an abrupt shortening of the action potential, but, in the continuous presence of ACh, action potential duration slowly recovers. This partial recovery in the presence of ACh can be explained by the decay of i_{K-ACh} as a result of desensitization (fig. 6). Fig. 8A also shows the changes in the contraction on addition of ACh. They are a mirror image of the changes in action potential duration: on addition of ACh there is a marked decrease in contraction, but then the contraction slowly recovers during the remainder of the exposure to ACh. Two experiments show that the changes in contraction in ferret ventricular cells are primarily the result of the changes in action potential duration. Fig. 8B shows that, when the contraction is triggered by voltage-clamp pulses of constant

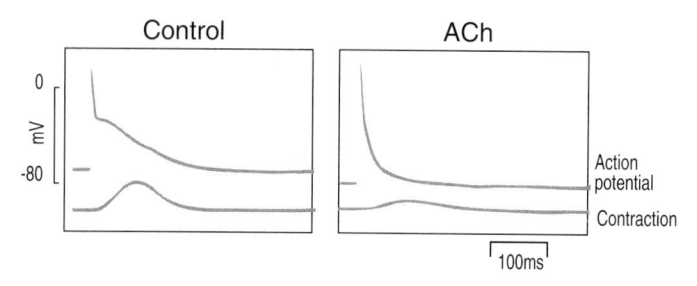

Fig. 7. Effect of 2 μM ACh on the action potential (top) and contraction (bottom) of a dog atrial fibre. Under control conditions the action potential has a characteristic low plateau, but after the application of ACh the action potential is reduced in duration. In the presence of ACh, the contraction is also smaller than under control conditions. The contractions are upward-going deflections. Taken from [5] with permission

duration, ACh has minimal effect on the contraction, the small decrease could be the result of a decrease in basal i_{Ca}. Fig. 8C shows that in the absence of ACh, when the duration of voltage-clamp pulses is altered to mimic the changes in action potential duration seen with ACh, changes in contraction are obtained similar to those observed with ACh. Similar results have been obtained by Ten Eick et al. [5] from atrial muscle. They concluded that with ACh concentrations sufficient to decrease contraction by 30 - 40%, the negative inotropic effect is exclusively the result of the activation of the muscarinic K$^+$ channel (and the consequent action potential shortening), whereas a decrease of basal i_{Ca} contributes to the inotropic effect at higher concentrations.

The muscarinic K$^+$ channel is only expressed in ventricular myocytes of the ferret. In the ventricular muscle of other species, ACh has no inotropic effect under basal conditions and only decreases the contraction after the tissue has first been stimulated by a ß-agonist. This inotropic effect is presumably exclusively the result of the inhibition of i_{Ca}.

Adenosine can also decrease contraction by activating the muscarinic K$^+$ channel

Adenosine, which is released by heart cells, is well known to have potent cardiovascular actions, including a negative inotropic effect in both atrial and ventricular muscle [6]. The mechanisms underlying the inotropic effect of adenosine are similar to those underlying the effect of ACh. In the heart, adenosine binds to the A1 receptor. On binding of an agonist, the A1 receptor, like the muscarinic receptor, inhibits adenylate cyclase and activates muscarinic K$^+$ channels via G proteins. In atrial but not ventricular muscle, adenosine produces a small decrease in basal i_{Ca}, which may or may not be the result of the inhibition of adenylate cyclase [7]. In all cardiac tissues, including ventricular muscle, after adenylate cyclase (and i_{Ca}) has been stimulated by a ß-agonist, adenosine produces a decrease in i_{Ca} by inhibiting adenylate cyclase [6].

In atrial muscle, adenosine produces a marked shortening of the action potential. Visentin et al. [7] attribute this to the activation of muscarinic K$^+$ current, rather than to the small decrease of basal i_{Ca}, because the shortening is well correlated with the increase of muscarinic K$^+$ current, but poorly correlated with the decrease of i_{Ca}. A decrease in i_{Ca}, simi-

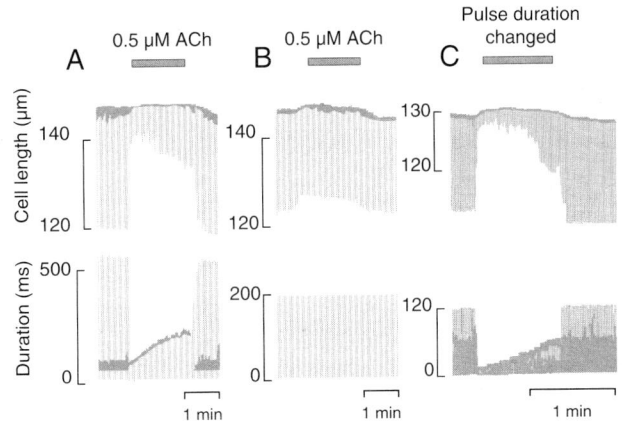

Fig. 8 A-C. The negative inotropic effect of ACh is the result of the ACh-induced shortening of the action potential. The contractions of ferret ventricular cells (**A** and **B** from the same cell, **C** from another cell) stimulated at 0.5 Hz are shown in the top traces and the duration of action potentials or voltage-clamp pulses are shown in the lower traces. **A** Effect of 0.5 µM ACh on contractions triggered by action potentials. On application of ACh, there is an abrupt decrease in action potential duration and contraction, but in the continued presence of ACh these effects decay as a result of desensitization. **B** Effect of 0.5 µM ACh on contractions triggered by voltage-clamp pulses of constant duration (as shown by the lower trace). On application of ACh, there is little decrease in the contraction as compared with the control in **A**. **C** Simulation of the effect of ACh on contraction. In the absence of ACh the voltage-clamp pulse duration (shown in the lower trace) was altered to simulate the changes in action potential duration in **A** on application of ACh. The resultant changes in the contraction are similar to the changes in the contraction in **A** on application of ACh. **B** and **C**: holding potential, -80 mV; pulse potential, 0 mV. The contractions are downward-going deflections. Modified from [49] with permission

lar to that caused by adenosine, brought about by lowering bathing $[Ca^{2+}]$ has relatively little effect on action potential duration as compared to that caused by adenosine. The shortening of the action potential is presumed to be responsible for the negative inotropic effect on atrial muscle, although this has not been proven, as it has in the case of ACh (fig. 8).

The negative inotropic effect of adenosine on ventricular muscle is only observed after the tissue has been stimulated by a ß-agonist and is presumably the result of the inhibition of adenylate cyclase and the subsequent decrease in i_{Ca} [6].

The role of transient outward current channels in modulating contraction

Transient outward current channels can be an important determinant of action potential duration

Fig. 9B shows the transient outward current of a rat ventricular cell during a 500 ms voltage-clamp pulse from -80 to +20 mV. On depolarization the outward current rapidly activates and then it inactivates (decays) over several hundred ms. The transient outward current is

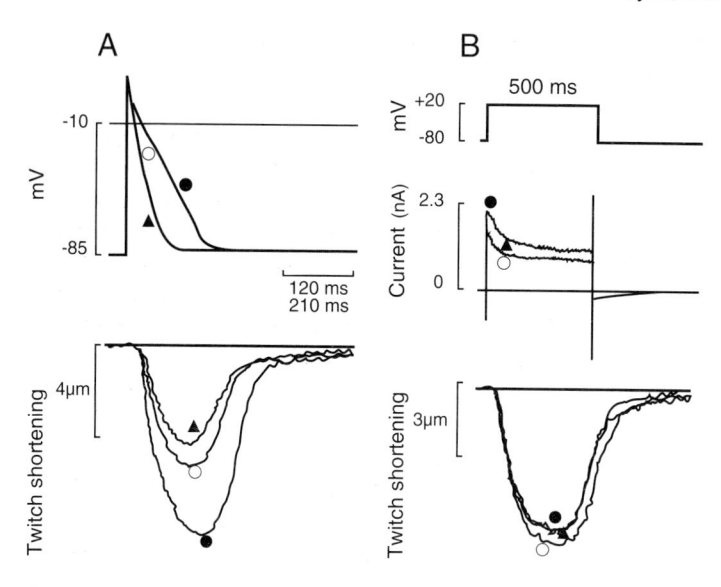

Fig. 9 A, B. The positive inotropic effect of α-adrenergic stimulation is the result of an inhibition of transient outward current channels and the consequent prolongation of the action potential. **A** Effect of 100 μM methoxamine on contractions triggered by action potentials. Action potentials (top) and contractions (bottom) of a rat ventricular cell under control conditions *(filled triangle)*, in the presence of 100 μM methoxamine *(filled circle)* and after washout *(open circle)* are shown. Methoxamine causes a prolongation of the action potential and an increase in the contraction. **B** Effect of 100 μM methoxamine on contractions triggered by voltage-clamp pulses of constant duration. The voltage-clamp pulse *(top)*, transient outward currents *(middle)* and contractions *(bottom)* under control conditions *(filled triangle)*, in the presence of 100 μM methoxamine *(filled circle)* and after washout *(open circle)* are shown. Methoxamine causes decreases in the transient outward current and contraction. Modified from [9] with permission

principally carried by K^+, shows voltage-dependent activation and inactivation analogous to the voltage-dependent activation and inactivation of the Na^+ current and is blocked completely by 4-aminopyridine. This current is present in atrial tissue, including human atrial tissue, Purkinje fibres and ventricular muscle. In most cases, block of the transient outward current by 4-aminopyridine results in a substantial prolongation of the action potential, demonstrating that this current is an important determinant of action potential duration.

The positive inotropic effect of α-adrenergic agonists is the result of the inhibition of transient outward current channels

In mammalian atrial and ventricular muscle, including that of the human, α-receptor stimulation produces a positive inotropic effect. The mechanisms responsible for the inotropic effect are controversial. Endoh and Blinks [8] suggested that the positive inotropic effect in rabbit ventricular muscle, in part at least, is the result of an increase in the Ca^{2+} sensitivity of the myofilaments. However, Fedida and Bouchard [9] found no evidence of an increase in Ca^{2+} sensitivity in rat ventricular cells and suggested that the positive inotropic effect must be the result of another mechanism.

The positive inotropic effect of α-stimulation is accompanied by an increase in action potential duration. An example is shown in fig. 9A. In this case, on addition of 100 μM methoxamine (an α-agonist) there is a substantial prolongation of the action potential and increase of the contraction. α-stimulation has no effect on i_{Ca} and the prolongation of the action potential is the result of an inhibition of the transient outward current. An example of the decrease in the transient outward current caused by 100 μM methoxamine in another rat ventricular cell is shown in fig. 9B. The pathway by which α-stimulation results in a decrease in the transient outward current is not fully understood, but involves a pertussis toxin insensitive G protein and a soluble second messenger [10]. Two experiments carried out by Fedida and Bouchard [9] demonstrate that the positive inotropic effect of α-stimulation is the result of the prolongation of the action potential. Firstly, they blocked the transient outward current by 4-aminopyridine; as expected this results in a similar prolongation of the action potential and increase in contraction to those observed on application of an α-agonist. Furthermore, when an α-agonist is then applied in the presence of 4-aminopyridine there is no further increase of action potential duration or contraction (as expected, because the transient outward current is already completely inhibited). Secondly, as shown in fig. 9B, when the contraction is triggered by voltage-clamp pulses of constant duration, on addition of the α-agonist there is a decrease, rather than an increase, in the contraction.

The role of ATP-sensitive K⁺ channels in modulating contraction

Intracellular levels of ATP in heart muscle modulate another class of K⁺ channel (the ATP-sensitive K⁺ channel). Many of the early studies in this area concentrated on the electrophysiology of cardiac muscle during metabolic inhibition. Metabolic inhibition can be induced in isolated cardiac preparations by blocking both anaerobic and aerobic glycolysis with 2-deoxyglucose and cyanide (CN^-), respectively. Metabolic blockade is of interest because it simulates the effects of hypoxia and ischaemia during which both aerobic and anaerobic glycolysis can be compromised. Early studies [11,12] revealed that during metabolic inhibition, the cardiac action potential shortens and this is accompanied by an increase in outward K⁺ current. ATP-sensitive K⁺ channels were first found in guinea-pig cardiac muscle [13]. These channels are closed as long as intracellular ATP, $[ATP]_i$, remains high. However, during interventions that decrease $[ATP]_i$ (e.g. metabolic inhibition, hypoxia or ischaemia), the blockade of these channels by ATP is reduced, K⁺ ions leave the cell and, as discussed earlier in this chapter, induce a shortening of the action potential and a consequent decrease in the strength of contraction. This section will deal with the influence of metabolic blockade and hypoxia on the strength of contraction and will summarise the evidence which suggests that ATP-sensitive K⁺ channels are important in controlling contractility under these conditions.

Metabolic inhibition leads to a reduction in action potential duration and contraction

Fig.10A illustrates the effect of metabolic inhibition on the action potential and contraction of an isolated guinea-pig ventricular cell. Addition of 2 mM CN^- (in the presence

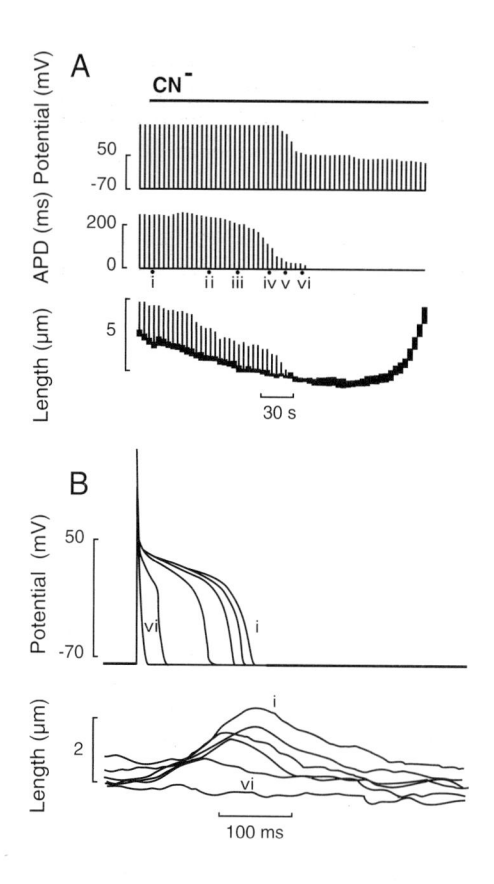

Fig. 10. A Action potentials (top), action potential duration (at -60 mV, middle) and contractions (bottom) of a guinea-pig ventricular cell during complete metabolic blockade. 2 mM CN⁻ was applied during the period represented by the bar. **B** Fast time base recordings of action potentials (top) and contractions (bottom) taken at the points i to vi marked in panel **A**. Exposure to 2 mM CN⁻ in the presence of 2-deoxyglucose led to a reduction of both action potential duration and contraction. Each action potential is preceded by a large stimulus artefact. From point vi onwards the action potential is abolished and only the stimulus artefact is seen. Taken from [26] with permission of the American Heart Association Inc

of 10 mM 2-deoxyglucose) leads to a progressive reduction in the duration of the action potential which fails completely 2 min after the addition of CN⁻. Associated with this decrease in action potential duration is a similar progressive decline in the strength of the contraction which fails when the action potential fails. Later the cell goes into a maintained contracture presumably as a result of the depletion of $[ATP]_i$ and the development of rigor.

Reduction of action potential duration during metabolic inhibition is associated with increased K^+ conductance

Lederer et al. [14] carried out experiments on rat ventricular cells to assess whether the shortening of the action potential is due to an increase in outward K^+ current or a reduction of the L-type Ca^{2+} current, both of which could induce a reduction of action potential duration and therefore, contraction. Under voltage-clamp, induction of metabolic inhibition (as above) led to a very large increase in outward current which reached its maximum before the cell went into contracture. The measured current reversed between -75 and -80 mV (near E_K) which suggests it was carried mainly by K^+. As explained below,

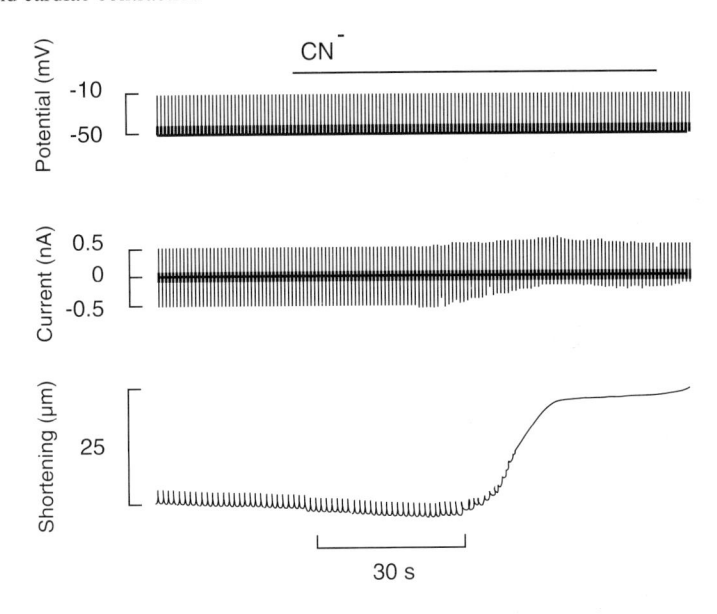

Fig. 11. Membrane potential (top), membrane current and contractions (bottom) of a voltage-clamped rat ventricular cell during metabolic inhibition. The contractions were triggered by 100 ms voltage-clamp pulses from -50 mV to -10 mV at 1 Hz. 2 mM CN^- was applied during the period represented by the bar. K^+ channel blockers were present in this experiment (see text for details). Under these conditions the contraction does not fail until the development of the contracture. Taken from [14] with permission

there is relatively little reduction of i_{Ca} during metabolic inhibition and, therefore, the action potential shortening under these conditions must be largely the result of the activation of the outward K^+ current (presumably the ATP-sensitive K^+ current).

The twitch does not fail under voltage clamp during metabolic inhibition

Lederer et al. [14] demonstrated that if the contractions of rat ventricular cells are triggered by voltage clamp pulses of constant duration during metabolic inhibition, there is only a small decline in the size of the contraction (see fig. 11), although the contraction does eventually fail once the contracture develops (compare fig. 10 with fig. 11). In this experiment K^+ channels were blocked by the addition of 2 mM tolbutamide, a blocker of ATP-sensitive K^+ channels, to the bathing and pipette solution, the addition of 1 mM tetrabutylammonium to the pipette solution and the replacement of K^+ in the pipette by N-methyl glucosamine to allow i_{Ca} to be recorded. Fig. 11 illustrates that there is a relatively small but progressive decrease in inward current during metabolic inhibition. The small decline in the contraction under these conditions could be the result of the gradual decline in i_{Ca} [14]. This experiment suggests that when the contraction is triggered by action potentials (fig. 10), the rapid failure of the contraction during metabolic inhibition is the result of the reduction in action potential duration.

An increase in [ATP]$_i$ during metabolic inhibition can restore the action potential and the contraction

In an elegant experiment, Lederer et al. [14] illustrated the importance of [ATP]$_i$ in the phenomenon of contractile failure during metabolic inhibition. The premise was that if the failure of the action potential (and the consequent failure of the twitch) resulted from activation of ATP-sensitive K$^+$ channels, it should be possible to reverse these effects by increasing [ATP]$_i$ during metabolic inhibition. To test this hypothesis, a rat ventricular cell was stimulated in the presence of 10 mM 2-deoxyglucose and 2 mM CN$^-$. The contraction failed in the characteristic manner (see fig. 10). A patch pipette, filled with 10 mM ATP was then brought up to the cell and a gigaseal formed. The patch of membrane under the electrode was then broken allowing the ATP access to the cell interior. Within 30 s, contractions in response to stimulation were observed. As expected, the action potential was also partially restored. In a similar experiment, Taniguchi et al. [15] reported a reduction of outward current (presumably K$^+$ current through ATP-sensitive K$^+$ channels) upon microinjection of ATP into a metabolically inhibited cell.

Glibenclamide reduces the shortening of the action potential during no-flow ischaemia

Recently, it has been reported that sulphonylurea antidiabetic agents such as tolbutamide and glibenclamide (glyburide) can inhibit ATP-sensitive K$^+$ channels in cardiac muscle (e.g. [16]). As a result, these agents would be expected to reduce the extent of the shortening of the action potential during ischaemia. Cole et al. [17] investigated the influence of glibenclamide on the electrical and mechanical activity of the guinea-pig perfused right ventricular wall during no-flow ischaemia. Fig. 12A illustrates the influence of 20 min of ischaemia followed by 60 min of reperfusion on the action potential and the contraction of this preparation. Ischaemia results in the expected shortening of the action potential and decrease in contraction. Fig. 12B illustrates that pretreatment of the preparation with 10 µM glibenclamide reduces the shortening of the action potential during the 20 min exposure to ischaemia. This is consistent with ATP-sensitive K$^+$ channels being responsible for or contributing to the action potential shortening during ischaemia. However, fig. 12B illustrates that glibenclamide does not prevent the failure of the contraction and the onset of contracture during ischaemia. Furthermore, in the glibenclamide-treated preparation, although the action potential recovers fully upon reperfusion, the contraction does not (the contraction recovers more fully in the absence of glibenclamide; see fig. 12A). In the presence of glibenclamide, because the action potential does not shorten during ischaemia, Ca^{2+} influx into the cells remains high during ischaemia. Because the cells are ischaemic, they cannot tolerate this relatively high Ca^{2+} influx and they become Ca^{2+} overloaded. Ca^{2+} overload is known to lead to cell damage and this probably explains the poor recovery of contraction in the glibenclamide treated preparation. Therefore, the action potential shortening during ischaemia, in the absence of glibenclamide, is a protective mechanism for the myocardium. Whether agents such as glibenclamide should be used clinically is, therefore, in some doubt. While these drugs reduce the incidence of ischaemic arrhythmias by maintaining the action potential [17] the degree of reperfusion damage following ischaemia is more pronounced if these drugs are used.

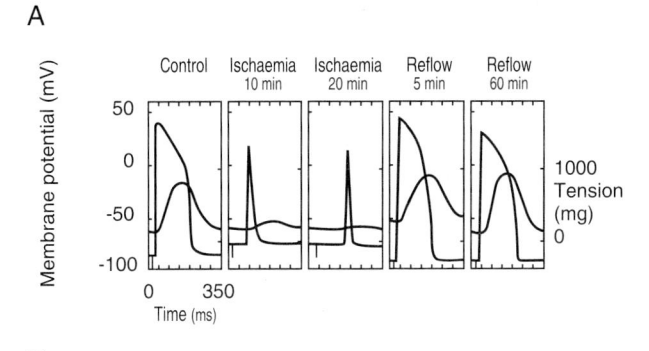

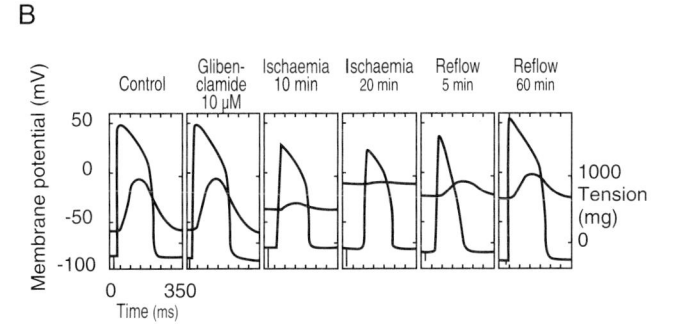

Fig. 12 A, B. Action potentials and contractions of a guinea-pig right ventricular wall preparation during a 20 min exposure to ischaemia followed by 60 min reperfusion (reflow). The preparation was made ischaemic by stopping the perfusion. **A** Control period of ischaemia and reperfusion. **B** Action potentials and contractions during ischaemia and reperfusion in the presence of 10 μM glibenclamide. Small ticks on time base represent 50 ms intervals. Adapted from [17] with permission of the American Heart Association Inc

Agents that open ATP-sensitive K+ channels mimic the effects of metabolic inhibition

The addition of 100 μM pinacidil (in the presence of 0.3 mM ATP) to an open cell attached patch preparation from guinea-pig ventricular muscle leads to an increase in the open probability (P_o) of the ATP-sensitive K+ channels [16]. Pinacidil and other agents which open ATP-sensitive K+ channels, such as nicorandil and cromakalim, should mimic the effects of metabolic inhibition or ischaemia. Fig. 13 illustrates that the addition of 100 μM pinacidil to a normally perfused guinea-pig papillary muscle markedly shortens the action potential and abolishes the contraction. This inhibition of the action potential and contraction can be reversed by the addition of tolbutamide or glibenclamide (both inhibitors of ATP-sensitive K+ channels).

Fig. 14 illustrates that pinacidil (10 μM) reduces action potential duration and the strength of contraction in a perfused right ventricular wall preparation from the guinea-pig. When the preparation is made ischaemic by stopping perfusion, action potential duration shortens significantly faster in the presence of pinacidil compared with control ischaemia (see fig. 12A). However, what is of note is that upon reperfusion in preparations pretreated with pinacidil, both the action potential and the contraction are fully restored. This reinforces the point that the shortening of the action potential during ischaemia is important in reducing reperfusion damage.

Ripoll et al. [18] reported that agents such as cromakalim which open ATP-sensitive

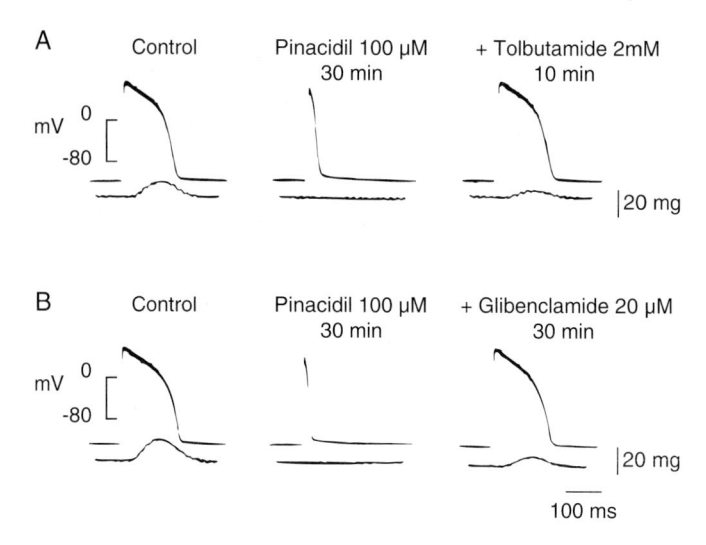

Fig. 13 A, B. Effect of tolbutamide (2 mM) and glibenclamide (20 µM) on the pinacidil (100 µM) induced shortening of the action potential and decrease of contraction in guinea-pig papillary muscle. **A** Effect of tolbutamide. **B** Effect of glibenclamide. Each panel illustrates superimposed records of the action potential (above) and contraction (below). Taken from [16] with permission

K$^+$ channels, do so by increasing the intracellular ATP concentration necessary to inhibit the activity of the channels. For example, 40 µM cromakalim increases the [ATP]$_i$ required for half maximal inhibition of the channels (K$_i$) from 80 to 150 µM ATP. One important question that now arises is that if the [ATP]$_i$ needs to fall to approximately 100 µM to open 50% of the ATP-sensitive K$^+$ channels, does [ATP]$_i$ fall sufficiently during metabolic inhibition or ischaemia to induce significant ATP-sensitive K$^+$ channel opening?

Metabolic inhibition leads to only a modest reduction in intracellular ATP

During metabolic inhibition, the [ATP]$_i$ level has been reported to fall by as much as 50% over a 10 min period [19], although other reports suggest more modest falls than this, e.g. 20% [20]; 25% [21]. Under normal conditions [ATP]$_i$ is approximately 5 mM and therefore severe ischaemia or complete metabolic inhibition is expected only to reduce [ATP]$_i$ to 2.5 mM over the first 10 min, the period of time when the action potential can shorten very dramatically. However, studies on the activity of ATP-sensitive K$^+$ channels in membrane patches have suggested that ATP needs to fall as low as 1 mM (K$_i$ is ~100 µM and see above) to give any significant change in the open probability of the ATP-sensitive K$^+$ channels [22]. How can these facts be reconciled? Either other cellular factors, for example [H$^+$]$_i$, [ADP]$_i$, [P]$_i$, or the ADP/ATP ratio, which change during metabolic blockade, are able to open ATP-sensitive K$^+$ channels or the opening of only a small number of the ATP-sensitive K$^+$ channels (or a small increase in the P$_o$) is required to give the dramatic changes in the duration of the action potential seen during metabolic inhibition.

Fig. 14. Action potentials and contractions of a guinea-pig right ventricular wall preparation during a 30 min exposure to ischaemia followed by 60 min reperfusion (reflow). The preparation was pretreated with 10 µM pinacidil (for 15 min) before ischaemia was induced by stopping the perfusion. Pinacidil was not present during reperfusion. Small ticks on time base represent 50 ms intervals. Taken from [17] with permission of the American Heart Association Inc

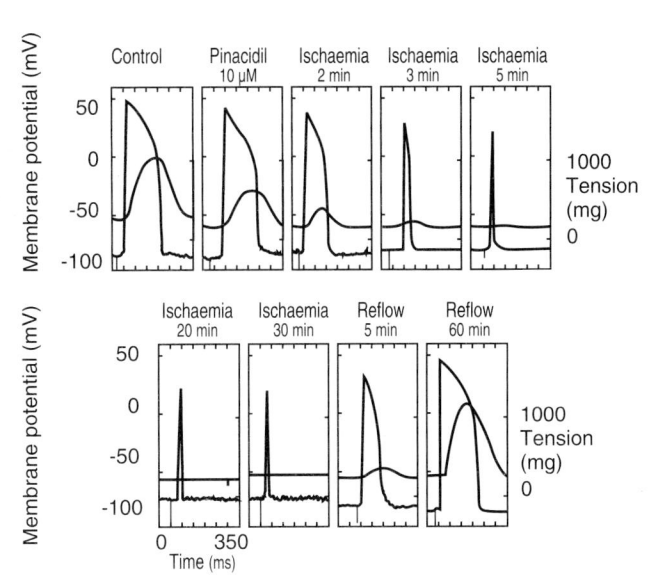

Only a small number of ATP-sensitive K⁺ channels need to open to significantly shorten the action potential

Recent studies in which ATP-sensitive K^+ channel activity was measured in membrane patches from rat ventricular cells suggest that these channels show a range of sensitivities to $[ATP]_i$ [23]. Fig. 15 illustrates that at zero ATP, the majority of the channels in the patch are open (P_o = 0.82). These channels have a slope conductance of approximately 25-35 pS [18, 24, 25] in a physiological extracellular K^+ concentration (5.4 mM). As $[ATP]_i$ is increased, channel activity decreases. However, even at millimolar concentrations of ATP, some channels still display openings. The mean K_i in these experiments of 37 µM ATP (range between 9 and 580 µM) is close to the values found by other experimenters (e.g. [13, 18, 25]). Findlay and Faivre [23] suggest that only a very small proportion (1%) of the ATP-sensitive K^+ channels need to be activated to significantly shorten the action potential because the channels have a large conductance. Furthermore, because of the range of sensitivities to $[ATP]_i$, activation of 1% of the ATP-sensitive K^+ channel conductance is expected to occur when $[ATP]_i$ falls from about 5 to 2.5 mM, as is known to happen during ischaemia.

This conclusion is supported by Nichols et al. [26]. They mimicked the opening of ATP-sensitive K^+ channels in a normal guinea-pig ventricular cell by injecting hyperpolarizing current into the cell. As progressively more current was injected into the cell, action potential duration and contraction fell together in a similar fashion to that observed experimentally during metabolic inhibition. Nichols et al. [26] also measured the maximal ATP-sensitive K^+ channel conductance in a single cell to be 195 nS. They then calculated that only 1.3 nS (0.7% of this conductance) has to be activated to shorten the action potential to 50% of control. This level of ATP-sensitive K^+ channel activity would occur at an $[ATP]_i$ of 1.4 mM. The action potential would be expected to shorten by 10% if $[ATP]_i$ fell to 3.1 mM. These calculations were based on a K_i of 114 µM [26].

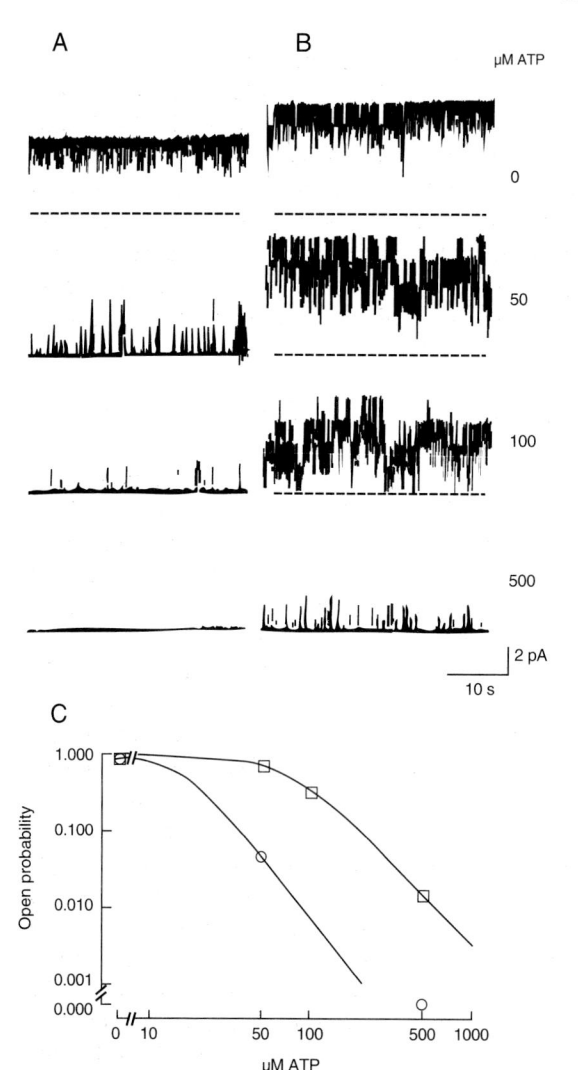

Fig. 15 A-C. Influence of the intracellular [ATP] on ATP sensitive K$^+$ channels. **A, B** Single ATP sensitive K$^+$ channel currents. In **A** the patch contained 3 channels and, in **B** the patch contained 5 channels. As the intracellular [ATP] increases (see concentrations to right of figure), channel activity decreases. Dotted lines indicate zero current level. **C** Log-log plot of open probability and [ATP] showing the dose-dependence of channel activity (circles represent the patch in **A**, squares the patch in **B**). Taken from [23] with permission

Therefore, the results of both Findlay and Faivre [23] and Nichols et al. [26] show that even a slight fall in [ATP]$_i$ would have a dramatic effect on action potential duration and contraction.

The role of Na$^+$-activated K$^+$ channels in modulating contraction

A high concentration of Na$^+$ is required to open Na$^+$-activated K$^+$ channels

Kameyama et al. [27] first reported the presence of Na$^+$-activated K$^+$ channels in mammalian cardiac muscle. This channel, which is not gated by intracellular Ca^{2+}, has a very large unitary conductance of 207 pS (conditions: $[K^+]_o$ 150 mM, $[K^+]_i$ 49 mM), approximately 20 times that of ATP-sensitive K^+ channels. They showed that the channel is activated when intracellular Na^+ rises to levels above 20-30 mM with a half activation constant of 66 mM Na^+. More recently, the channel conductance has been investigated under more physiological K^+ concentrations [28]. When intracellular $[K^+]$ exceeds extracellular $[K^+]$ the current displays outward rectification and reverses close to the equilibrium potential for K^+. Under these conditions the single channel conductance is 132 pS. To achieve a high P_o (0.8), $[Na^+]_i$ has to be above 70 mM. As $[Na^+]_i$ is reduced, P_o decreases though in some cases channel activity is still observed at $[Na^+]_i$ of 30 mM and below [28].

Are Na^+-activated K^+ channels important in cardiac physiology?

There is controversy about the role played by this channel in normal cardiac physiology. Under physiological conditions cytoplasmic $[Na^+]_i$ is closely regulated between 5 and 10 mM (e.g. [29]) and therefore Na^+-activated K^+ channels are likely to remain closed. However, a recent concept concerning the cellular distribution of ions suggests that the role of these (and most certainly other) channels may indeed be more important than first thought. This concept recently termed "the fuzzy space" (e.g. [30]) suggests that intracellular ions (and also compounds such as ATP) may be unevenly distributed within the cell. For example, the concentrations of ions such as Na^+ and Ca^{2+} may be higher closer to the inside face of the sarcolemma (due to the transmembrane fluxes of these ions) where the regulatory sites of transmembrane proteins, such as ion channels, are located. As a result, conclusions concerning the physiological role of channels based on the level of regulatory ions/compounds measured in the bulk cytoplasm may be spurious.

Furthermore, it has been suggested that under pathological conditions, for example during either severe ischaemia or blockade of the Na/K pump (the main mechanism by which $[Na^+]_i$ is regulated), $[Na^+]_i$ can rise sufficiently at the inner face of the membrane to induce a proportion of the Na^+-activated K^+ channels to open. It should be noted here that the argument put forward to explain the effect of ATP-sensitive K^+ channels on the duration of the action potential holds in this case too, i.e. as the Na^+-activated K^+ channels have such a large conductance (many times that of ATP-sensitive K^+ channels) only a very small change in the P_o of these channels is required to markedly alter the duration of the action potential.

Activation of Na^+-activated K^+ channels is thought to reduce action potential duration during exposure to cardiac glycosides

Upon exposure of ventricular myocytes to cardiac glycosides the duration of the action potential initially increases due to the inhibition of the outward Na/K pump current. However, during continued exposure, action potential duration progressively shortens to below control levels [31]. This is associated with a marked increase in $[Na^+]_i$. Activation of Na^+-activated K^+ current under these conditions may contribute to this phenomenon [32].

Activation of this current has also been implicated in the outward shift of current observed during exposure to Ca^{2+} and Mg^{2+}-free solutions [33] and was identified as such due to its dependence upon [Na⁺]ᵢ, its sensitivity to caesium and TEA and the reversal of the current at -92 mV.

Class III antiarrhythmic agents and contraction

Class III antiarrhythmic agents are discussed in detail in chapter 11. In this section the effect of class III agents on contraction is discussed. In order to discuss their actions specific to contraction, it is necessary to re-iterate some details of their general mode of action.

Class III antiarrhythmics prolong the action potential

The definition of a class III antiarrhythmic agent is one which prolongs the action potential. It can therefore be predicted that these agents should have a positive inotropic effect. The combination of an antiarrhythmic action with positive inotropy could be clinically useful as hearts in need of antiarrhythmic therapy are often also failing.

Compounds such as DPI 201-106 prolong the action potential by enhancing the inward, depolarizing Na⁺ current [34] and thus fall outside the scope of this chapter. However, clinical agents such as sotalol and amiodarone prolong the action potential by reducing outward K⁺ currents, principally the delayed rectifier (i_K) [35, 36]. The fact that positive inotropy is not universally associated with these agents probably results from their secondary modes of action. Sotalol and Amiodarone will be discussed as examples.

Sotalol and amiodarone can produce positive and negative inotropy

As well as prolonging the action potential, sotalol has ß-adrenergic receptor blocking properties which are more pronounced with the l-isomer than the d-isomer form of the drug. Thus any positive inotropic effect resulting from a prolongation of the action potential will be countered by a reduction in the ß-stimulated i_{Ca}. The interaction of these mechanisms was demonstrated by Tande and Refsum [37]. Using threshold stimulation of rat atrial muscle they found a positive inotropic response to d-sotalol (7.5 x 10⁻⁵ M) but not to the l-isomer. The positive inotropic response was presumably due to the prolongation of the action potential. When suprathreshold stimuli were applied in the absence of sotalol there was an increase in force thought to result from the stimulation of ß-receptors on the surface of the muscle cells by noradrenaline released from nerve terminals in the muscle. Under these conditions both isomers caused a negative inotropic response, the greater being caused by l-sotalol. The negative inotropic response was probably the result of the inhibition of ß-stimulated i_{Ca}.

Although known as a class III antiarrhythmic agent, amiodarone can claim membership to all four classes of anti-arrhythmic drugs. It reduces Na⁺ channel activity (Class I [38]), acts as a non-competitive ß-blocker (Class II [39]), prolongs the action potential by reducing i_K (class III [36]) and inhibits i_{Ca} (Class IV [40]). Class I, II and IV actions are all negatively inotropic. Amiodarone has been shown to have a transient positive inotropic effect on rat atrial muscle at 10⁻⁴ M [41]. This effect is perhaps a result of prolongation of the action potential which is then countered by a decrease in i_{Ca}.

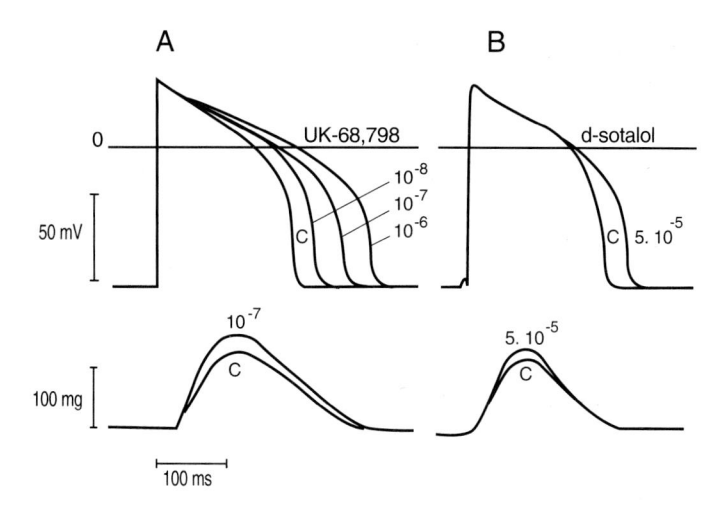

Fig 16 A, B. Comparison of the effects of (**A**) UK-68,798 (10^{-8}, 10^{-7} and 10^{-6} M) and (**B**) d-sotalol (5×10^{-5} M) on the action potential shape (upper traces) and contraction (lower traces) of guinea pig papillary muscle. Control records indicated by c. The prolongation of the action potential was greater for UK-68,798. Taken from [43] with permission

New class III agents have increased selectivity for K⁺ channels

Because existing class III agents do not invariably cause a positive inotropic effect and because the combination of anti-arrhythmic and positive inotropic properties are potentially useful, there has been a search for new class III agents which can prolong the action potential in the absence of secondary effects. Several agents have been produced that reduce i_K (and thus prolong the action potential) at concentrations several orders of magnitude less than sotalol and which do not have associated secondary actions; for example: E-4031 [42], UK-68,798 [43], and Compound II [44]. All these agents produce positive inotropic responses. A comparison of one of these agents (UK-68,798) and sotalol is shown in fig. 16. The positive inotropic effects are presumably the result of the changes in the shape of the action potential.

References

1. Morad M, Goldman Y (1973) Excitation-contraction coupling in heart muscle: Membrane control of development of tension. Prog Biophys Mol Biol 27: 257-313
2. Terrar DA, White E (1989) Mechanism of potentiation of contraction by depolarization during action potentials in guinea-pig ventricular muscle. Q J Exp Physiol 74: 355-358
3. Bers DM (1991) Excitation-contraction coupling and cardiac contractile force. Kluwer Academic Publishers, Dordrecht, The Netherlands
4. Zang WJ, Yu XJ, Honjo H, Kirby MS, Boyett MR (1993) On the role of G protein activation and phosphorylation in desensitization to acetylcholine in guinea-pig atrial cells. J Physiol (Lond) 464: 649-679

5. Ten Eick R, Nawrath H, McDonald TF, Trautwein W (1976) On the mechanism of the negative inotropic effect of acetylcholine. Pflügers Arch 361: 207-213
6. Belardinelli L, Linden J, Berne RM (1989) The cardiac effects of adenosine. Prog Cardiovascular Diseases 32: 73-97
7. Visentin S, Wu S-N, Belardinelli L (1990) Adenosine-induced changes in atrial action potential: contribution of Ca^{2+} and K^+ currents. Am J Physiol 258: H1070-H1078
8. Endoh M, Blinks JR (1988) Actions of sympathomimetic amines on the Ca^{2+} transients and contractions of rabbit myocardium: reciprocal changes in myofibrillar responsiveness to Ca^{2+} mediated through α- and ß-adrenoceptors. Circ Res 62: 247-265
9. Fedida D, Bouchard RA (1992) Mechanisms for the positive inotropic effect of $α_1$-adrenoceptor stimulation in rat cardiac myocytes. Circ Res 71: 673-688
10. Braun AP, Fedida D, Clark RB, Giles WR (1990) Intracellular mechanisms for $α_1$-adrenergic regulation of the transient outward current in rabbit atrial myocytes. J Physiol (Lond) 431: 689-712
11. Carmeliet E (1978) Cardiac transmembrane potentials and metabolism. Circ Res 42: 577-587
12. Vleugels A, Vereecke J, Carmeliet E (1980) Ionic currents during hypoxia in voltage-clamped cat ventricular muscle. Circ Res 47: 501-508
13. Noma A (1983) ATP regulated K^+ channels in cardiac muscle. Nature 305: 147-148
14. Lederer WJ, Nichols CG, Smith GL (1989) The mechanism of early contractile failure of isolated rat ventricular myocytes subjected to complete metabolic inhibition. J Physiol 413: 329-349
15. Taniguchi J, Noma A, Irisawa H (1983) Modification of the cardiac action potential by intracellular injection of adenosine triphosphate and related substances in guinea-pig ventricular cells. Circ Res 53: 131-139
16. Nakaya H, Takeda Y, Tohse N, Kanno M (1991) Effects of ATP-sensitive K^+ channel blockers on the action potential shortening in hypoxic and ischaemic myocardium. Br J Pharmacol 103: 1019-1026
17. Cole WC, McPherson CD, Sontag D (1991) ATP-regulated K^+ channels protect the myocardium against ischaemia/reperfusion damage. Circ Res 69: 571-581
18. Ripoll C, Lederer WJ, Nichols CG (1990) Modulation of ATP-sensitive K^+ channel activity and contractile behaviour in mammalian ventricle by the potassium channel openers cromakalim and RP49356. J Pharmacol Exp Ther 255: 429-435
19. Allen DG, Morris PG, Orchard CH, Pirolo JS (1985) A nuclear magnetic resonance study of metabolism in the ferret heart during hypoxia and inhibition of glycolysis. J Physiol (Lond) 361: 185-204
20. Elliot AC, Smith GL, Allen DG (1989) Simultaneous measurement of action potential duration and intracellular ATP in isolated ferret hearts exposed to cyanide. Circ Res 64: 583-591
21. Deutsch N, Klitzner TS, Lamp ST, Weiss JN (1991) Activation of cardiac ATP-sensitive K^+ current during hypoxia: e. c. Correlation with tissue ATP levels. Am J Physiol 261: H671-H676
22. Noma A, Shibasaki T (1985) Membrane current through adenosine-triphosphate-regulated potassium channels in guinea-pig ventricular cells. J Physiol (Lond) 363: 463-480
23. Findlay I, Faivre J-F (1991) ATP-sensitive K^+ channels in heart muscle: Spare channels. FEBS Lett 279: 95-97
24. Kakei M, Noma A, Shibasaki T (1985) Properties of adenosine-triphosphate-regulated potassium channels in guinea-pig ventricular cells. J Physiol (Lond) 363: 441-462
25. Nichols CG, Lederer WJ (1990) The regulation of ATP-sensitive K^+ channel activity in intact and permeabilized rat ventricular cells. J Physiol (Lond) 423: 91-110
26. Nichols CG, Ripoll C, Lederer WJ (1991) ATP-sensitive potassium channel modulation of the guinea-pig ventricular action potential and contraction. Circ Res 68: 280-287
27. Kameyama M, Kakei M, Sato R, Shibasaki T, Matsuda H, Irisawa H (1984) Intracellular Na^+ activates a K^+ channel in mammalian cardiac cells. Nature 309: 354-356

28. Luk H-N, Carmeliet E (1990) Na⁺-activated K⁺ current in cardiac cells: e. c. Rectification, open probability, block and role in digitalis toxicity. Pflügers Arch 416: 766-768
29. Harrison SM, McCall E, Boyett MR (1992) The relationship between contraction and intracellular sodium in rat and guinea-pig ventricular myocytes. J Physiol (Lond) 449: 517-550
30. Carmeliet E (1992) A fuzzy subsarcolemmal space for intracellular Na⁺ in cardiac cells? Cardiovasc Res 26: 433-442
31. Levi AJ (1991) The effect of strophanthidin on action potential, calcium current and contraction in isolated guinea-pig ventricular myocytes. J Physiol (Lond) 443: 1-24
32. Rodrigo GC (1990) The relationship between intracellular sodium activity (a^i_{Na}) and action potential duration in guinea-pig ventricular myocytes exposed to strophanthidin. J Physiol (Lond) 423: 61P
33. Rodrigo GC, Chapman RA (1990) A sodium-activated potassium current in intact ventricular myocytes isolated from the guinea-pig heart. Exp Physiol 75: 839-842
34. Kohlhardt M, Frobe U, Herzig W (1987) Removal of inactivation and blockade of cardiac Na channels by DPI 201-106: different voltage-dependencies of the drug action. Naun Schmeid Archiv Pharmacol 335: 183-188
35. Carmeliet E (1985) Electrophysiologic and voltage clamp analysis of the effects of sotalol on isolated cardiac muscle and Purkinje fibres. J Pharmacol Exp Ther 232: 817-825
36. Balser JR, Bennett PB, Hondeghem LM, Roden DM (1991) Suppression of time-dependent outward current in guinea-pig ventricular myocytes, action of quinidine and amiodarone. Circ Res 69: 519-529
37. Tande PM, Refsum H (1988) Class III antiarrhythmic action linked with positive inotropy: effects of the d- and l-isomer of sotalol on isolated rat atria at threshold and suprathreshold stimulation. Pharmacol Toxicol 62: 272-277
38. Follmer CH, Aomine M, Yeh JZ, Singer DH (1987) Amiodarone-induced block of sodium current in isolated cardiac cells. J Pharmacol Exp Ther 243: 187-194
39. Polster P, Broeckhuysen J (1976) The adrenergic antagonism of amiodarone. Biochem Pharmacol 25: 133-140
40. Nashimura M, Follmer CH, Singer DH (1989) Amiodarone blocks calcium current in single guinea-pig ventricular myocytes. J Pharmacol Exp Ther 251: 650-659
41. Tande PM, Refsum H (1990) Class III antiarrhythmic action linked with positive inotropy: acute electrophysiological and inotropic effects of amiodarone in vitro. Pharmacol Toxicol 66: 18-22
42. Wettner E, Scholtysik G, Schaad A, Himmel H, Ravens U (1991) Effects of the new class III antiarrhythmic drug E-4031 on the myocardial contractility and electrophysiological parameters. J Cardiovasc Pharmacol 17: 480-487
43. Tande PM, Bjornstad H, Refsum H (1990) Rate-dependent class III antiarrhythmic action, negative chronotropy and positive inotropy of a novel i_K blocking drug, UK-68,798: potent in guinea-pig but no effect in rat myocardium. J Cardiovasc Pharmacol 16: 401-410
44. Conners SP, Gill EW, Terrar DA (1992) Action and mechanism of action of novel analogues of sotalol on guinea-pig and rabbit ventricular cells. Br J Pharmacol (in press)
45. Antoni RJ, Kaufmann R (1963) Mechanishe Reaktionen des Frosch und Saugetiermyokards bei Veranderung der Aktionspotential-Dauer durch konstante Gleichstromimpulse. Pflügers Arch 306: 33-57
46. Terrar DA, White E (1989) Mechanism and significance of calcium entry at positive membrane potentials in guinea-pig ventricular muscle cells. Q J Exp Physiol 74: 121-139
47. White E, Terrar DA (1991) Action potential duration and the inotropic response to reduced extracellular potassium in guinea-pig ventricular myocytes. Exp Physiol 76: 705-716
48. Neer E, Clapham DE (1988) Roles of G-protein subunits in transmembrane signalling. Nature 333: 129-134
49. Boyett MR, Kirby MS, Orchard CH, Roberts A (1988) The negative inotropic effect of acetylcholine on ferret ventricular myocardium. J Physiol (Lond) 404: 613-635

Physiology
The vasculature

CHAPTER 6
The role of potassium channels in the regulation of peripheral resistance

MT Nelson

Small arteries exist in a partially contracted state from which they can constrict further or dilate depending upon the demand for blood. A significant portion of this arterial tone is caused by transmural pressure and has been termed myogenic tone. Arterial smooth muscle tone is the main determinant of peripheral vascular resistance and so of blood pressure. Potassium channels in arterial smooth muscle cells are important regulators of arterial smooth muscle tone. Activation of arterial smooth muscle K^+ channels leads to vasodilation whereas inhibition of K^+ channels causes vasoconstriction. Defects in potassium channel function may lead to vasoconstriction or vasospasm as well as compromise the ability of an artery to dilate. Thus, alteration in smooth muscle K^+ channel function could be involved in pathological conditions of the vasculature such as vasospasm, hypertension, ischaemia, hypotension following sepsis, and diabetes.

The goal of this chapter is to provide an overview of the roles of potassium channels in the control of vascular tone. Potassium channels regulate arterial smooth muscle function through controlling membrane potential. Therefore, as background, the membrane potential and its relationship to smooth muscle function will be discussed. Three distinct types of potassium channels have been identified in arterial smooth muscle: 1) calcium-activated potassium channels; 2) delayed rectifier K^+ channels; 3) ATP-sensitive potassium channels. A fourth type (inward rectifier) of potassium channel is probably also important in some types of artery. Current information on the properties and function of these four types of potassium channels will be addressed.

Patch-clamp, membrane potential, and force measurements give information about K^+ channels of arterial smooth muscle

Our knowledge of potassium channels in arterial smooth muscle has increased significantly over the last ten years, owing largely to two technical advances: 1) improved techniques to isolate single smooth muscle cells from arteries and, in particular, small arteries; 2) the patch-clamp technique which permitted the measurement of whole-cell and single potassium channel currents in small cells such as arterial smooth muscle cells (see chapter 1). Using these techniques, investigators have been able to measure whole cell and single channel currents through Ca^{2+}-activated K^+ channels, ATP-sensitive K^+ channels, and voltage-dependent K^+ channels in single smooth muscle cells from arteries.

The functional manifestations of K^+ channel activation or inhibition can also be monitored as changes in membrane potential (measured with conventional microelectrodes) and as changes in the isometric force of ring segments of arteries. K^+ channel opening drugs have often been screened using force measurements. A compound that relaxes an artery constricted by modest elevations in extracellular K^+ (e.g. 20 mM) but not an artery constricted with high potassium (e.g. 80-100 mM) would be considered as a possible K^+ channel opener. When external potassium is 20 mM, the potassium equilibrium potential (E_K) (about -52 mV) is more negative than the smooth muscle membrane potential (about -40 mV). Under this condition, a K^+ channel opener would still hyperpolarize and relax a blood vessel. However, E_K and the membrane potential are very close together with external K^+ as high as 80-100 mM, thus preventing any membrane hyperpolarization and relaxation to a K^+ channel opening drug.

Many studies of K^+ channel openers have involved the investigation of the effects of these drugs on isometric force of large conduit arteries. Large arteries differ in many important ways from small resistance arteries. Recent technical improvements have enabled several groups to investigate the role of K^+ channels in the control of membrane potential and arterial diameter in small blood vessels. To understand fully the role of K^+ channels in the regulation of peripheral resistance, it will be important to focus on these channels in resistance arteries.

The membrane potential of arterial smooth muscle regulates vascular tone

A prerequisite to understanding the role of potassium channels in arterial smooth muscle is knowledge of the physiological range of membrane potentials exhibited by resistance arteries that have tone. Smooth muscle cells in arteries and arterioles, in vitro, have stable membrane potentials when subjected to normal levels of intravascular pressure between -40 and -50 mV [1,2]. Membrane potentials measured in vivo are in the range of -40 to -55 mV [1,3]. Thus, potassium permeability does not completely dominate the membrane conductance, since the membrane potential is considerably more positive than the potassium equilibrium potential (about -85 mV). The membrane potential of arterial smooth muscle may be positive to E_K because the chloride permeability is relatively high [3]. The chloride equilibrium potential is about -25 mV.

The membrane potential of smooth muscle cells in the arterial wall appears to be an important regulator of vascular tone. The relationship between smooth muscle membrane potential and arterial tone is very steep, so that even membrane potential changes of a few millivolts cause significant changes in blood vessel diameter [1, 2, 4]. The effect of this membrane potential dependence of arterial tone on blood flow is further amplified by the fourth power relationship between blood vessel radius and blood flow.

Membrane potential of smooth muscle primarily regulates muscle contractility through alterations in calcium influx through voltage-dependent calcium channels [1]. The relationship between Ca^{2+} influx through voltage-dependent Ca^{2+} channels and membrane potential can be very steep, with 3 mV depolarization or hyperpolarization increasing or decreasing Ca^{2+} influx as much as twofold [1,4]. Any physiological or pharmacological agent that alters membrane potential should cause a significant change in blood vessel diameter. Thus, it is not surprising that membrane hyperpolarization through activation of K^+ channels is a powerful mechanism to lower blood pressure through vasodilation. Further, many types of vasoconstrictors may act through membrane depolarization.

Table 1. Major types of potassium channels in arterial smooth muscle

Type of K⁺ channel	Possible Function	Pharmacological activators	Pharmacological external blockers
Ca^{2+}-activated K^+	Negative feedback control of membrane potential in response to increases in intracellular calcium		TEA, Charybdotoxin, iberiotoxin
ATP-sensitive K^+	Membrane potential control in responses to metabolic changes	Levcromakalim, diazoxide, minoxidil sulphate, nicorandil, pinacidil, RP 49356	Glibenclamide, tolbutamide, low concentrations of external barium
Delayed rectifier	Control membrane potential in response to changes in depolarizing currents		4-aminopyridine
Inward rectifier	Control membrane potential in response to changes in extracellular K^+		Low concentrations of external barium

Ca^{2+}-activated K^+ channels provide negative feedback on depolarization and are inhibited by some vasoconstrictors

Large conductance potassium channels that are activated by intracellular calcium and membrane depolarization have been found in virtually every type of smooth muscle [5]. These potassium channels permit potassium to pass readily through them and thus, have often been called "Big" Ca^{2+}-activated K^+ channels or "Maxi" Ca^{2+}-activated K^+ channels. The fraction of time that a Ca^{2+}-activated K^+ channel spends in an ion conducting state, in other words its open probability (P_o) increases with membrane depolarization (2.7-fold increase per 12-14 mV depolarization). Elevations of intracellular calcium through the physiological range also cause substantial increases in P_o. Thus, Ca^{2+}-activated K^+ channels should be activated by the membrane depolarization and elevation in intracellular calcium that occurs during vasoconstriction.

Pharmacology of Ca^{2+}-activated K^+ channels

Agents that either activate or block potassium channels are both useful tools for exploring the role of a particular K^+ channel and are potentially important therapeutic agents. Ca^{2+}-activated K^+ channels are blocked by external tetraethylammonium ions (TEA) (concentration for half-block, K_i, about 200 µM), charybdotoxin (K_i, 10 nM), and ibe-

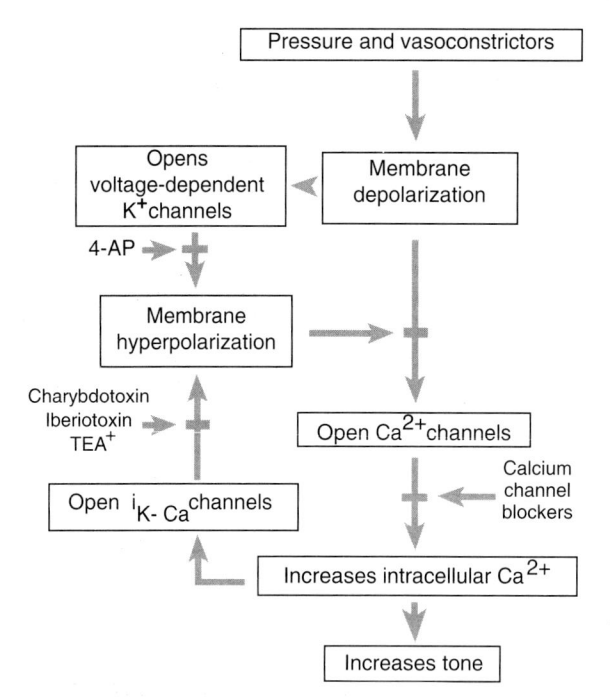

Fig. 1. Proposed mechanism for the roles of voltage-dependent potassium channels and Ca^{2+}-activated K^+ channels in the regulation of arterial tone

riotoxin ($K_i < 10$ nM) [1, 2, 5] (see Table 1) (figs. 1, 2). External TEA below 1 mM appears to be selective for Ca^{2+}-activated K^+ channels, since no other known K^+ channel in arterial smooth muscle is significantly blocked by TEA in this concentration range. Although charybdotoxin, a peptide produced by the scorpion *Leirus quinquestriatus*, has been shown to block some other types of K^+ channels in other tissues, this peptide appears to be selective for Ca^{2+}-activated K^+ channels in arterial smooth muscle. Iberiotoxin from the venom of the scorpion *Buthus tamulus* also blocks Ca^{2+}-activated K^+ channels in arterial smooth muscle and is thought to be highly selective for Ca^{2+}-activated K^+ channels. Ca^{2+}-activated K^+ channels of smooth muscle are not inhibited by apamin, which blocks low-conductance calcium-activated potassium channels in other tissues [6], by glibenclamide, which blocks ATP-sensitive K^+ channels [1, 7, 8], and by low concentrations (100 µM) of external barium, which blocks ATP-sensitive K^+ channels [1, 7] and inward rectifier K^+ channels [3].

The physiological role of Ca^{2+}-activated K^+ channels

Elevation of intravascular pressure depolarizes smooth muscle cells in resistance arteries and causes vasoconstriction. This tone has been referred to as "myogenic tone" and is a major contributor to peripheral resistance. Ca^{2+}-activated K^+ channels play an important role in the control of myogenic tone. It has been proposed that pressure-induced membrane depolarization and increases in intracellular calcium activate Ca^{2+}-activated K^+ channels [2]. Activation of Ca^{2+}-activated K^+ channels would increase potassium efflux,

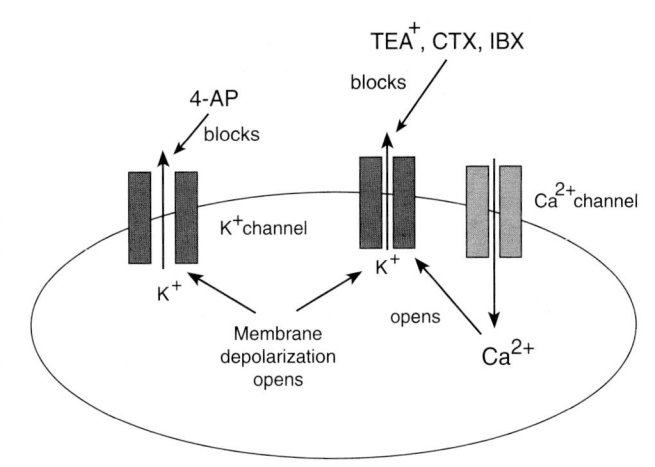

Fig. 2. Possible functions of calcium-activated potassium channels and voltage-dependent potassium channels in arterial smooth muscle. *TEA*, tetraethylammonium ions; *CTX*, charybdotoxin; *IBX*, iberiotoxin; *4-AP*, 4-aminopyridine

which would counteract the depolarization and constriction caused by pressure and vasoconstrictors (figs. 1, 2). This mechanism would predict that blockers of Ca^{2+}-activated K^+ channels would depolarize and constrict arteries with tone. This is, indeed, the case, in that TEA, charybdotoxin and iberiotoxin depolarized and constricted myogenic cerebral, coronary and skeletal muscle arteries [2]. Lowering intravascular pressure or blocking calcium channels, which would decrease intracellular calcium and thus decrease active tone, as predicted from the scheme in fig. 1, greatly attenuated the effects of Ca^{2+}-activated K^+ channel blockers on membrane potential and arterial tone [2]. Thus, Ca^{2+}-activated K^+ channels appear to serve a dynamic role in the control of arterial smooth muscle membrane potential by serving as a negative feedback pathway to regulate the degree of membrane depolarization and hence vasoconstriction caused by pressure as well as possibly other vasoconstrictors. Recent evidence that a number of vasoactive substances may act in part through modulation of Ca^{2+}-activated K^+ channels will be described.

Regulation by endogenous vasoactive substances

Most vasoconstrictors (e.g. noradrenaline, angiotensin II, endothelin, and serotonin) depolarize vascular smooth muscle [1]. It is conceivable that inhibition of Ca^{2+}-activated K^+ channels contributes to this membrane depolarization. Recently, angiotensin II and a thromboxane A_2 agonist (U46619) have been shown to inhibit Ca^{2+}-activated K^+ channels from coronary artery smooth muscle [5].

Activation of Ca^{2+}-activated K^+ channels would tend to hyperpolarize smooth muscle and lead to muscle relaxation. ß-adrenergic stimulation activates Ca^{2+}-activated K^+ channels in airway smooth muscle cells and thus may contribute to ß-adrenergic bronchodilation. This activation of Ca^{2+}-activated K^+ channels in airway smooth muscle appears to be caused by phosphorylation mediated by a cAMP-dependent protein kinase [9]. It is possible that cAMP-related arterial dilations may also involve activation of Ca^{2+}-activated K^+ channels.

Delayed rectifier K^+ channels may contribute to vascular membrane potential

Potassium channels activated by membrane depolarization have been identified in a variety of cell types [10] including arterial smooth muscle [1, 3] (figs. 1, 2). K^+ efflux through these channels increases steeply with membrane depolarization. After a depolarizing voltage step, these channels tend to turn on slowly and inactivate slowly. These channels do exhibit inactivation so it is important to consider the relationship between steady-state K^+ efflux through these channels and membrane potential. Steady-state K^+ efflux increases very steeply with membrane depolarization over the physiological range of membrane potentials in arterial smooth muscle and then decreases at positive potentials as steady-state inactivation becomes more significant. These channels are not activated by intracellular Ca^{2+} or directly regulated by intracellular ATP.

Pharmacology

Delayed rectifier K^+ channels in smooth muscle are blocked by 4-aminopyridine (0.5-2 mM) and by relatively high concentrations of TEA (>10 mM) (Table 1) (figs. 1, 2). They are not inhibited by blockers of ATP-sensitive K^+ channels (e.g. glibenclamide) or Ca^{2+}-activated K^+ channels (e.g. charybdotoxin). Some types of voltage-dependent K^+ channels are blocked by a peptide from snake venom, dendrotoxin. It is unclear whether or not dendrotoxin has any effect on delayed rectifier K^+ channels in arterial smooth muscle.

Physiological role

Although the physiological role of these channels in resistance arteries has yet to be firmly established, recent work suggests that these channels play a vital role in the control of arterial smooth muscle membrane potential and hence arterial tone (fig. 1) [3, 11]. This proposal is based on the observations that K^+ efflux through these channels over the physiological membrane potential range is sufficient to regulate membrane potential and that blockers of these channels (4-aminopyridine) depolarize and constrict arteries. Thus, we propose that potassium efflux through delayed rectifier K^+ channels regulates arterial smooth muscle membrane potential and that these channels act as a negative feedback pathway to minimize membrane depolarization (and action potentials) and hence vasoconstriction (fig. 1).

Inward rectifier K^+ channels may be important in arterioles and small arteries

Inward rectifier potassium channels in other tissues permit K^+ to enter cells more rapidly than they allow K^+ to leave. These potassium channels are activated by elevations in extracellular K^+ and usually are blocked by low concentrations of external barium

(<100 µM) [3] (Table 1). These potassium channels tend to stabilize membrane potential at negative values. Some but not all vascular beds exhibit properties consistent with the existence of inward rectifier K$^+$ channels [3]. Voltage-clamped small segments of submucosal and cerebral arterioles have current-voltage relationships that show inward rectification. This inward rectification is blocked by external Ba^{2+} (0.25 to 1 mM). The inward rectifier in these small cerebral arteries and arterioles differs from the type measured in heart and skeletal muscle in that it is partly activated at potentials substantially positive to E$_K$, so that it can contribute significant K$^+$ conductance to the normal resting potential. Whole cell or single channel inward rectifier potassium currents have not been reported for any type of smooth muscle.

Inward rectifier K$^+$ channels may play an important role in the regulation of the membrane potential of smooth muscle cells in some arterioles. Their existence may explain the observations that resistance-sized cerebral arteries dilate and hyperpolarize to small elevations in external K$^+$ and these effects are blocked by Ba^{2+} [3, 12]. In noninnervated cerebral arterioles, activation of the inward rectifiers by an elevation in extracellular K$^+$ may underlie the autoregulatory relaxation that occurs in response to raised extracellular K$^+$.

ATP-sensitive K$^+$ channels can mediate relaxation to hypoxia and hyperpolarizing vasodilators

Potassium channels closed by intracellular adenosine-5'-triphosphate (called ATP-sensitive or ATP-dependent K$^+$ channels) were first identified in cardiac muscle [13]. Since then, they have been found in skeletal muscle, pancreatic ß-cells and certain types of neurons [8]. ATP-sensitive K$^+$ channels have recently been identified in smooth muscle [1, 5, 7]. Whereas Ca^{2+}-activated K$^+$ channels respond to changes in membrane potential and intracellular Ca^{2+}, ATP-sensitive K$^+$ channels respond to changes in the cellular metabolic state as well as to a number of endogenous vasodilators. ATP-sensitive K$^+$ channels in different tissues exhibit considerable variation in their single channel properties as well as in their sensitivity to K$^+$ channel-opening drugs and to the inhibitory action of sulphonylurea drugs such as glibenclamide (also called glyburide) and tolbutamide [8]. ATP-sensitive K$^+$ channels in pancreatic ß-cells appear to be the targets of these anti-diabetic sulphonylurea drugs [8]. The drugs are thought to increase insulin release by causing by inhibition of ATP-sensitive K$^+$ channels in pancreatic ß-cells and so leading to membrane depolarization. Glibenclamide appears to be selective for ATP-sensitive K$^+$ channels and does not block a variety of other ion channels, including calcium channels, inward rectifying and delayed rectifying K$^+$ channels, and Ca^{2+}-activated K$^+$ channels [1, 8]. Glibenclamide has also become an important tool in the investigation of the role of ATP-sensitive K$^+$ channels in cellular function.

Evidence that ATP-sensitive K$^+$ channels exist in smooth muscle came first from both direct single channel measurements of these channels in smooth muscle [7] and the observation that vasodilating actions of diazoxide which had been shown to activate of ATP-sensitive K$^+$ channels in pancreatic ß-cells [8] were inhibited by glibenclamide [7]. Whole cell and single channel currents through ATP-sensitive K$^+$ channels have now been identified in a number of different types of smooth muscle [1, 5, 7].

ATP-sensitive K$^+$ channels in cardiac muscle, skeletal muscle and pancreatic ß-cells respond to changes in cellular metabolism. ATP-sensitive K$^+$ channels in smooth muscle are no exception. Metabolic poisons can lead to arterial dilation that is blocked by inhibitors of ATP-sensitive K$^+$ channels such as glibenclamide [14]. I will next discuss the evidence that ATP-sensitive K$^+$ channels in arterial smooth muscle are the targets of a variety of synthetic and endogenous vasodilators (fig. 3).

Pharmacology of ATP-sensitive K$^+$ channels

ATP-sensitive K$^+$ channels in smooth muscle are inhibited by the anti-diabetic sulphonylurea drugs such as glibenclamide and tolbutamide and external barium [1, 5, 7] (see Table 1). ATP-sensitive K$^+$ channels are not blocked by inhibitors of Ca^{2+}-activated K$^+$ channels such as charybdotoxin and iberiotoxin, and are relatively insensitive to external TEA (half-block constant > 5 mM) [1, 5, 7].

ATP-sensitive K$^+$ channels are activated by a variety of synthetic substances including cromakalim, pinacidil, minoxidil sulphate and diazoxide. These drugs appear to be relatively selective for ATP-sensitive K$^+$ channels and have significant clinical applications including as anti-hypertensive drugs.

ATP-sensitive K$^+$ channels are targets of anti-hypertensive drugs

A number of anti-hypertensive drugs appear to act through potassium channel activation and have collectively been referred to as potassium channel openers. This class of anti-hypertensive drugs include compounds that have been in use for some time as anti-hypertensives (e.g. minoxidil sulphate and diazoxide) and newer drugs such as pinacidil, nicorandil, cromakalim (or its more active isomer levcromakalim), and RP 49356. Vasodilation to all of these compounds is blocked by glibenclamide. The actions of these vasodilators are not affected by blockers of Ca^{2+}-activated K$^+$ channels (charybdotoxin, iberiotoxin, and low concentrations of TEA) [1, 5] (Table 1). Cromakalim (or levcromakalim) and pinacidil have been directly shown to activate ATP-sensitive K$^+$ channels in vascular smooth muscle [1, 5, 7] and cardiac muscle [8, 15]. These studies led to the proposal [1, 7] that ATP-sensitive K$^+$ channels (fig. 3) and not Ca^{2+}-activated K$^+$ channels are the targets of K$^+$ channels openers such as diazoxide, cromakalim, and pinacidil.

The physiological role of ATP-sensitive K$^+$ channels

A number of endogenous vasodilators act at least in part through membrane hyperpolarization caused by K$^+$ channel activation [1, 5]. Recent evidence suggests that many of these vasodilators hyperpolarize by activating ATP-sensitive K$^+$ channels in arterial smooth muscle [1, 5]. Hyperpolarizations and part of the vasorelaxations to the neuropeptides, calcitonin gene-related peptide (CGRP) [16] and vasoactive intestinal peptide (VIP) [7] were blocked by inhibitors (glibenclamide and barium) of ATP-sensitive K$^+$ channels and not affected by blockers of Ca^{2+}-activated K$^+$ channels. Vasorelaxations and hyperpolarizations to CGRP and VIP appear to involve activation of peptide receptors on the vascular smooth muscle cells and subsequent second messenger activation of ATP-sensitive K$^+$ channels [1, 5, 16]. The nature of the second messenger system remains to be elucidated.

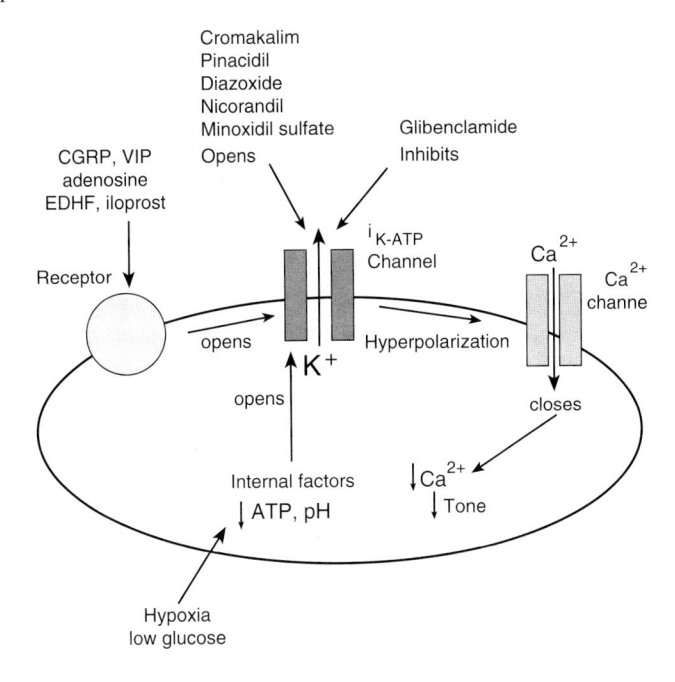

Fig. 3. Possible functions of ATP-sensitive K⁺ channels in arterial smooth muscle

Other endogenous vasodilators such as adenosine diphosphate (ADP) and acetylcholine (ACh) act through releasing factors from the vascular endothelium. Some of these factors (called endothelium-derived hyperpolarizing factors or EDHF) hyperpolarize arterial smooth muscle. Endothelial-dependent hyperpolarizations in middle cerebral arteries [5, 7, 17], skeletal muscle and mesenteric arteries [18] were inhibited by glibenclamide and barium. One possible candidate for one of the EDHFs is prostacyclin which can be released from the vascular endothelium by ACh, since Iloprost (a stable prostacyclin) causes a glibenclamide-sensitive hyperpolarization and vasorelaxation of mesenteric arteries [5]. However, indomethacin, which blocks the synthesis of prostanoids, did not prevent ACh-induced hyperpolarizations in other vascular preparations [5]. Another candidate for an EDHF is nitric oxide (NO). Exogenous nitric oxide has been shown to hyperpolarize several different types of arteries [5, 19]. Glibenclamide blocked the hyperpolarization of some but not all arterial preparations to NO [5]. However, NO, at concentrations that caused marked vasodilation, failed to hyperpolarize arteries in a number of other vascular preparations [5, 17]. In summary, there appear to be several hyperpolarizing factors from the endothelium, including NO and prostacyclin, and some, but not all, of these hyperpolarizing factors work through activation of ATP-sensitive K⁺ channels.

Recent evidence also suggests that ATP-sensitive K⁺ channels in coronary artery cells mediate the hypoxia-induced dilation of coronary arteries, since glibenclamide completely blocks this response in isolated, intact hearts [14] (see chapter 7). Hypoxia could activate ATP-sensitive K⁺ channels in smooth muscle cells of coronary arteries through a reduction in intracellular ATP. However, ATP-sensitive K$^+$ channels could also be activated by elevation in intracellular ADP or by intracellular acidification that may accompany hypoxia [5, 14]. Hypoxia also causes the release of adenosine from cardiac myocytes. Adenosine is a potent dilator of coronary arteries. Glibenclamide blocks part of the adenosine-induced dilations in vitro [14] and in vivo [5], suggesting that adenosine can activate ATP-sensitive K$^+$ channels in coronary artery smooth muscle cells. Reactive hyperaemia is also attenuated by glibenclamide, suggesting a role of ATP-sensitive K$^+$ channels in this phenomenon [5]. Further, recent evidence suggests that activation of ATP-sensitive K$^+$ channels may play an important role in endotoxic hypotension and that block of ATP-sensitive K$^+$ channels with sulphonylurea drugs (e.g. glibenclamide and tolbutamide) during septic shock may be an effective therapeutic approach [20].

Although it appears that activation of ATP-sensitive K$^+$ channels can lead to vasodilation, it is unclear whether or not ATP-sensitive K$^+$ channels contribute to arterial smooth muscle membrane potential in the absence of pharmacological activators of this channel. Glibenclamide by itself is usually without effect on blood pressure [5], suggesting a minimal tonic role of ATP-sensitive K$^+$ channels. However, recent evidence supports the idea that ATP-sensitive K$^+$ channels may indeed play an essential role in the maintainence of tone in certain vascular beds such as the coronary circulation [21]. Further, the aforementioned results on endogenous vasodilators suggest that ATP-sensitive K$^+$ channels may regulate the membrane potential of cells that are exposed to endogenous vasodilators such as CGRP, VIP and adenosine. It is also possible that some vasoconstrictors may depolarize arteries through inhibition of ATP-sensitive K$^+$ channels [5].

Distribution of ATP-sensitive K$^+$ channels

Cromakalim causes glibenclamide-sensitive hyperpolarizations and relaxations in a wide variety of vascular beds (e.g. coronary, mesenteric, skeletal muscle) as well as in a variety of non-vascular smooth muscle (urinary bladder, colonic, airway) [1, 5], suggesting that ATP-sensitive K$^+$ channel distribution is widespread in smooth muscle. However, the presence of ATP-sensitive K$^+$ channels in smooth muscle does not appear to be universal. Small resistance-sized cerebral arteries from rats and rabbits do not relax in response to cromakalim and pinacidil [12] or hyperpolarize to endogenous, glibenclamide-sensitive, hyperpolarizing vasodilators such as ADP [18], whereas larger cerebral arteries from the same animals respond to these vasodilators. These results suggest that ATP-sensitive K$^+$ channels are not expressed in small cerebral arteries from rat and rabbit. It has been suggested [12] that this apparent lack of ATP-sensitive K$^+$ channels in small cerebral arteries may be related to the degree of innervation since small cerebral arteries are scantily innervated. Understanding the regulation of the expression of these channels in smooth muscle should be an important area of future investigation.

Summary of the role of K$^+$ channels in vascular tone

Activation of potassium channels in arterial smooth muscle cells can increase blood flow and lower blood pressure through vasodilation. A variety of anti-hypertensive drugs (e.g. minoxidil sulphate, diazoxide, lemakalim, pinacidil) act through activation of ATP-sensitive potassium channels. A number of endogenous vasodilators (adenosine, hypoxia, CGRP, VIP, and EDHF) also appear to act partly through activation of ATP-sensitive K^+ channels. Pathological conditions such as hypotension associated with septic shock may involve excessive activation of ATP-sensitive K^+ channels. In this case, ATP-sensitive K^+ channel inhibitors such as glibenclamide and tolbutamide may be useful in restoring normal blood pressure.

Ca^{2+}-activated K^+ channels in the membranes of arterial smooth muscle cells respond to changes in intracellular calcium to regulate membrane potential. Ca^{2+}-activated K^+ channels appear to play a fundamental role in regulating the degree of intrinsic tone in resistance arteries. These channels help regulate arterial responses to pressure (figs. 1, 2) and vasoconstrictors. Inhibition of Ca^{2+}-activated K^+ channels could contribute to vasoconstriction, whereas activation of Ca^{2+}-activated K^+ channels would tend to cause vasodilation. Defects in Ca^{2+}-activated K^+ channels could lead to or contribute to pathological conditions that are characterized by highly constricted arteries (vasospasm).

Voltage-dependent (delayed rectifier) potassium channels and inward rectifier potassium channels appear to be important in the regulation of membrane potential in a number of vascular beds. Inward rectifier potassium channels may be important in the regulation of blood flow in response to changes in extracellular potassium.

In summary, potassium channels in arterial smooth muscle are important modulators of blood vessel diameter. Many of the functional and molecular aspects of the regulation of these channels in smooth muscle remain to be investigated.

Supported by the NIH (HL44455), NSF (DCB-8702476 and DCB-90195663), and NATO.

References

1. Nelson MT, Patlak JB, Worley JF, Standen NB (1990) Calcium channels, potassium channels and the voltage dependence of arterial smooth muscle tone. Am J Physiol 259: C3-C18
2. Brayden JE, Nelson MT (1992) Regulation of arterial tone by activation of calcium-dependent potassium channels. Science 256: 532-535
3. Hirst GDS, Edwards FR (1989) Sympathetic neuroeffector transmission in arteries and arterioles. Physiol Rev 69: 546-604
4. Nelson MT, Standen NB, Brayden JE, Worley JF (1988) Noradrenaline contracts arteries by activating voltage-dependent calcium channels. Nature 336: 382-385
5. Nelson MT (1993) Ca^{2+}-activated potassium channels and ATP-sensitive potassium channels as modulators of vascular tone. Trends Cardiovasc Med. 3: 54-60
6. Blatz AL, Magleby KL (1987) Calcium-activated potassium channels. Trends Neurosci 10: 463-467
7. Standen NB, Quayle JM, Davies NW, Brayden JE, Huang Y, Nelson MT (1989) Hyperpolarizing vasodilators activate ATP-sensitive K^+ channels in arterial smooth muscle. Science 245: 177-180.
8. Ashcroft SJH, Ashcroft FM (1990) Properties and functions of ATP-sensitive K-channels. Cell Sign 2: 197-214
9. Kume H, Takai A, Tokuno H, Tomita T (1989) Regulation of Ca^{2+}-dependent K^+-channel activity in tracheal myocytes by phosphorylation. Nature 341: 152-154
10. LY Jan & YN Jan (1990) How might the diversity of potassium channels be generated? Trends Neurosci 13: 415-419

11. Knot HJ, Robertson BE, Brayden JE, Nelson MT (1993) Voltage-dependent K$^+$ channels regulate arterial smooth muscle membrane potential and tone. Biophys J 64: A225
12. McCarron JG, Quayle JM, Halpern W, Nelson MT (1991) Cromakalim and pinacidil dilate small mesenteric arteries but not small cerebral arteries. Am J Physiol 261: H287-H291
13. Noma A (1983) ATP-regulated K$^+$ channels in cardiac muscle. Nature 305: 147-148
14. Daut J, Maier-Rudolph W, Von Beckerath N, Mehrke G, Günther K, Goedel-Meinen L (1990) Hypoxic dilation of coronary arteries is mediated by ATP-sensitive potassium channels. Science 247: 1341-1344
15. Nichols CG, Lederer WJ (1991) Adenosine triphosphate-sensitive potassium channels in the cardiovascular system. Am J Physiol 261: H1675-H1686
16. Nelson MT, Huang Y, Brayden JE, Hescheler J, Standen NB (1990) Activation of K$^+$ channels is involved in arterial dilations to calcitonin gene-related peptide. Nature 344: 770-773
17. Brayden JE (1990) Membrane hyperpolarization is a mechanism of endothelium-dependent cerebral vasodilation. Am J Physiol 259: H668-H673
18. Brayden JE (1991) Hyperpolarization and relaxation of resistance arteries in response to adenosine diphosphate. Circ Res 69: 1415-1420
19. Tare M, Parkington HA, Coleman HA, Neild TO, Dusting GJ (1990) Hyperpolarisation and relaxation of arterial smooth muscle caused by nitric oxide derived from the endothelium. Nature 346: 69-71
20. Landry DW, Oliver JA (1992) The ATP-sensitive K$^+$ channel mediates hypotension in endotoxemia and hypoxic lactic acidosis in dog. J Clin Invest 89: 2071-2074
21. Samaha FF, Heineman FW, Ince C, Fleming J, Balaban RS (1992) ATP-sensitive potassium channel is essential to maintain basal coronary vascular tone in vivo. Am J Physiol 262: C1220-C1227

CHAPTER 7
The role of potassium channels in the regulation of coronary blood flow

J Daut

One of the most striking characteristics of the coronary circulation is the interdependence between the work of the heart and the supply of oxygen and metabolic substrates through the coronary arteries. When the oxygen tension in the coronary circulation falls as a result of increased oxygen consumption of the working heart the coronary resistance arteries dilate considerably and the total coronary blood flow can be increased by a factor of 5 or more. This is called hypoxic vasodilation or metabolic regulation of coronary blood flow. Following a period of cardiac ischaemia, i.e. interruption of coronary blood flow with concomitant intravascular hypoxia and accumulation of metabolites, the coronary arteries also dilate maximally. This is called post-ischaemic vasodilation or reactive hyperaemia. Conversely, the amount of oxygen and substrate supplied through the coronary arteries can profoundly influence myocardial oxygen consumption by altering the contractility of cardiac muscle cells and thus the amount of blood ejected during each heart beat. Hypoxia can cause a marked decrease in the contractility of cardiac muscle within a few seconds. The mechanisms underlying this "early contractile failure" are complex and incompletely understood [1]. Myocardial contractility can also be changed by altering the amount and type of substrate supplied to the heart. For example, increasing the pyruvate concentration to which the cardiac muscle cells are exposed causes a marked increase in contractility. This phenomenon is called substrate dependence of cellular energy metabolism [2].

The mechanisms underlying the apparent cross-talk between cardiac muscle cells and coronary arteries are still poorly understood. As early as 1929, Drury and Szent-György [3] proposed that adenine nucleotides may be involved in the regulation of coronary blood flow, and about 30 years ago Berne [4] and Gerlach et al. [5] proposed that adenosine released from cardiomyocytes may play a crucial role in metabolic regulation of coronary blood flow. The sequence of events leading to hypoxic dilation of coronary arteries has been controversial ever since [6, 7]. However, there is some agreement that these adaptive processes are primarily mediated by local, non-neural mechanisms [8], which means that they can also be studied in isolated perfused hearts.

One of the main reasons why these mechanisms are so elusive lies in the complexity of the coronary microvasculature. Only about one third of the cells in the heart are cardiomyocytes, the rest consists of endothelial cells, coronary smooth muscle cells, pericytes, nerve cells and fibroblasts. Every cardiomyocyte makes contact with three to four capillaries running along its longitudinal axis, and the diffusion distance for oxygen is usually lower than 10 µm [8, 9]. The largest pressure drop in the coronary vascular tree and the most important regulatory processes reside in the terminal arterioles, where at least four types of cells are closely apposed: cardiomyocytes, vascular smooth muscle cells, endothelial cells and perivascular nerve cells (fig. 1). The endothelial cells are relatively impermeable to water-soluble substances. Thus local messengers or so-called

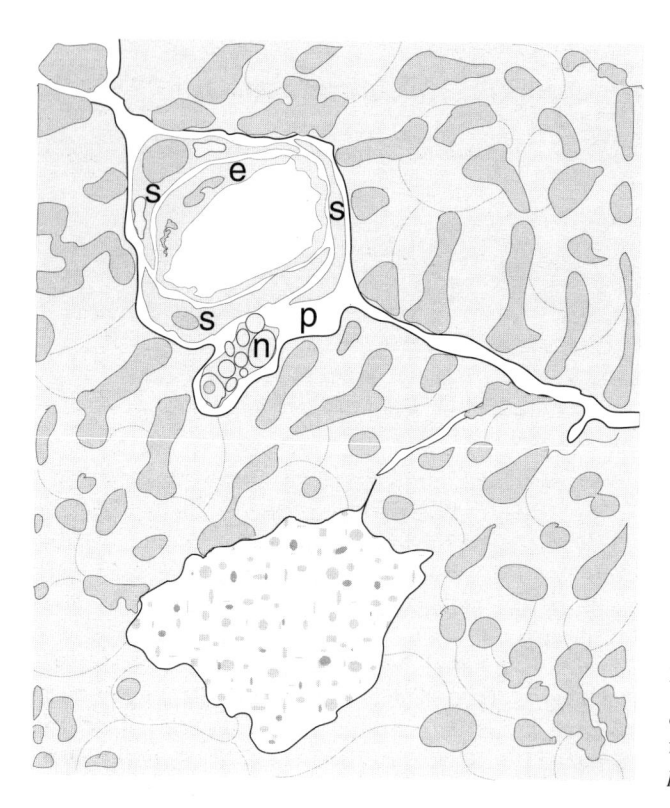

Fig. 1. Schematic diagram of a terminal arteriole in the heart (constructed using electron micrographs). A cross section of two cardiomyocytes with numerous mitochondria, thin filaments and a nucleus (at the bottom) can be seen; *e*, endothelium; *s*, smooth muscle cells; *n*, nerve; *p*, perivascular space

autacoids can accumulate in the the perivascular space (denoted p in fig. 1) [10, 11] and establish chemical communication between cardiomyocytes, smooth muscle cells, perivascular nerves, and the abluminal side of the endothelium. The cross-talk between different cell types is responsible for interdependence of cardiac work and coronary blood flow. These considerations illustrate that the microvasculature and the surrounding cardiac muscle cells form not only a structural but also a functional unit.

K^+ channels play a major role in the chemical communication between the different cell types in the terminal arterioles because they are part of the cascade of events underlying stimulus-contraction coupling and stimulus-secretion coupling. In principle, the function of the K^+ channels is to transduce a chemical stimulus into a change of membrane potential, which in turn influences transmembrane ion fluxes (mainly Ca^{2+} influx). The resulting change in intracellular calcium then influences the tone of smooth muscle cells or the secretion of vasoactive compounds from the endothelium. The two most important types of K^+ channels involved in the inter-cellular communication in the heart are ATP-sensitive potassium channels and Ca^{2+}-activated potassium channels.

ATP-sensitive potassium channels modulate the tone of coronary smooth muscle cells

ATP-sensitive potassium channels were first described in cardiac muscle cells by Noma [12] and have since been found in numerous other cells types including pancreatic

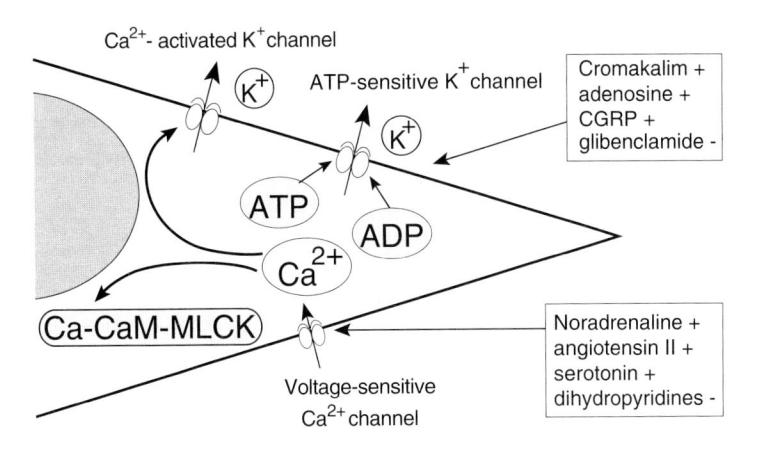

Fig. 2. Simplified scheme of the mechanism of action of some hyperpolarizing vasodilator agonists. The ATP-sensitive K⁺ channels in coronary smooth muscle cells can be activated by K⁺ channel openers *(cromakalim)*, adenosine, calcitonin gene related peptide *(CGRP)*, and can be inhibited by antidiabetic sulphonylureas *(glibenclamide)*. Voltage-sensitive (L-type) Ca^{2+} channels can be activated by noradrenaline, angiotensin II and serotonin, and can be inhibited by dihydropyridines

ß-cells, central neurons, skeletal muscle cells, and vascular smooth muscle cells [13-16]. ATP-sensitive K⁺ channels are activated by a fall in intracellular ATP and/or by a rise in intracellular ADP, as occurs during hypoxia or ischaemia. Thus these channels form a link between energy metabolism and electrical activity of the cells. In addition, the open-state probability (P_o) of these channels can be modulated by various endogenous agonists, either directly via receptor-coupled G-proteins or via intracellular second messengers. P_o can also be increased by K-channel opening drugs like cromakalim.

The mechanism by which modulation of the open-state probablity of ATP-sensitive K⁺ channels in coronary smooth muscle cells influences vascular tone is outlined in Fig. 2. The membrane potential of vascular smooth muscle cells in vivo is probably around -40 to -55 mV. Isolated arteries in vitro have a more negative resting potential, but applying intravascular pressure causes depolarization and brings the membrane potential close to the range observed in vivo. Thus the difference between the potassium equilibrium potential and the membrane potential is larger in vascular smooth muscle than in cardiac muscle or in skeletal muscle. Under physiological conditions the open-state probablity of ATP-sensitive K⁺ channels in smooth muscle is very low. When P_o of the ATP-sensitive K⁺ channels is elevated the membrane potential moves closer to the K⁺ equilibrium potential, i.e. the smooth muscle cells hyperpolarize. Due to the high input resistance of coronary smooth muscle cells (around 5-10 G Ω) the opening of one or two channels at a time is sufficient to hyperpolarize the cells by several mV.

The second element necessary for transducing a chemical stimulus into a change of vascular tone is the steep voltage dependence of the gating of calcium channels. The activation properties of single Ca^{2+} channels in arterial smooth muscle cells have been studied in systemic and coronary arteries [17, 18]. The open-state probability in the steady state is an exponential function of the membrane potential. In the physiological range a hyperpolarization of 2 mV is sufficient to decrease P_o by about 25%. As a consequence, calcium influx is reduced and the steady-state concentration of intracellular calcium

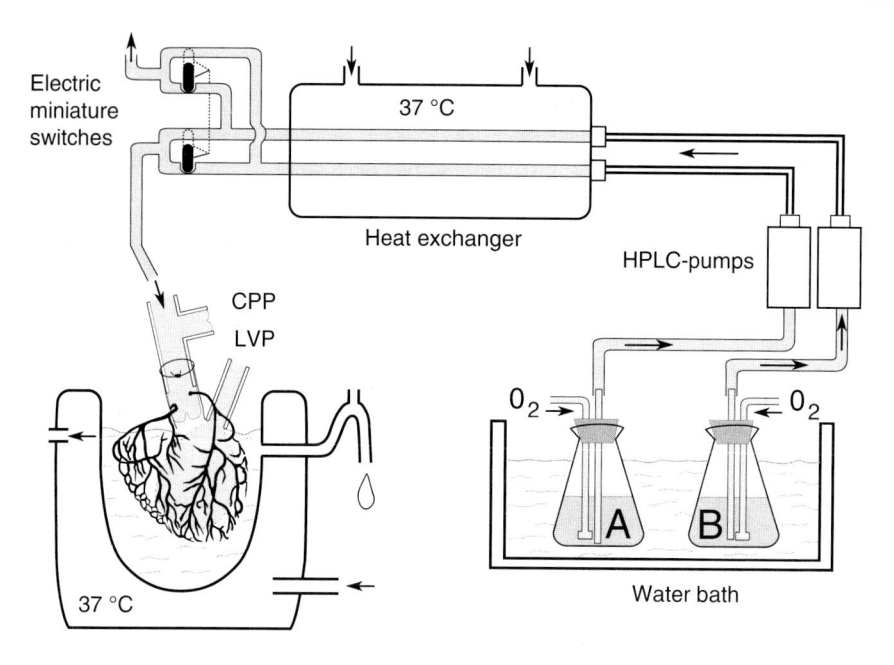

Fig. 3. Schematic diagram of a modified isolated perfused heart (Langendorff) preparation [19]. The heart is submerged in a small temperature-controlled bath and perfused at a constant rate using two HPLC pumps. The perfusate can be switched (between solution A and B) using two inert electric valves. For inducing hypoxia one of the perfusates is equilibrated with nitrogen instead of oxygen. Changes in coronary perfusion pressure *(CPP)* measured at the level of the aorta indicate changes in coronary resistance. Isovolumetric left ventricular pressure *(LVP)* can be measured with a Latex balloon

decreases because it depends on the balance between Ca^{2+} influx and efflux. The fall in intracellular Ca^{2+} reduces the activation of Ca^{2+}-calmodulin-dependent myosin light chain kinase and causes relaxation. Thus activation of ATP-sensitive K^+ channels eventually leads to a decrease in vascular resistance.

It is likely that the cascade of events outlined above, which has been deduced from experiments on different types of vascular smooth muscle, is also operative in coronary arterioles. However, since the terminal arterioles are rather inaccessible for patch-clamp studies, the evidence for the regulatory mechanisms in the intact heart is still rather indirect and has been derived mainly from studies on isolated perfused hearts. Fig. 3 illustrates the approach used in our laboratory to monitor changes in coronary vascular resistance [19]. Isolated guinea-pig hearts were perfused at a constant rate (between 5 and 10 ml/min) using two HPLC pumps and electronic microswitches to change solutions rapidly. The perfusion rate was adjusted to give a coronary perfusion pressure (CPP) of approximately 80 mmHg. CPP was measured with a piezoresistive pressure transducer. At constant flow, changes in perfusion pressure indicate changes in coronary vascular resistance.

Fig. 4 shows an experiment designed to elucidate the role of ATP-sensitive K^+ channels in the regulation of coronary vascular tone. When the perfusate was changed to hypoxic solution ($Po_2 < 10$ mmHg) a drop of CPP to less than 40 mmHg was observed,

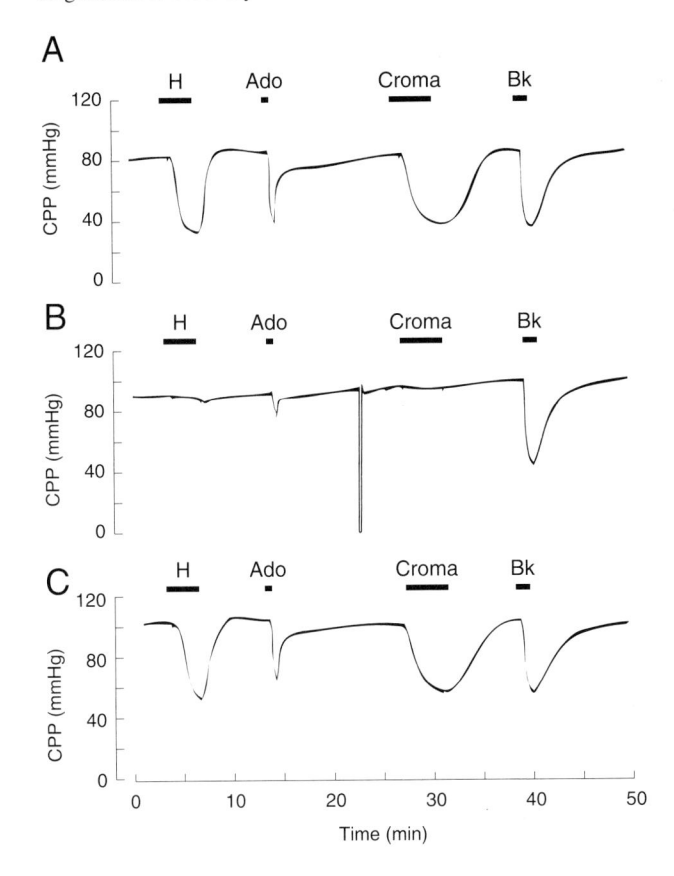

Fig. 4 A-C. The effects of glibenclamide on dilation of coronary arteries in an isolated guinea-pig heart, which was arrested by increasing extracellular K^+ to 15 mM. Vasodilation was induced by switching to a perfusate containing: *(H)* hypoxic solution, *(Ado)* 1 μM adenosine, *(Croma)* 200 nM cromakalim, or *(BK)* 500 pM bradykinin. **A** Control; **B** in the presence of 2 μM glibenclamide; **C** 1 hour after washout of glibenclamide. In the middle trace the perfusion of the heart was briefly interrupted (ischaemia) to check the zero-pressure level

indicating a maximal dilation of the coronary resistance arteries. The remaining resistance resides in the capillaries and in the coronary conduit arteries. Short-lasting intravascular application of adenosine produced a similar vasodilation which was rapidly reversed after wash-out. The K^+ channel opener cromakalim and the endogeneous vasoactive peptide bradykinin also produced a maximal vasodilation.

The middle trace of fig. 4 shows what happened when the same experimental procedure was repeated in the presence of the sulphonylurea derivative glibenclamide, a blocker of ATP-sensitive K^+ channels. The vasodilation induced by hypoxia and cromakalim was prevented completely by 2 μM glibenclamide [19, 20], whereas the vasodilation induced by bradykinin was almost unchanged. The lower trace in fig. 4 shows that the effects of glibenclamide were reversed after washing out the drug for 1 hour.

The lack of effect of glibenclamide on the vasodilatory effect of bradykinin was to be expected because bradykinin is thought to cause dilation by eliciting release of endothelium-derived relaxing factor (EDRF). This is a useful check for any non-specific effects of glibenclamide on the responsiveness of vascular smooth muscle cells to vasodilatory stimuli. Our results suggest that the effects of 2 μM glibenclamide were almost entirely attributable to the reduction of the open-state probability of ATP-sensitive K^+ channels [19, 20]. This interpretation is supported by our recent finding that glibenclamide concentrations as low as 5 nM partially inhibit the vasodilation induced by hypoxia or cromakalim [21].

Our current working hypothesis is that during hypoxia the submembrane concentrations of adenine nucelotides change, ATP decreases slightly and ADP increases, and that this change causes a small change in the steady-state open probability of ATP-sensitive K$^+$ channels, which is sufficient to hyperpolarize the cell membrane by a few mV. The interdependence between energy metabolism and the activation of ATP-sensitive K$^+$ channels is even more obvious in experiments where extracellular glucose was removed from the perfusate. Fig. 5A shows that removal of glucose (replaced by deoxy-glucose, which blocks glycolysis) in the oxygenated heart caused a maximal dilation of coronary arteries that could be inhibited by glibenclamide. Removal of glucose alone, without addition of deoxyglucose, caused a similar vasodilation, although with a somewhat longer delay. We conclude from these experiments that ATP-sensitive K$^+$ channel activity in coronary smooth muscle cells is linked to energy metabolism in a way similar to that observed with ATP-sensitive K$^+$ channels in pancreatic ß-cells [13].

The vasodilator effect of adenosine was only partly blocked by 2 µM glibenclamide (figs. 4 and 6). This can be explained by the fact that adenosine has two effects on the coronary microvasculature: (i) it induces the opening of ATP-sensitive K$^+$ channels in coronary smooth muscle cells, which can be prevented by glibenclamide, and; (ii) it induces release of EDRF from the microvascular endothelium. This effect is not related to ATP-sensitive K$^+$ channels (see below) and is insensitive to glibenclamide. Adenosine is continuously released even in the fully oxygenated heart and is taken up again by a specific adenosine transporter, which means that interstitial adenosine is in a dynamic equilibrium. During hypoxia, cardiomyocytes release more adenosine, which accumulates in the perivascular space. Thus the experiment shown in fig. 4 is not a very good model for what happens during hypoxia because there can be large differences between the adenosine concentration in the blood and in the lymph that drains the interstitial space [10, 11].

However, it is possible to mimic perivascular accumulation of adenosine pharmacologically by applying dipyridamole. Dipyridamole inhibits adenosine re-uptake and has been shown to produce a marked perivascular accumulation of adenosine [11]. Fig. 5B shows that 0.5 µM dipyridamole produced a maximal vasodilation in the isolated perfused guinea-pig heart. This vasodilation was strongly inhibited when the experiment was repeated in the presence of 2 µM glibenclamide, as can be seen in fig. 5C. The dilation of coronary arteries induced by dipyridamole could also be blocked by the adenosine antagonist 8-phenyl-theophylline, indicating that it was indeed mediated by perivascular accumulation of adenosine [19].

These results suggest that adenosine induces the opening of ATP-sensitive K$^+$ channels in the coronary resistance arteries. In cardiac muscle cells, adenosine has been shown to open ATP-sensitive K$^+$ channels via a G-protein-mediated mechanism, without involvement of intracellular second messengers [22]. Therefore, the simplest explanation for our results is that a similar adenosine-activated channel exists in the membrane of smooth muscle cells of coronary resistance vessels. Adenosine has recently been shown to acivate a glibenclamide-sensitive K$^+$ current in isolated porcine coronary arterial myocytes [23]. Recent studies on open-chest dogs confirmed the conclusions drawn from our studies on isolated perfused hearts: reactive hyperaemia after 30 or 60 s coronary occlusion and the vasodilation following intracoronary application of adenosine were significantly inhibited by glibenclamide infusion [24, 25].

Regulation of coronary blood flow

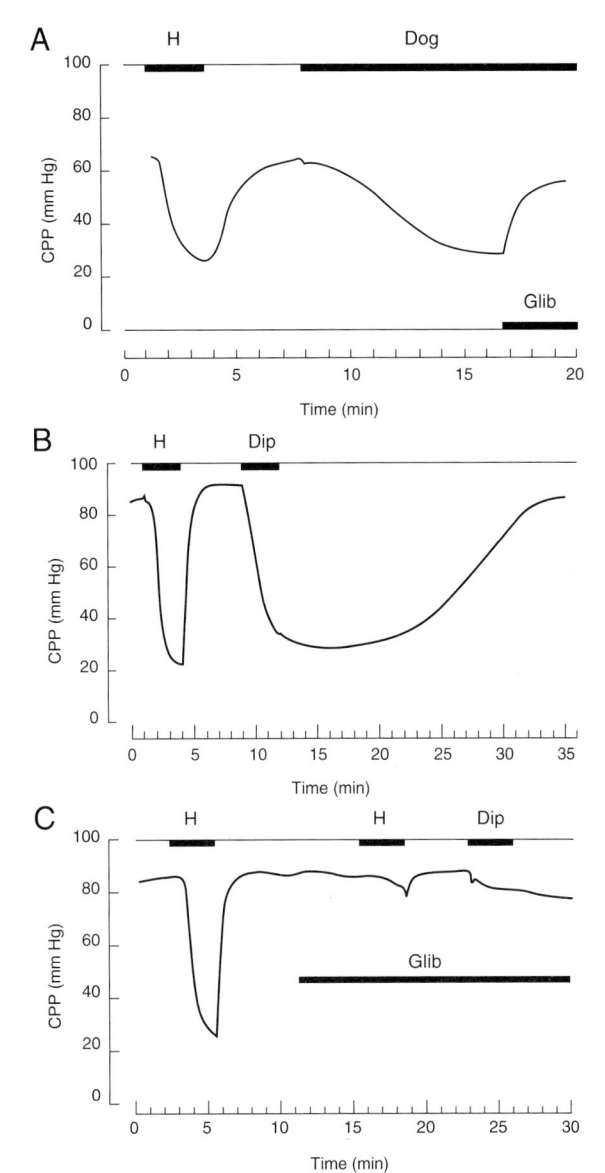

Fig. 5. A The effect of replacing 10 mM glucose with 10 mM deoxyglucose in the perfusate of an arrested guinea-pig heart. On the left, a hypoxic vasodilation (*H*) is shown. On the right, 2 µM glibenclamide was added in the presence of 10 mM deoxyglucose (*Dog*). **B** The effect of hypoxia (*H*) and 500 nM dipyridamole (*Dip*) on coronary perfusion pressure (*CPP*). Dipyridamole induces perivascular accumulation of adenosine by inhibiting adenosine re-uptake from the interstitial space. **C** The effect of 2 µM glibenclamide on hypoxic vasodilation and on the vasodilation induced by 500 nM dipyridamole (records **A**, **B** and **C** are taken from [19])

Ca^{2+}-activated K^+ channels provide a negative feedback for elevated vascular tone

Calcium activated potassium channels have been found in many different cell types including neurons, hepatocytes, vascular smooth muscle cells and endothelial cells [26-29]. They are activated by depolarization and intracellular calcium and have been subdivided somewhat arbitrarily into three classes: (i) large-conductance Ca^{2+}-activated K^+ channels (BK channels, 100-250 pS), that are activated by intracellular calcium and membrane depolarization; (ii) small-conductance Ca^{2+}-activated K^+ channels (SK channels, 6-22 pS) that are less voltage sensitive, and; (iii) intermediate-conductance Ca^{2+}-activated K^+ channels (about 40 pS). All of the conductances given here apply to symmetrical high K^+ solution inside and outside the patch. BK channels (and some intermediate-type Ca^{2+}-activated K^+ channels) are blocked by low concentrations of charybdotoxin, SK channels are blocked by low concentrations of apamin. Large-conductance Ca^{2+}-activated K^+ channels have been found in most types of vascular smooth muscle. A rise in intracellular calcium sufficient to activate the contractile proteins causes a substantial increase in the open-state probability (P_o) of BK channels.

Recent experimental evidence suggests that Ca^{2+}-activated K^+ channels are involved in the regulation of vascular tone [30]. The tone of vascular smooth muscle cells in intact coronary arteries is very sensitive to transmural pressure: elevation of transmural pressure causes resistance arteries to depolarize and constrict. This so-called "myogenic response" or "autoregulation" is an intrinsic property of arterial smooth muscle and is accompanied by a rise in intracellular Ca^{2+}. In the mammalian heart, myogenic tone seems to be higher than in other organs with a similar rate of oxygen consumption per gram tissue, for example brain or kidney. As a result, basal coronary resistance is high and the venous oxygen tension is lower in the coronary sinus than the venous Po_2 in any other organ. Thus the regulation of coronary blood flow is mainly mediated by vasodilatory mechanisms superimposed on a high basal tone. Nevertheless, the high basal tone can be increased even further by endogenous pressor agents like catecholamines or angiotensin II or by an increase in transmural pressure.

It has recently been proposed that the tone of arteries is limited by calcium-activated potassium channels, i.e. if the rise in intracellular Ca^{2+} exceeds a certain limit, the resulting activation of Ca^{2+}-activated K^+ channels causes a hyperpolarization that reduces Ca^{2+} influx [30]. This means that vascular smooth muscle cells possess two negative feedback mechanisms that limit increases in tone. The first one signals inadequate oxygen supply to the terminal arterioles and operates through ATP-sensitive K^+ channels, and the second negative feedback mechanism signals Ca^{2+} overload and operates through Ca^{2+}-activated K^+ channels. The latter mechanism is not directly related to changes in oxygen tension, although it is not completely independent of energy metabolism because energy deprivation will eventually lead to an increase in intracellular Ca^{2+}. Interestingly, both negative feedback mechanisms converge to a common element: membrane hyperpolarization. Thus both mechanisms utilize the steep voltage dependence of the open-state probablity of Ca^{2+} channels. More details on the role of Ca^{2+}-activated K^+ channels can be found in chapter 6.

Ca^{2+}-activated potassium channels modulate stimulus-secretion coupling in the endothelium

Coronary endothelium releases a number of vasoactive autacoids, for example "endothelium derived relaxing factor" (NO or a related compound), "endothelium derived hyperpolarizing factor" (an as yet chemically unidentified compound), prostacyclin, endothelin and adenosine. These autacoids play an important role in regulating coronary blood flow, especially in terminal arterioles, where only one or two layers of smooth muscle cells overlie the endothelium. The release of EDRF and prostacyclin is regulated

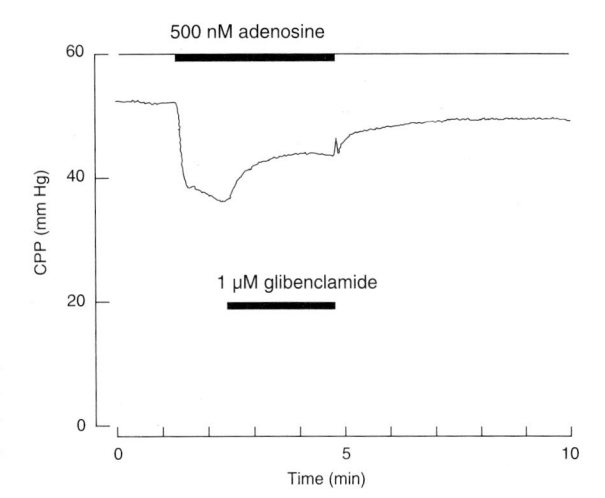

Fig. 6. The effect of 500 nM adenosine and 1 µM glibenclamide on coronary perfusion pressure (CPP) in an arrested isolated guinea-pig heart (Cyrys and Daut, unpublished experiment)

by the level of intracellular Ca^{2+} in the endothelium [31], and can be modulated over a wide range by a host of endogenous vasoactive compounds, for example braykinin, histamine, thrombin, ADP, ATP and adenosine [32]. This means that Ca^{2+} regulation in the endothelium is as important as in smooth muscle. However, the signal transduction mechanisms are quite different in the two cell types. Some substances that cause vasoconstriction when applied directly to smooth muscle cells cause vasodilation when applied to the endothelium (for example, acetylcholine). In other cases vasoactive autacoids have synergistic effects on endothelial and smooth muscle cells. The vasodilation induced by adenosine, for example, is mediated by both endothelial and smooth muscle cells. Fig. 6 shows that the effects of intracoronary application of adenosine consisted of a fast and a slow phase, and that only one of the phases was blocked by glibenclamide. The glibenclamide-sensitive part of the response was presumably mediated by ATP-sensitive K^+ channels in coronary smooth muscle cells, whereas the glibenclamide-insensitive part of the vasodilator response was related to the release of autacoids from the endothelium.

Fluorimetric studies have shown that application of vasoactive agonists increases intracellular Ca^{2+} via two pathways: (i) release of Ca^{2+} from intracellular stores; (ii) Ca^{2+} influx through the cell membrane. Ca^{2+} influx is abolished in the absence of extracellular Ca^{2+}, whereas Ca^{2+} release from intracellular stores continues in the absence of extracellular Ca^{2+} as long as the stores are full [33, 34]. This is illustrated in the schematic diagram of fig. 7. The release of Ca^{2+} from intracellular stores is mediated by the G-protein - phospholipase C - inositoltrisphosphate cascade. The Ca^{2+} influx through the cell membrane is mediated by channels that are linked to the state of the intracellular Ca^{2+} stores (Ca^{2+} release activated channels, or refill channels), and probably by receptor operated channels (ROCs, fig. 7). The receptor operated Ca^{2+} permeable channels may be activated via a G-protein or by direct binding of the agonist to the channel. It should be noted, though, that the existence of receptor operated channels in the endothelium has not yet been demonstrated unequivocally [34]. The changes in intracellular Ca^{2+} are usually accompanied by changes in membrane potential: an increase in intracellular Ca^{2+} activates Ca^{2+}-activated K^+ channels and hyperpolarizes the cells.

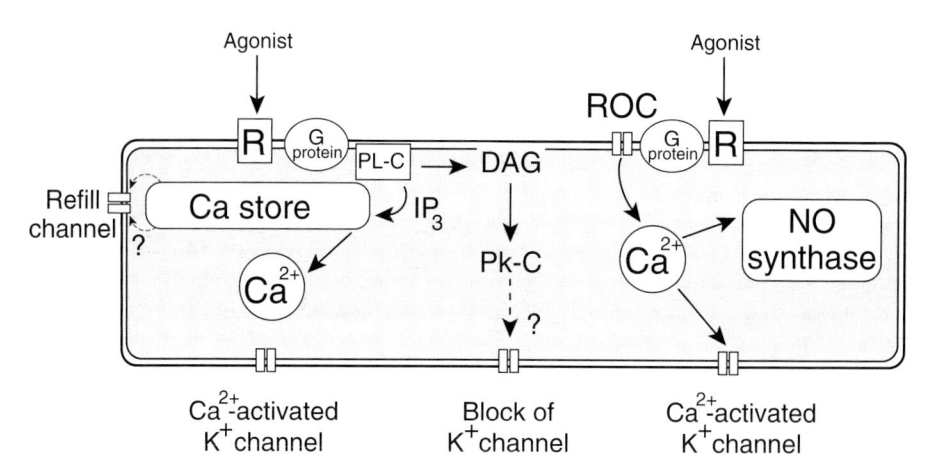

Fig. 7. Simplified scheme of the hypothetical mechanisms underlying agonist-induced Ca^{2+} movements in endothelial cells. The mechanism by which the state of the intracellular Ca^{2+} stores influences transmembrane Ca^{2+} influx, and the mechanism by which activation of protein kinase C may induce depolarization are still unclear

Fig. 8A illustrates how the potential changes in monolayers of cultured coronary endothelial cells can be recorded with the patch-clamp technique [32]. One of the cells is approached with a patch electrode, and after formation of a gigaseal and rupturing of the patch the membrane potential of the entire monolayer can be recorded in the current-clamp mode of the patch clamp-technique. This is so because the individual cells in the monolayer are tightly coupled by gap junctions and form an electrical syncytium, like cardiac muscle cells. The entire monolayer is virtually isopotential because the lumped resistance of the gap junctions of a given cell is very low compared with the membrane resistance of that cell. The large membrane capacitance of the monolayer filters all rapid fluctuations of membrane potential caused by the opening of single channels. As a result, agonist-induced changes in the membrane potential of confluent monolayers can be recorded for hours with a high signal-to-noise ratio.

Fig. 8B shows a typical record of the membrane potential of a monolayer of coronary endothelial cells. During application of the vasoactive peptide bradykinin a transient hyperpolarization was observed which turned into a sustained depolarization in the continued presence of the peptide. The transient hyperpolarization was caused by the opening of Ca^{2+}-activated K^+ channels. The Ca^{2+}-activated K^+ channels in the endothelial membrane have a lower conductance than the channels in vascular smooth muscle cells, and probably a different pharmacology [32, 35]. The ion channel underlying the depolarization is still unknown. Since the depolarization can be mimicked by activators of protein kinase C it has been speculated that it may be attributable to the closing of K^+ channels [32], mediated by diacylglycerol (DAG), activation of protein kinase C and subsequent phosphorylation of channel proteins (fig. 7). Most vasoactive agonists, for example ATP, bradykinin and histamine, induce a biphasic membrane potential change in coronary endothelial cells. However, under some experimental conditions adenosine can also evoke a sustained hyperpolarization [33], as illustrated in fig. 8C. This hyperpolarization is dependent on the presence of extracellular Ca^{2+} and l-arginine [37] and is

Fig. 8. A Schematic diagram of patch-clamp recording in confluent monolayers of cultured coronary endothelial cells. The cells are about 1 µm thick, the bottom of the Petri dish is proportionally much thicker than shown here. The patch pipette (filled with "intracellular" solution) is lowered vertically (and carefully) towards the endothelium. After establishing a gigaseal contact with the cell membrane in the region of the nucleus (where the cells are somewhat thicker) the membrane patch is ruptured with negative current pulses. In the current-clamp mode (zero current, i.e. membrane potential recording) the average membrane potential of thousands of cells in the monolayer is recorded. Analysis of the cable properties showed that the cells are electrically coupled by gap junctions and are virtually isopotential [38]. **B** The effect of 10 nM bradykinin on the membrane potential of a monolayer of coronary endothelial cells. The transient hyperpolarization is followed by a depolarization in the continued presence of bradykinin [from 32]. **C** The effect of 2 µM adenosine on the membrane potential of coronary endothelial cells in the presence of 2 mM arginine [from 36]

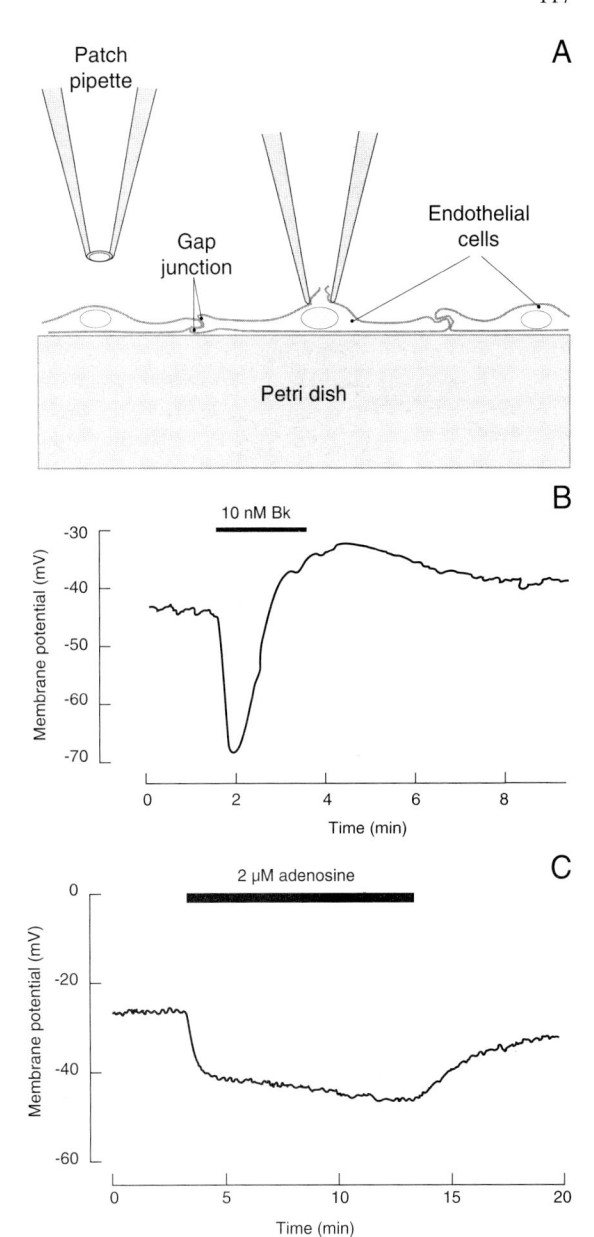

probably related to a transmembrane Ca^{2+} influx that activates Ca^{2+}-activated K^+ channels. It is not yet clear whether the Ca^{2+} influx is mediated by receptor operated channels or by second-messenger operated channels. The hyperpolarization is insensitive to glibenclamide and is not related to ATP-sensitive K^+ channels, which are apparently not expressed in the endothelium [38].

A complicating factor in the analysis of Ca^{2+} movements in non-excitable cells is the finding that Ca^{2+} influx (which serves to re-fill the intracellular Ca^{2+} stores) appears to depend on the degree of filling of the Ca^{2+} stores [39-41] (fig. 7). A current activated by emptying the intracellular stores (Ca^{2+} release activated current) has recently been recorded in mast cells [42]. The experiment illustrated in fig. 9 shows that such a current is also present in the endothelium. In this experiment the Ca^{2+} stores of a monolayer of coronary endothelial cells were emptied by application of 200 nM bradykinin in Ca^{2+}-free solution. The transient hyperpolarization was entirely due to release of Ca^{2+} from intracellular stores and the subsequent opening of Ca^{2+}-activated K^+ channels. A second application of bradykinin had almost no effect because the Ca^{2+} released into the cytoplasm had been pumped out of the cell so that the stores remained empty. Subsequently, the cells were superfused with Ca^{2+} containing solution for 3 min to re-fill the Ca^{2+} stores. When the application of bradykinin was repeated in Ca^{2+} free solution the transient hyperpolarization was again observed. During application of external Ca^{2+} the endothelial monolayer depolarized. This was probably due to inward current flowing through calcium-release activated Ca^{2+} channels. The mechanism that couples Ca^{2+} influx to the state of the Ca^{2+} stores is still obscure [39, 40].

In endothelial cells hyperpolarization increases transmembrane calcium influx, i.e. the effect of membrane potential on Ca^{2+} influx is in the direction opposite to that in vascular smooth muscle cells. The reason for this is that the open-state probablity of the endothelial Ca^{2+} channels is either potential independent [34] or increased by hyperpolarization [41]. Even with a potential-independent open-state probability Ca^{2+} influx would increase with hyperpolarization due to the increased Ca^{2+} driving force. The "inverse" voltage dependence of the endothelial Ca^{2+} influx has important functional consequences: first, it implies that the hyperpolarization induced by vasoactive agonists (caused by the opening of Ca^{2+}-activated K^+ channels) modulates the agonist-induced rise in intracellular Ca^{2+}, and thus the release of EDRF and prostaglandins. Second, rise in intracellular Ca^{2+} induced by the hyperpolarization constitutes a positive feedback loop: it leads to a larger hyperpolarization, and, in turn, to a still higher Ca^{2+} influx, etc. In other words, the influx of Ca^{2+} could represent a regenerative process, once a certain threshold of intracellular Ca^{2+} is crossed. The resulting wave of intracellular Ca^{2+} and hyperpolarization could propagate in the endothelium along the vessel wall. Propagated responses to local application of vasodilator agonists have been measured in various vascular preparations [43, 44]. It should be noted, however, that the hypothetical mechanism outlined here needs to be verified (or falsified) by direct Ca^{2+} and potential measurements in intact arterioles.

Coronary smooth muscle and endothelial cells form a myoendothelial regulatory unit

The data reviewed show that the membrane potential of both coronary smooth muscle cells and coronary endothelial cells contributes to the regulation of coronary resistance. In both cell types the effect of membrane potential on Ca^{2+} permeability is an essential element of the control mechanism. In smooth muscle cells hyperpolarization decreases Ca^{2+} influx and thus initiates vasorelaxation. In the endothelium hyperpolarization increases Ca^{2+} influx and thus stimulates release of vasoactive autacoids, which also leads to vasorelaxation. Furthermore, there is a growing body of indirect evidence that in

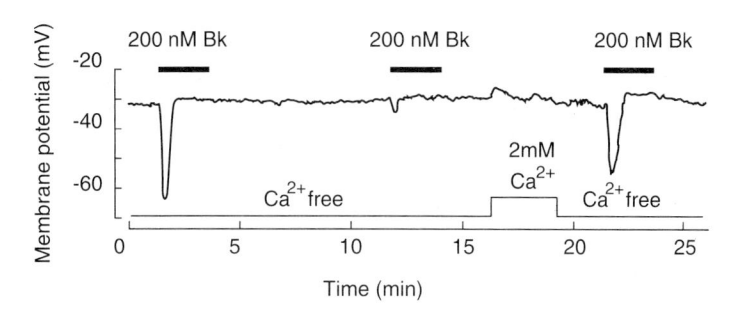

Fig. 9. Experiment designed to study the emptying and re-filling of intracellular Ca^{2+} stores in coronary endothelial cells. The monolayer was superfused with Ca^{2+}-free solution except for the brief period indicated at the bottom. Note the absence of effect of the second application of 200 nM bradykinin and the transient depolarization observed during the increase in external Ca^{2+} (Mehrke and Daut, unpublished experiment)

the terminal arterioles the endothelium is electrically coupled to the underlying smooth muscle cells via gap junctions [45-47]. Hence agonist-induced hyperpolarization of the endothelium might induce vasodilation via two different pathways: (i) by increasing Ca^{2+} influx into the endothelium, which augments release of EDRF; (ii) by hyperpolarizing the smooth muscle cells of the terminal arterioles via myo-endothelial gap junctions.

Since both endothelial cells and smooth muscle cells have a very high input resistance the existence of myo-endothelial gap junctions implies that both cell types are virtually isopotential in the steady state. Thus electrical coupling of endothelial cells and smooth muscle cells would confer to the terminal arterioles the properties of a "myo-endothelial regulatory unit". The electrical signals generated in the capillaries could be propagated in the endothelium against the direction of blood flow and converge onto the terminal arteriole. There the hyperpolarization generated in different capillaries could be summed and transmitted to the vascular smooth muscle cells. The result might be an integrated response of the myo-endothelial regulatory unit.

The functional consequences of electrical coupling between endothelium and smooth muscle cells are illustrated in the schematic diagram of fig. 10, using the effects of adenosine as an example. Adenosine released by hypoxic cardiomyocytes hyperpolarizes both endothelial and smooth muscle cells, although via different mechanisms. This is expected to induce a substantial hyperpolarization of the myo-endothelial regulatory unit and, as a result, a strong local vasodilation. Thus hypoxic regions of the heart could adjust their own local blood supply by inducing dilation of their feed arteriole. This type of local regulation of blood flow can only be achieved by means of propagation of electrical signals, because humoral substances released in the capillary bed diffuse in the downstream direction and cannot influence the upstream arterioles.

In conclusion, the microarchitecture of terminal arterioles provides the structural basis for chemical communication between cardiac muscle cells, coronary smooth muscle cells, endothelial cells and perivascular nerves (see fig. 1). The cells in the microcirculation form a functional unit which communicates both electrically and via vasoactive compounds released locally. K^+ channels in endothelial and smooth muscle cells serve to adapt coronary blood flow to the changing energy requirements of the cardiomyocytes. ATP-sensitive K^+ channels in cardiac muscle cells adapt action potential dura-

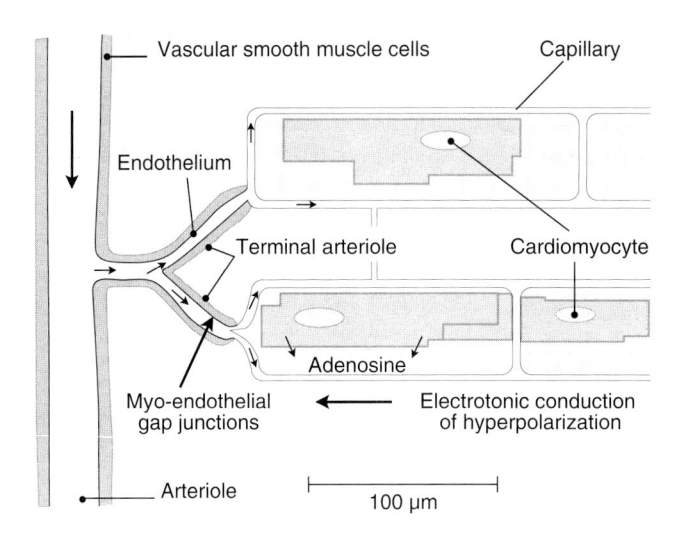

Fig. 10. Simplified schematic drawing of the myo-endothelial regulatory unit. Blood flow is from left to right, whereas the the electrical signal in the endothelium (induced by perivascular accumulation of adenosine) is assumed to be propagated from right to left (against the direction of blood flow) towards the terminal arterioles

tion and cardiac work to the energy supply via the coronary arteries (see chapters 3 and 5). Thus K^+ channels endow the heart with the ability to produce an integrated response which adapts the function of the different cell types to the wide range of work loads encountered under physiological conditions and also to more extreme situations such as ischaemia and hypoxia.

References

1. Allen DG, Orchard CH (1987) Myocardial contractile function during ischaemia and hypoxia. Circ Res 60: 153-168
2. Daut J, Elzinga G (1989) Substrate dependence of energy metabolism in isolated guinea-pig cardiac muscle: a microcalorimetric study. J Physiol (Lond) 413: 379-397
3. Drury AN, Szent-Györgyi A (1929) The physiological activity of adenine compounds with special reference to their action upon the mammalian heart. J Physiol (Lond) 68: 213-237
4. Berne RM (1963) Cardiac nucleotides in hypoxia: possible role in regulation of coronary blood flow. Am J Physiol 204: 317-322
5. Gerlach E, Deuticke B, Dreisbach RH (1963) Der Nucleotidabbau im Herzmuskel bei Sauerstoffmangel und seine mögliche Bedeutung für die Coronardurchblutung. Naturwissenschaften 50: 228-229
6. Berne RM (1980) The role of adenosine in the regulation of coronary blood flow. Circ Res 47: 807-813
7. Olsson RA, Bünger R, Spaan JAE (1991) Coronary circulation. In: Fozzard HA (ed) The Heart and Cardiovascular System, Vol. 2. Raven Press, New York, pp. 1293-1426
8. Berne RM, Rubio R (1979) Coronary circulation. In: Berne RM (ed) Handbook of Physiology, section 2, vol. I, The Heart. American Physiological Society, Bethesda, MD, pp. 873-952
9. Rakusan KJ, Moravec J, Hyatt PY (1980) Regional capillary supply in the normal and hypertrophied heart. Microvasc Res 20: 319-326
10. Decking UKM, Jüngling E, Kammermeier H (1988) Interstitial transudate concentration of adenosine and inosine in rat and guinea-pig hearts. Am J Physiol 254: H1125-H1132
11. Wangler RD, Gorman MW, Wang MCY, DeWitt DF, Chan IS, Bassingthwaite JB, Sparks HV (1989) Transcapillary adenosine transport and interstitial adenosine concentration in guinea-pig hearts. Am J Physiol 257: H89-H106

12. Noma A (1983) ATP-regulated K^+ channels in cardiac muscle. Nature 305: 147-148
13. Ashcroft SJH, Ashcroft FM (1990) Properties and functions of ATP-sensitive K-channels. Cell Signalling 2: 197-214
14. Standen NB, Quayle JM, Davies NW, Brayden JE, Huang Y, Nelson MT (1989) Hyperpolarising vasodilators activate ATP-sensitive K^+ channels in arterial smooth muscle. Science 245: 177-180
15. Inoué I, Nakaya S, Nakayama Y (1990) An ATP-sensitive K^+ channel activated by extracellular Ca^{2+} and Mg^{2+} in primary cultured arterial smooth muscle cells. J Physiol (Lond) 430: 123P
16. Kajioka S, Kitmura K, Kuriyama H (1991) Guanosine diphosphate activates an adenosine-5'-triphosphate-sensitive K^+ channel in the rabbit portal vein. J Physiol (Lond) 444: 397-418
17. Nelson MT, Patlak JB, Worley JF, Standen NB (1990) Calcium channels, potassium channels, and voltage dependence of arterial smooth muscle tone. Am J Physiol 259: C3-C18
18. Ganitkevich VY, Isenberg G (1990) Contribution of two types of calcium channels to membrane conductance of single myocytes from guinea-pig coronary artery. J Physiol (Lond) 426: 19-42
19. von Beckerath N, Cyrys S, Dischner A, Daut J (1991) Hypoxic vasodilation in isolated perfused guinea-pig heart: an analysis of the underlying mechanisms. J Physiol (Lond) 442: 297-319
20. Daut J, Maier-Rudolph W, von Beckerath N, Mehrke G, Günther K, Goedel-Meinen, L (1990) Hypoxic dilation of coronary arteries is mediated by ATP-sensitive potassium channels. Science 247: 1341-1344
21. Cyrys S, Daut J (1991) Hypoxic vasodilatation in isolated guinea-pig heart is inhibited by nanomolar concentrations of the antidiabetic sulphonylurea glibenclamide. Pflügers Arch 419: (Suppl 1) R107
22. Kirsch GE, Codina J, Birnbaumer L, Brown AM (1990) Coupling of ATP-sensitive K^+ channels to A_1 receptors by G proteins in rat ventricular myocytes. Am J Physiol 259: H820-H826
23. Dart C, Standen NB (1993) The potassium current activated by adenosine in isolated coronary arterial myocytes of the pig. J Physiol (Lond) (in press)
24. Aversano T, Ouyang P, Silverman H (1991) Blockade of ATP-sensitive potassium channel modulates reactive hyperemia in the canine coronary circulation. Circ Res 69: 618-622
25. Clayton FC, Hess TA, Smith MA, Grover GJ (1992) Coronary reactive hyperemia and adenosine-induced vasodilation are mediated partially by a glyburide-sensitive mechanism. Pharmacol 44: 92-100
26. Blatz AL, Magleby KL (1987) Calcium-activated potassium channels. Trends Neurosci 10: 463-467
27. Langton PD, Nelson MT, Huang Y, Standen NB (1991) Block of calcium-activated potassium channels in mammalian arterial myocytes by tetraethylammonium ions. Am J Physiol 260: H927-H934
28. Van Renterghem C, Lazdunski M (1992) A small-conductance charybdotoxin-sensitive, apamin-resistant Ca^{2+}-activated K^+ channel in aortic smooth muscle cells (A7r5 line and primary culture). Pflügers Arch 420: 417-423
29. Rusko J, Tanzi F, van Breemen C, Adams DJ (1992) Calcium-activated potassium channels in native endothelial cells from rabbit aorta: conductance, Ca^{2+} sensitivity and block. J Physiol (Lond) 455: 601-621
30. Brayden JE, Nelson MT (1992) Regulation of arterial tone by activation of calcium-dependent potassium channels. Science 256: 532-535
31. Moncada S, Palmer RMJ, Higgs EA (1991) Nitric oxide: physiology, pathophysiology, and pharmacology. Pharmacol Rev 43: 109-142
32. Mehrke G, Daut J (1990) The electrical response of cultured coronary endothelial cells to endothelium-dependent vasodilators. J Physiol (Lond) 430: 251-272
33. Hallam TJ, Jacob R, Merritt JE (1989) Influx of bivalent cations can be independent of receptor stimulation in human endothelial cells. Biochem J 259: 125-129

34. Adams DJ, Barakeh J, Laskey R, Van Breemen C (1989) Ion channels and regulation of intracellular calcium in vascular endothelial cells. FASEB J 3: 2389-2400
35. Daut J, Standen NB, Nelson MT (1993) The role of the membrane potential of endothelial and smooth muscle cells in the regulation of coronary blood flow. J Cardiovasc Electrophysiol (in press)
36. Seiss-Geuder M, Mehrke G, Daut J (1992) Sustained hyperpolarization of cultured guinea-pig endothelial cells induced by adenosine. J Cardiovasc Pharmacol 20 (Suppl 12): S97-S100
37. Mehrke G, Seiss-Geuder M, Daut J (1991) Sustained hyperpolarization of cultured coronary endothelium induced by adenosine. Pflügers Archiv 418 (Suppl 1): R104
38. Mehrke G, Pohl U, Daut J (1991) Effects of vasoactive agonists on the membrane potential of cultured bovine aortic and guinea-pig coronary endothelial cells. J Physiol (Lond) 439: 277-299
39. Putney JW (1990) Capacitative calcium entry revisited. Cell Calcium 11: 611-624
40. Jacob R (1990) Agonist-stimulated divalent cation entry into single cultured human umbilical vein endothelial cells. J Physiol (Lond) 421: 55-77
41. Penner R, Neher E (1988) The role of calcium in stimulus-secretion coupling in excitable and non-excitable cells. J Exp Biol 139: 329-345
42. Hoth M, Penner R (1992) Depletion of intracellular calcium stores activates a calcium current in mast cells. Nature 355: 353-356
43. Segal SS, Duling BR (1986) Communication between feed arteries and microvessels in hamster striated muscle: segmental vascular responses are functionally coordinated. Circ Res 59: 283-290
44. Segal SS, Damin DN, Duling BR (1989) Propagation of vasomotor responses coordinates arteriolar resistances. Am J Physiol 256: H832-H837
45. Davies, PF, Olesen SP, Clapham D, Morrel EM, Schoen FJ (1988) Endothelial communication. Hypertension 11: 563-572
46. Bény, JL (1990) Endothelial and smooth muscle cells hyperpolarized by bradykinin are not dye coupled. Am J Physiol 258: H836-H841
47. Bürrig KF, Decher CU, Jurokowa, Z (1992) Häufigkeit myo-endothelialer Verbindungen in Arterien und Arteriolen. Anat Anz 174 (Suppl 1): 204

CHAPTER 8
The regulation of blood flow to skeletal muscle during exercise

F Karim and D Cotterrell

The blood flow to skeletal muscle increases with the increase in its metabolic activity during exercise, so increasing the supply of oxygen and other nutrients and providing for the removal of waste products of metabolism. It appears that several factors act in combination to bring about the rapid increase in muscle blood flow. These factors are myogenic, neural, hormonal and local chemical factors that include potassium, adenosine, phosphate ions, lactic acid, hydrogen ions (pH), hyperosmolarity, decreased oxygen level, increased carbon dioxide level, prostaglandins and histamine. Possible roles of endothelium-derived relaxing and hyperpolarizing factors remain to be clarified. The relative contribution of the different factors outlined above varies with the period and type of exercise, as well as between different muscle types. Thus myogenic and neural mechanisms and K^+ appear to be involved in the initial stages of exercise, while adenosine, phosphate, lactate and H^+ contribute to sustained exercise hyperaemia. This chapter will evaluate the current evidence for the role of the individual factors and their mode of action, including the possible contribution of K^+ channels at the cellular level in the regulation of skeletal muscle blood flow during exercise.

Multiple factors are involved in exercise hyperaemia

The blood flow to the human forearm and calf muscles is about 4 ml 100 g^{-1} tissue min^{-1} at rest (fig.1) and that in non-contracting skeletal muscles of most animals under anaesthesia is about 10 ml 100 g^{-1} min^{-1}. The flow increases with the increase in exercise intensity to meet the changing metabolic demands, and may be as high as 15-20 times the resting value in heavy exercise (fig. 1). This suggests that the resistance vessels of the resting muscles of the limbs, which constitute about 40% of the body mass, have a very high basal constrictor tone [1]. The blood flow through contracting muscle is closely correlated with its metabolic activity and hence with its oxygen consumption. Control of the blood flow is exerted at the level of the tone of smooth muscle present in small arteries, arterioles and pre-capillary sphincters. The rapid and dramatic decrease in vascular resistance with increased metabolic demand occurs through dilation of these resistance vessels and through an increase in the number of blood-perfused capillaries [2]. Recent evidence from direct observation of the microcirculation and from histology suggests that about 50% of capillaries are open at rest and that they all open fully during rhythmic muscle contraction at 4 Hz. At rest about 20% of the cardiac output goes to muscles: the total blood flow to skeletal muscle of an average adult is about 1.2 litres min^{-1}. During heavy isotonic or dynamic exercise this may reach 20-25 litres min^{-1}, so that about 80% of the cardiac output may be directed to skeletal muscle under these conditions.

The mechanism of exercise hyperaemia is still not completely understood. Current

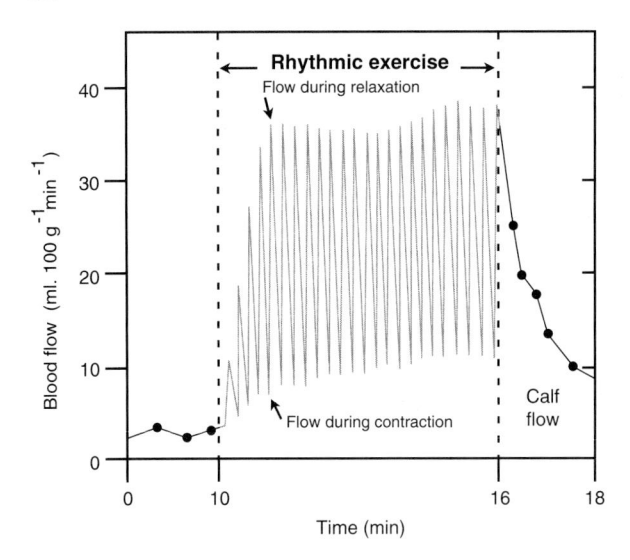

Fig. 1. Effects of rhythmic muscle contraction and relaxation on human calf muscle blood flow. The blood flow was low during muscle contraction due to mechanical compression of the vessels and high during muscle relaxation due to release of compression. From [Barcroft H and Swan HJC (1953) Sympathetic Control of Human Blood Vessels. Arnold, London. Chapter 5, p 48]

evidence suggests that no single mechanism can explain the rapid onset of the increase in blood flow with exercise (within 0.5 s), the steady relationship of blood flow to workload, and the recovery phase following cessation of exercise. In fact each stage appears to involve several factors. These include myogenic, neural, hormonal and local factors including K^+, adenosine, decreased oxygen level, H^+, phosphate, lactic acid, carbon dioxide, hyperosmolarity, histamine, prostaglandins and endothelium-derived relaxing factor.

Myogenic regulation may be involved in initial hyperaemia

According to the myogenic theory, the smooth muscle of small arteries and arterioles contracts strongly when the intravascular pressure (or stretch on the wall) increases, resulting in a reduction in the radius of the vessels [3]. Therefore, when intravascular pressure increases there is an immediate passive distension of the vessels and an increase in blood flow, which is followed by an active contraction of the smooth muscle which reduces the diameter (fig. 2A) and returns the flow to a value close to its previous level. Conversely, when the intravascular pressure falls, blood flow decreases immediately. However, since the smooth muscle then relaxes, the vessel radius increases, restoring the blood flow to near the control level [3, 4]. Such a myogenic autoregulatory response has been observed in the vessels of various organs, including kidneys, intestine and skeletal muscle.

In skeletal muscle, it has been suggested that the myogenic mechanism is involved in producing initial hyperaemia during exercise [5]. In dog muscle the myogenic response can account for about one third of the vasodilation produced by a 1-second tetanus [5]. However, the involvement of a myogenic mechanism in twitch exercise is not so clear. When skeletal muscles contract, the vessels within them are compressed, so passively decreasing their calibre. This is expected to reduce stretch to the smooth muscle, and thus produces relaxation of the vessel wall. Subsequent relaxation of the skeletal muscle would release the vessels from compression, resulting in an increase in the calibre of the vessels and so an immediate reduction of the vascular resistance and an increase in blood

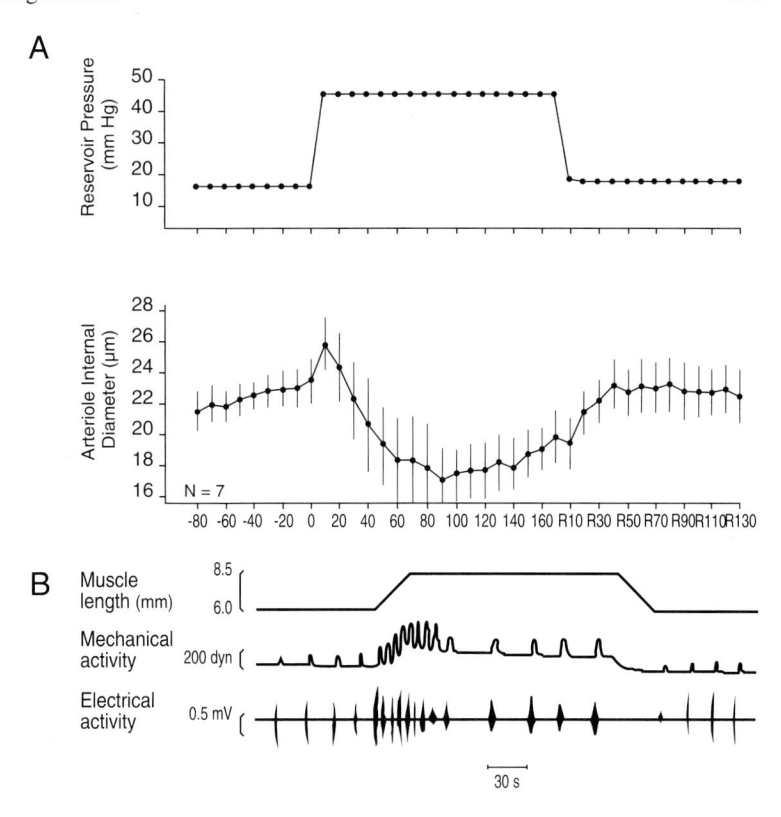

Fig. 2. A Decrease in arteriolar diameter in response to a step change in static intravascular pressure. Studies were done while blood flow was arrested in the isolated cat mesenteric preparation and arterial and venous pressure were increased simultaneously. From [Johnson P C and Intagleitta M (1976) Am J Physiol 231: 1686-1696]. **B** Effect of stretch on mechanical and electrical activity of guinea-pig portal vein. Both mechanical and electrical activity increased with dynamic stretch and then declined during steady stretch to lower levels, but still higher than control levels. From [Johanson B, Mellander S (1975) Circ Res 36: 76-83]

flow (fig. 1). In recent years, the myogenic hypothesis has received strong experimental support and been used to explain the normal basal vascular tone, autoregulatory responses, and the rapid onset of exercise hyperaemia [3, 6]. The pressure-sensitive myogenic mechanism is capable of overriding local metabolic control [7].

The mechanism by which a change in the intravascular pressure causes a myogenic response is not yet fully understood [8]. However, there is a general agreement that the phenomenon originates within the smooth muscle itself and does not seem to be mediated by neurohumoral mechanisms, although it can be modified by them [3, 6, 8]. Thus it has been suggested that the vascular renin-angiotensin system and endothelin, a vasoconstrictor released from the endothelium, may be involved, and that their actions may be modulated by endothelium-derived relaxing factor [8]. When stretched, the cell membrane of smooth muscle is likely to change its ionic permeability, and there is evidence for a change in membrane potential prior to the mechanical constrictor response (figs. 2A, B) [6]. However, it might be expected that once the muscle cells shorten, and remain shortened

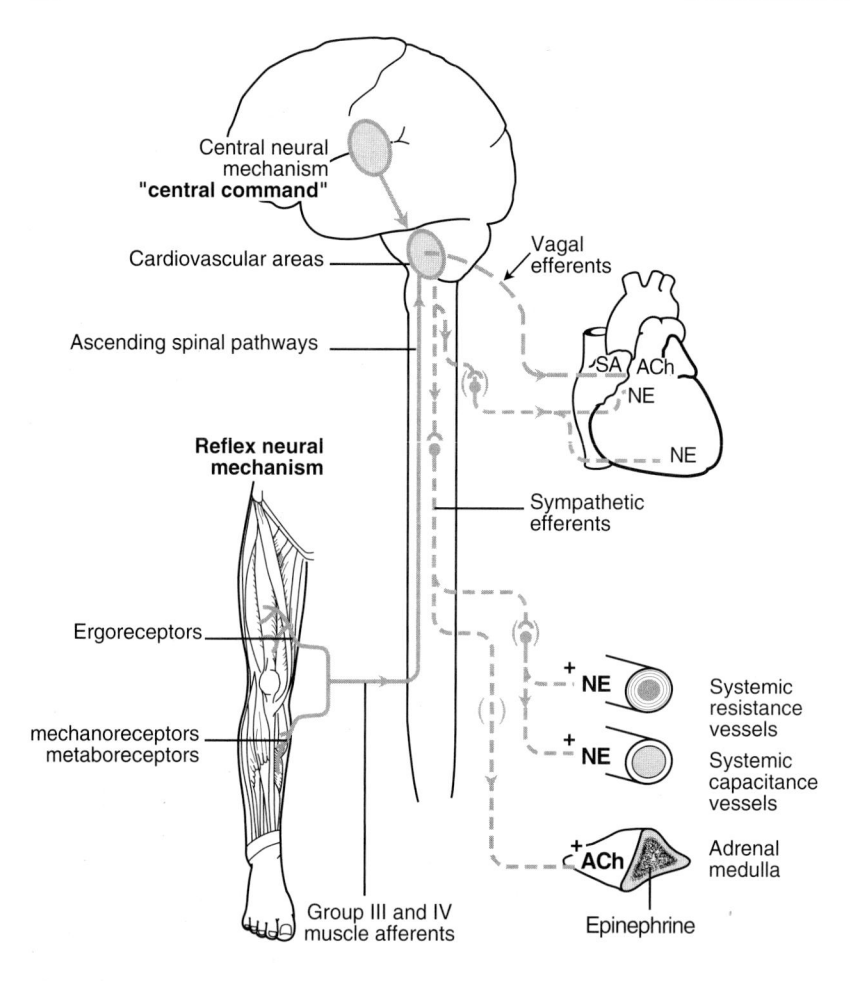

Fig. 3. Neural control of the circulation during exercise. From [12]

for a whole period of elevated pressure, the membrane permeability would go back to its original level, so that the change in the electrical activity that occurs initially would disappear [6]. However, there is evidence that the electrical activity remains higher than control during maintained stretch to the smooth muscle (fig. 2B), and that the vessel diameter remains decreased when the pressure to wall is maintained (fig. 2A). Therefore it has been postulated that the circular smooth muscle contains a sensor which monitors tension even after contraction has occurred. It has also been suggested that distension of proximal vessels gives rise to a propagated signal that maintains increased contraction of distal vessels [9, 10]. It is not yet certain whether stretch sensitive ion channels in smooth muscle cells are involved in the myogenic response to alteration of intravascular pressure. Whatever the molecular mechanism of the myogenic response may be, there is some evidence for its enhancement by catecholamines and by sympathetic nerve stimulation, and for its inhibition by local vasodilator metabolites and endothelium-derived relaxing factor [8].

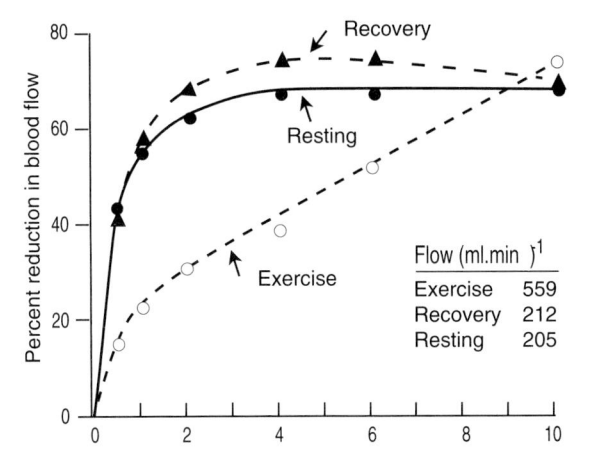

Fig. 4. Effect of electrical stimulation of the lumbar sympathetic chain, with increased frequencies, on the dog limb flow during rest and exercise. At rest there was marked flow reduction, but during exercise sympathetic stimulation was less effective due to effects of metabolites. From [Donald DE, Rowlands DJ, Ferguson D (1970) Circ Res 26: 185-189]

Flow (ml.min)^1
Exercise 559
Recovery 212
Resting 205

Neural factors regulate muscle blood flow at rest and at the start of exercise

The autonomic nervous system controls the cardiovascular responses during exercise. During different types of exercise, there is an overall increase in the activity of the sympathetic nervous system and a decrease in the activity of the parasympathetic nervous system (fig. 3). The increased sympathetic activity increases the plasma concentration of catecholamines, increases the rate and force of contraction of the heart, and decreases the calibre of the resistance and capacitance vessels [11]. The decreased parasympathetic activity primarily increases heart rate and the force of atrial contraction. It is generally accepted that during exercise two mechanisms are involved in changing the activities of the autonomic nervous system: a) a central neuronal mechanism, or "central command", originating in the brain and; b) a reflex neuronal mechanism originating in the contracting muscles themselves from stimulation of sensory fibres by muscle metabolites such as K^+ (fig. 3). Group III and IV nerve fibres carry the afferent impulses to the brain [12]. Both central and peripheral signals activate the cardiovascular control areas of the medulla [12]. In humans, an increased sympathetic nerve activity is responsible for the redistribution of blood flow away from the splanchnic area, the kidneys and resting muscles to the active muscles [13]. The effects of local vasodilators are considerably reduced by this increased sympathetic constrictor effect during muscular exercise (fig. 4), and therefore a serious fall in peripheral vascular resistance and so in the arterial blood pressure is prevented.

During rest, the autonomic nervous system is the primary regulator of skeletal muscle blood flow, giving rise to a tonic vasoconstriction. The extent of this vasomotor tone can be seen by blocking the nerves by local anaesthetics [1] or by denervating the muscle (fig. 5); muscle blood flow increases by 2-3 fold following nerve blockade or section. However, during exercise local factors (both myogenic and metabolic) can quickly override the effects of both strong sympathetic activity and of elevated concentrations of vasoconstrictor hormones (adrenalin, noradrenaline, angiotensin and vasopressin) on the skeletal muscle microcirculation [8]. These local factors can increase muscle blood flow by 20-30 fold during heavy exercise. During non-exercise situations vasodilation can be achieved either passively, by decreasing impulse activity in constrictor fibres, or actively,

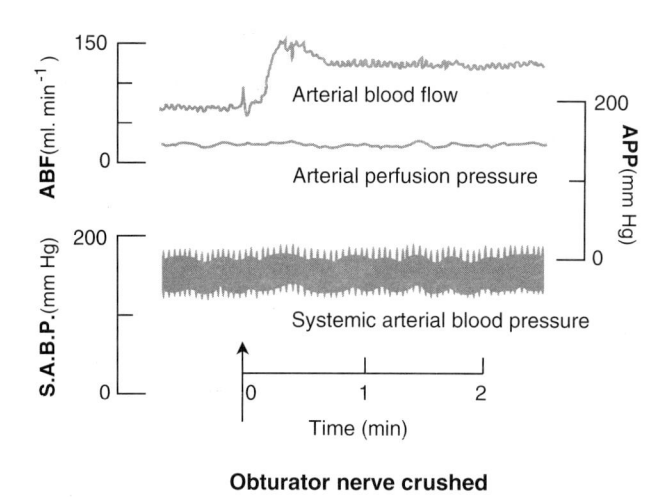

Fig. 5. The effect of denervation of the gracilis muscle on the gracilis arterial blood flow in anaesthetized dogs. *ABF*, mean arterial blood flow; *APP*, mean arterial perfusion presssure; *SABP*, systemic arterial blood pressure. Note the immediate increase in the blood flow following crushing the nerve supply to the muscle showing a basal sympathetic constrictor tone. From [Ballard H, Cotterrell D, Karim F, unpublished]

by stimulation of sympathetic cholinergic or ß-adrenergic vasodilator nerve fibres. It has been suggested that the cholinergic vasodilation occurs mainly from cerebral activity under conditions such as emotional reactions and exercise. However, there is no clear evidence that cholinergic vasodilation occurs in human skeletal muscle during exercise.

Skeletal muscle blood flow sometimes increases at or before the start of exercise, an effect that is believed to be neurally mediated via sympathetic cholinergic and ß-adrenergic vasodilator fibres (see below). It is also believed that the impulses from the higher brain centres bypass the medullary cardiovascular control areas. It is not known whether there is any reduction of vasoconstrictor tone at the start of exercise contributing to a rapid initial hyperaemia. There is evidence, however, that during heavy exercise muscle blood flow is the net result of active metabolic vasodilation and the reflex vasoconstriction that competes with it. Rowell [14] suggested that the limited capacity of the human heart to pump blood is exceeded by the great capacity of skeletal muscle to cause vasodilation, so that the only way to maintain arterial pressure during severe whole-body exercise is to limit the vasodilation in active muscle by reflex vasoconstriction. Nevertheless, the role of autonomic nerves in the regulation of skeletal muscle blood flow during exercise is still far from clear.

The role of humoral factors

The arterioles of skeletal muscle contain sympathetic α, ß, and cholinergic receptors (fig. 6). The ß receptors in the dog, cat and pig are mainly $ß_2$ subtypes [15]. On stimulation the sympathetic nerve terminals normally release a large amount of noradrenaline and a small amount of adrenalin, whereas the adrenal medulla produces a large amount of adrenalin and a small amount of noradrenaline. Noradrenaline produces strong

Fig. 6 A-D. The effects of sympathetic nerve stimulation on blood flow in skeletal muscle shown diagramatically to illustrate the various sympathetic receptors. **A** Decreased flow due to strong α-receptor stimulation. **B** After α blockade, sympathetic nerve stimulation produces an increase in blood flow due to stimulation of cholinergic and ß adrenergic stimulation. **C** After α blockade and cholinergic blockade with atropine the vasodilator effect is markedly reduced. **D** Small sympathetic ß-receptor vasodilation (as seen in **C**) is blocked with a ß blocker. From [Ruch TC, Patton HD (1974) Physiology and Biophysics. WB Saunders, Philadelphia, p 125]

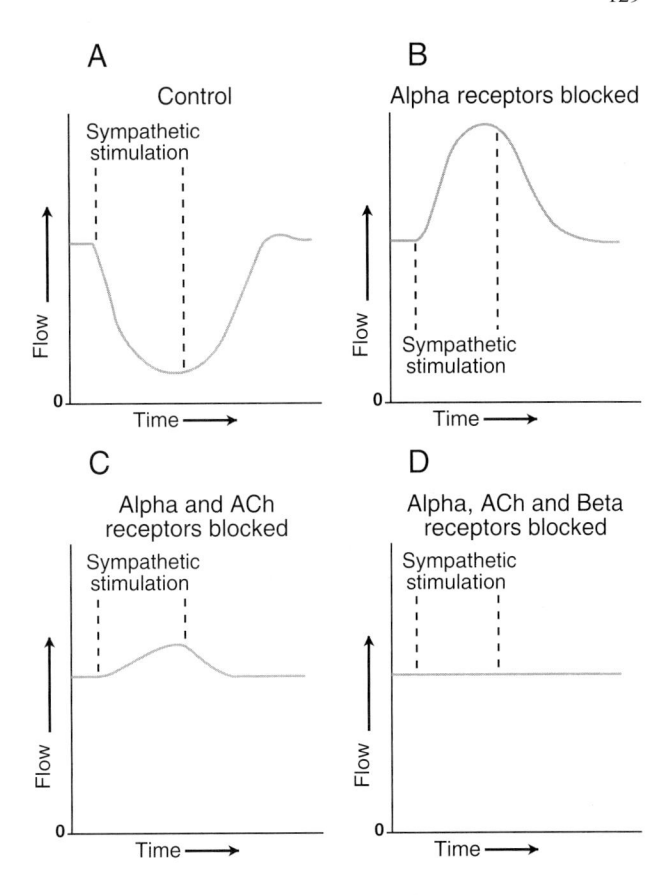

α-constrictor action and a small ß-dilator effect. On the other hand, adrenalin at a low concentration produces a large ß-dilator action in the vessels of skeletal muscle, which have a large number of ß$_2$-dilator receptors [15]. Therefore, the results of stimulation of the sympathoadrenal system, as occurs during exercise, which raises the plasma concentration of both noradrenaline and adrenalin would be a balance between the actions of these transmitters on the above receptors [16].

Patients who are under treatment with beta blockers having effects on both ß$_1$- and ß$_2$-adrenoceptors, such as propranolol, would not be expected to increase blood flow to the active muscle during exercise to the same extent as do a normal persons for the following main reasons: a) the dilator action of catecholamines via ß-adrenoceptors is eliminated; b) the sympathetically induced positive inotropic and chronotropic effects on the heart are greatly reduced, so limiting the rise in cardiac output and arterial blood pressure during exercise; c) there is also evidence that ß-adrenergic blockers can unmask α-adrenergic vasoconstrictor actions on the coronary vessels, so limiting the increase in coronary blood flow during exercise and therefore the cardiac output and the arterial blood pressure.

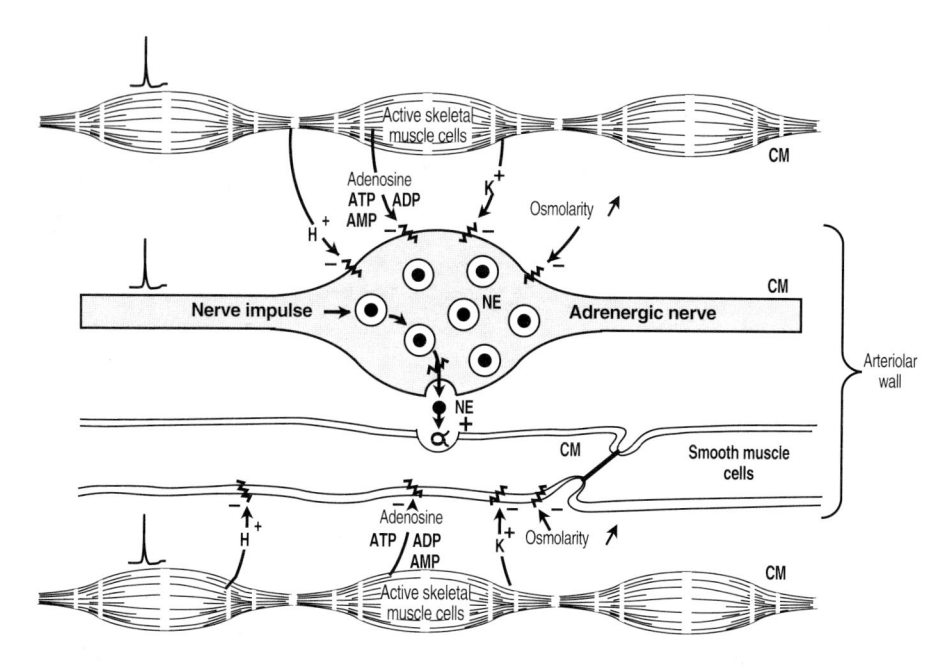

Fig. 7. Products of muscle contraction cause dilatation of arterioles because of their direct inhibitory effect on smooth muscle cells and because they interrupt vasoconstrictor impulses of sympathetic nerves. Arteriolar wall is shown as one layer of smooth muscle cells with adrenergic nerve and one of its varicosities containing adrenergic neurotransmitter norepinephrine *(NE)*. AMP, adenosine monophosphate; *ADP,* adenosine diphosphate; *ATP,* adenosine triphosphate; *CM,* cell membrane; +, activation; -, inhibition; α, α-adrenergic receptor. From [13]

Local metabolic factors sustain muscle vasodilation

Adenosine contributes to exercise hyperaemia

The vasodilator action of adenosine has been known for a long time (see also chapter 7). However, it was in 1964 that Berne and colleagues (see [17]) suggested, without real evidence, the adenosine hypothesis for the regulation of blood flow in skeletal muscle and the coronary circulation. According to their hypothesis, when oxygen demand exceeds oxygen supply during muscular exercise there is a net breakdown of ATP to adenosine (fig. 7). Adenosine then diffuses down its concentration gradient into the perivascular space and so into the venous blood (figs. 7, 8). During this passage it acts on the smooth muscle of the resistance vessels to bring about vasodilation, and thus increases blood flow to a level where oxygen supply matches oxygen demand. Conversely, if oxygen demand is decreased due to reduced activity of the tissues or the blood flow is increased to a level where oxygen supply exceeds oxygen demand, then little or no adenosine is produced, leading to a vasoconstriction and a reduction of the blood flow to a level where oxygen demand and supply are again equal. Thus blood flow is matched to the metabolic requirements of the muscle by a negative feedback mechanism. This may be the reason why several investigations failed to observe a significant increase in the concentration of venous adenosine during free-flow muscle contraction because of the dilution of adeno-

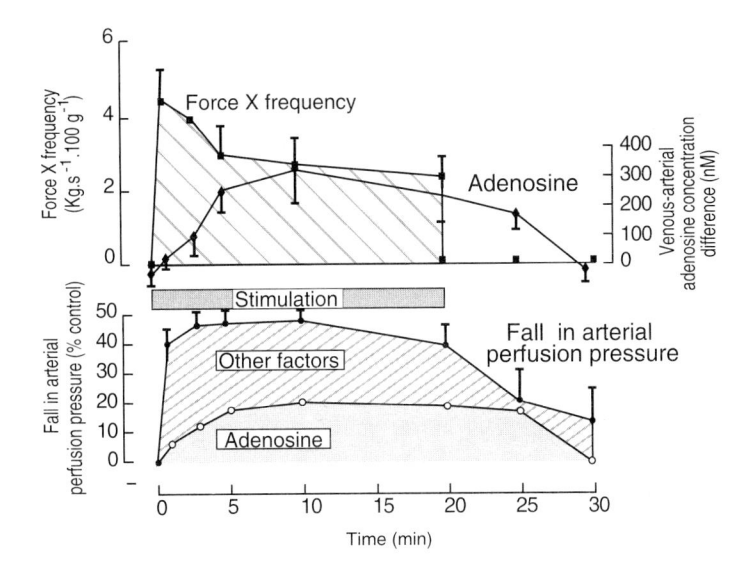

Fig. 8. A comparison between the time courses of the changes in arterial perfusion pressure, venous-arterial difference in plasma adenosine concentration, and the force x frequency of contractions during and after contractions of the high constant-flow perfused gracilis muscle for 20 min. Arterial and venous adenosine concentrations were measured by HPLC; the fall in arterial perfusion pressure caused by these adenosine concentrations was calculated from the vasodilation caused by intra-arterial adenosine infusion. The proportion of the total vasodilation during contractions that could be caused by the released adenosine is shown by the stippled areas, and the remainder of the vasodilation, which must have been caused by other factors, is shown by the hatched area. Values are the mean and SE of seven tests in six anaesthetized dogs. Note that the fall in arterial perfusion pressure is near maximum within 1 min of contractions, whereas release of adenosine in the venous plasma is very small. From [20]

sine by the increased blood flow. However, the release of adenosine into the venous blood increases along with the increase in blood flow as exercise proceeds [18].

Although numerous studies have been carried out over the past 30 years, the importance of the role of adenosine in skeletal muscle hyperaemia during exercise has only recently been clarified by studies using modern biochemical and pharmacological techniques [18, 19]. Ballard and colleagues [20] clearly demonstrated that adenosine is unlikely to initiate active hyperaemia (fig. 8). The contribution of adenosine to sustained exercise hyperaemia depends on the type of blood flow (restricted or free-flow) and the type of muscle fibres (glycolytic or oxidative): adenosine makes a smaller contribution to vasodilation in free-flow exercise (about 20%) than to high restricted flow exercise vasodilation (about 40%) and to post tetanic free-flow hyperaemia (about 30%). The adenosine-induced component of hyperaemia is greater in oxidative fibres (red muscle) than in glycolytic fibres (white muscle). Glycogen stores in red muscles are low and the increased blood flow brings more glucose to replenish glycogen in these muscles [17]. It has also been shown that adenosine is much more important in the regulation of myocardial active hyperaemia than in skeletal muscle active hyperaemia [21] (see chapter 7).

The mechanism of the vasodilator action of adenosine is not yet fully understood. Several mechanisms have been proposed to explain the relaxant effect of adenosine:

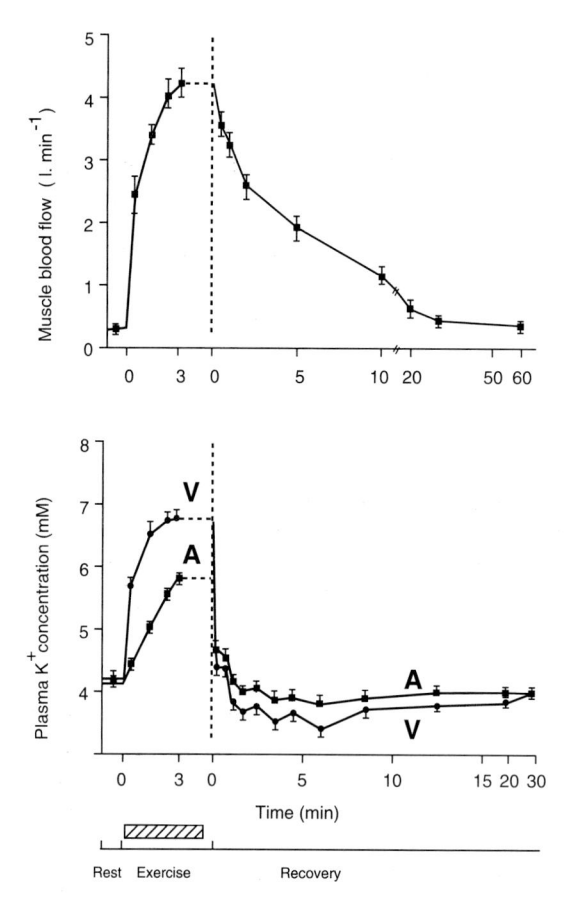

Fig. 9. Muscle blood flow and arterial *(A)* and venous *(V)* plasma K⁺ concentrations from human knee extensor muscle at rest, during exercise to exhaustion *(shaded bar)* and during recovery from exercise are shown over the same time course. Modified from [30]

a) inhibition of Ca^{2+} influx into smooth muscle; b) inhibition of noradrenaline release at the sympathetic nerve endings to the muscle vessels (fig. 7); and c) modulation of K^+ channels of the smooth muscle membrane. There is indirect experimental evidence that adenosine may produce its effect by opening the K^+ channels: Daut et al. [22] showed in isolated, perfused guinea-pig hearts that glibenclamide, a blocker of ATP-sensitive K^+ channels, can inhibit the coronary vasodilation induced by hypoxia, ischaemia or exogenous adenosine. In addition, Merkel et al. [23] have demonstrated that glibenclamide blocked the A_1 component of the adenosine-induced vasodilation in isolated segments of porcine coronary arteries. Therefore it is not unreasonable to speculate that adenosine may produce some of its vasodilatory effects in exercising skeletal muscle by acting via ATP-sensitive K^+ channels of the smooth muscle membrane of skeletal muscle arteries and arterioles. However, this has not yet been demonstrated directly.

K^+ Release is involved in exercise hyperaemia

During muscular exercise there is a large release of K^+ from the intracellular fluid of skeletal muscle into the interstitial fluid and plasma (fig. 7). Two types of K^+ channel in the skeletal muscle membrane are thought to be important in this release: delayed rectifier

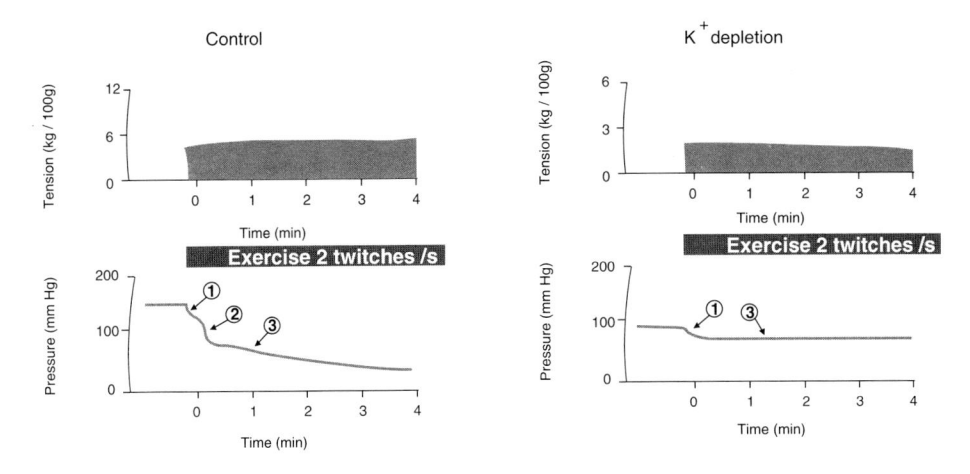

Fig. 10. Tension and perfusion pressure (at constant flow) of isolated skeletal muscle preparations from a control dog and a dog depleted of K^+ by dietary restriction. Note that muscle of K^+-depleted dog developed less tension and exhibited less vasodilation with elimination of second phase of initial vasodilator response. Circled numbers indicate phases of vasodilator response. From [Hazeyama Y, Sparks HV (1979) Am J Physiol 236: H480-H486]

and ATP-sensitive K^+ channels [24]. The delayed rectifier K^+ channel is mainly responsible for the repolarization phase of the skeletal muscle action potential, so that channel activity increases with increased firing of action potentials in the contracting muscle. Medbo and Sejersted [25] have calculated that the gross efflux of potassium via these channels is at least three times the minimum required to account for the observed rise in plasma potassium.

Secondly, some K^+ release may be through ATP-sensitive K^+ channels opening in response to metabolic changes in the skeletal muscle fibres. Local decreases in ATP concentration [24], the decrease in internal pH seen during exercise, and other metabolites act to open ATP-sensitive K^+ channels and so increase K^+ efflux [26]. Direct evidence for K^+ release through this channel in exercising skeletal muscle has been obtained by blocking with glibenclamide, indicating that the ATP-sensitive K^+ channels do contribute to net K^+ loss from muscles during fatiguing work [27].

The K^+ lost from muscles into the interstitial fluid rapidly equilibrates with the plasma and may increase the plasma K^+ concentration from its normal value of 4 mM to some 8-15mM (fig. 9). Several studies have shown that blood flow to skeletal muscles during exercise is related to the increase in plasma [K^+] released from contracting muscles [24]. For example, blood flow in the femoral artery in human subjects exercising their quadriceps to exhaustion increases some 10-fold and is accompanied by a rapid increase in venous plasma [K^+] from 4 to 7 mM (fig. 9).

During maximal exercise there is clear evidence that K^+ release increases linearly with both intensity and duration of exercise. However, during submaximal exercise K^+ release diminishes with duration of exercise indicating that its contribution may not be as important in sustained submaximal exercise. The K^+ release is not dependent on H^+ ion release: K^+ release has been shown to be proportional to workload in both acidotic and alkalotic subjects, in other words it is independent of acid-base status [24]. Depletion of skeletal muscle K⁺ by dietary restriction reduces the vasodilator response in contracting skeletal muscles (fig. 10). Although muscles from K⁺-depleted animals do not contract as forcibly as normal muscles, they exhibit far less vasodilation and the second initial phase of dilation is completely abolished (fig. 10). There is considerable evidence, therefore, to suggest that the raised extracellular [K⁺] correlates well with the initial rapid onset of exercise hyperaemia.

The mechanism of the vasodilator action of K⁺ is less clear, but inward rectifier K⁺ channels in the smooth muscle of arterioles may play a role in causing hyperpolarization and thus dilation of arterioles [28]. Unlike other tissues, in submucosal arterioles inward rectifier channels are activated at membrane potentials positive to the potassium equilibrium potential, E_K (see chapter 1), with normal internal and external K⁺ concentrations. As a consequence, virtually all the resting K⁺ conductance at normal resting potentials results from these inward rectifier K⁺ channels. When external [K⁺] is increased to the range seen during severe exercise (8-15 mM) E_K will become less negative, but the relative increase in K⁺ permeability due to increased activation of inward rectifiers will dominate the effect on the membrane potential and result in hyperpolarization. Thus the net result of an increase in external [K⁺] on the arteriole will be a suppression of contractile activity and the vessels will dilate [10].

K⁺ does not play a role in post-exercise hyperaemia

K⁺ release does not appear to correlate with the post-exercise hyperaemia seen during recovery from exercise. Blood flow remains high when venous plasma [K⁺] has fallen to its initial level or even to levels below arterial [K⁺] (fig. 9). The re-uptake of K⁺ during recovery is mediated by the Na/K pump in skeletal muscle membrane [25]. The pump is stimulated by a fall in intracellular [Na⁺] rather than an increase in external [K⁺] which could only increase pump rate by some 10%. Although training increases the re-uptake of K⁺ into muscles, there appears to be no induction of new Na/K pump sites as judged by ³H-ouabain binding [29]. Increased re-uptake after exercise may well be stimulated by catecholamines acting on α-adrenoceptors to increase Na/K pump activity [24].

K⁺ balance during exercise

Overall, during heavy sustained exercise, considerable amounts of K⁺ are released from skeletal muscle through K⁺ channels into the interstitium. The rise in [K⁺] may be related to the rapid initial vasodilation seen. 70% of the released K⁺ is rapidly taken up again into the muscles by the Na/K pump whilst about 30% is carried away in the plasma: both mechanisms reduce interstitial [K⁺] rapidly at the cessation of exercise [30], whereas hyperaemia persists for a period after exercise. Some of the released K⁺ may be taken up from the plasma into erythrocytes [31] as there is a transient rise in erythrocyte [K⁺], which falls, however, to control levels within 5 min of cessation of exercise. However, quite marked haemo-concentration occurs during strenuous exercise, causing erythrocyte shrinkage. When this is taken into account there appears to be no net change in erythrocyte K⁺ content indicating that plasma is the main route by which extracellular K⁺ is cleared from the active skeletal muscles [30]. Clinically, the raised extracellular potassium may produce changes in cardiac muscle function; high plasma [K⁺] is known to induce cardiac arrhythmias in some circumstances [29]. On the other hand a reduced plasma [K$^+$] seen during the recovery phase of exercise from stimulation of the Na/K pump may cause ventricular tachycardia. Similarly, a low [K$^+$] as seen in such neurological disorders as "grand-mal" seizures, (where intense muscular activity causes stimulation of the Na/K pump and thus accelerates re-uptake of K$^+$) may cause cardiovascular disturbances.

Lactate may be involved during recovery from exercise hyperaemia

Lactate has long been known as an effective dilator of arteriolar smooth muscle and has been proposed as a candidate for the mediation of part of the sustained exercise hyperaemia [2]. A large increase in muscle lactate is seen during strenuous exercise to exhaustion (fig.11). Lactate in muscle biopsy samples from quadriceps muscle rises from about 2 to 28 mmol kg^{-1} at the end of 3 min of exercise to exhaustion [30]. Intracellular lactate diffuses down its concentration gradient into the interstitial fluid and plasma, where a peak level of about 15 mM is reached from the resting level of 0.8 mM. The increase in lactate release and the increase in blood flow are similar whether the exercise is continuous or intermittent [32].

There is, however, some degree of correlation between lactate release and K$^+$ release from muscles during strenous exercise [29], but as expected K$^+$ release is more rapid than lactate release. In contrast to the very rapid return of plasma [K$^+$] to normal at the end of exercise (see above), both intramuscular and plasma lactate decline slowly during recovery, with approximately the same time course (10 - 15 min) as the return of blood flow to control levels [30]. The lactate released into plasma is partially taken up into erythrocytes by both the inorganic anion exchanger and by a specific carrier for monocarboxylates. Erythrocyte lactate levels rise to about 50% of the peak concentration in plasma [31]. Since the time course of release of lactate into the interstitial fluid is slower than either that of K$^+$ release or that of the increase in blood flow, it is unlikely that lactate plays a role in the initiation of vasodilation. However, the parallel decline in post-exercise hyperaemia and in plasma and intracellular lactate concentrations suggests that lactate may be an important mediator during this recovery phase.

The roles of H$^+$, Pco$_2$ and Po$_2$ are less clear

H$^+$ ions (and CO$_2$) dilate arterioles in skeletal muscle [2]. During short duration strenuous exercise, either continuous [30] or intermittent [2] venous pH falls very rapidly from 7.4 to 7.1 in line with increased blood flow (fig.11). Venous CO$_2$ rises equally rapidly from about 48 to 97 mm Hg. Both pH and CO$_2$ then stabilize until exhaustion, and may even decline with repeated bouts of exercise. Thus neither H$^+$ nor CO$_2$ mirror closely the changes in blood flow during sustained exercise or recovery. After exercise the return of Pco$_2$ to the control level is more rapid than that of H$^+$, but both return more rapidly than does lactate, which follows the restoration of blood flow as described above. In sustained exercise more H$^+$ ions than CO$_2$ are added to blood and pH changes may well be altered by Na/H exchange and CO$_2$/HCO$_3$ exchange as well as by released lactate. Overall, in exercise CO$_2$ seems to change with a time course which does not mirror initiation or recovery of blood flow, while H$^+$ movements are a result of a complex interaction of lactate release, CO$_2$ production and Na/H co-transport, making a direct comparison with blood flow difficult.

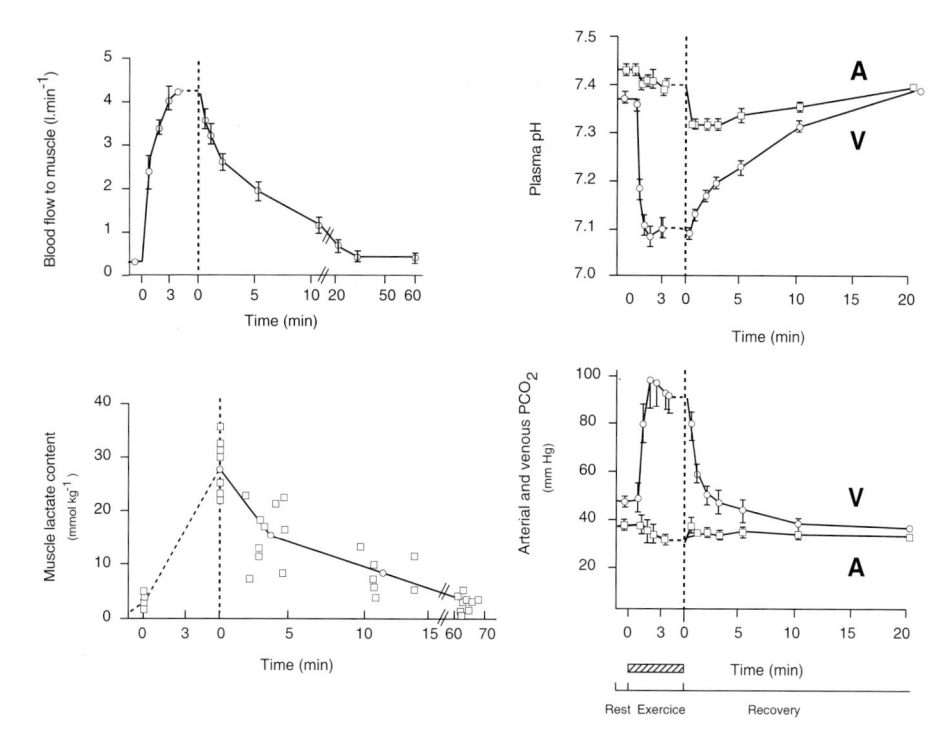

Fig. 11. Blood flow, muscle lactate, blood pH and P_{CO_2} from contracting human knee extensor muscles during rest, maximal exercise to exhaustion and recovery are shown over the same time course. From [30]

Undoubtedly oxygen demand is a very important determinant of vasodilation. However, P_{O_2} per se may not be a major determinant of active hyperaemia [2]. Although a decrease in muscle oxygen had been thought to be a primary stimulus for exercise hyperaemia [25], the following evidence suggests that its effect is indirect, occurring via metabolism [2]: a) the O_2 content of venous blood and tissues during recovery from exercise can be higher than those observed at rest, despite a high blood flow during the recovery phase; b) breathing pure O_2 at normal pressure does not reduce post-exercise blood flow as would be expected if P_{O_2} were a stimulus; c) perfusing resting muscles with blood deoxygenated to the same level as that seen during contractions only causes about one third of the vasodilation seen during contraction; d) restoration of tissue P_{O_2} by superfusion with O_2 only reduces by about half the increase in blood flow caused by contraction. Oxygen lack is therefore unlikely to be a direct stimulus for metabolic vasodilation [2].

Inorganic phosphate at normal pH does not cause vasodilation

In many forms of human forearm or quadriceps exercise, phosphate levels in venous plasma rise only slightly (about 10%) [33]. Nevertheless, phosphate release, particularly from oxidative muscle fibres, correlates with changes in blood flow in exercise of short duration. The mechanism of action at a cellular level is not clear. Acid phosphate applied by close arterial injection has marked vasodilator properties [34]. However, at physiolo-

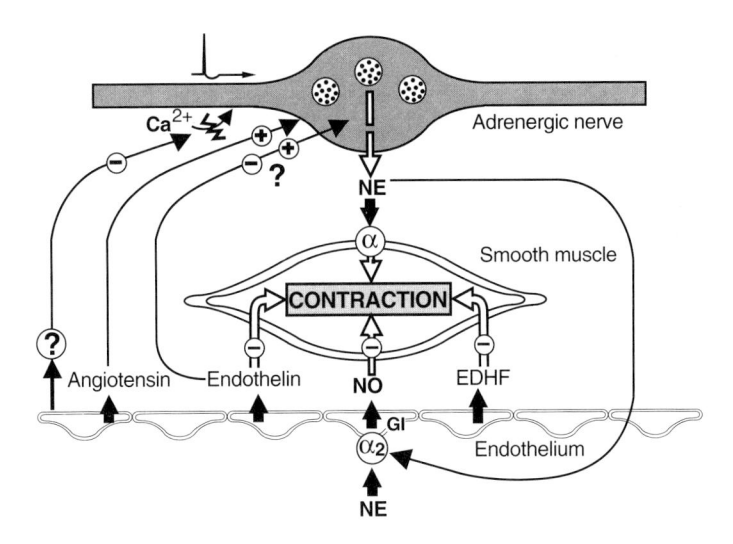

Fig. 12. Schematic diagram of possible interactions between substances released from adrenergic nerve endings and the endothelium. α_2, α_2-adrenergic receptor; Ca^{2+}, calcium; *EDHF*, endothelium-derived hyperpolarizing factor; *Gi*, guanine nuceotide regulatory protein sensitive to inhibition by pertussis toxin; *NE*, norepinephrine; *NO*, nitric oxide; +, stimulation; -, inhibition. From [Miller V M (1991), News Physiol Sci 6: 60-63]

gical pH its infusion does not cause vasodilation of the skeletal muscle vascular bed [34]. Two possible modes of action are by chelation of Ca^{2+} in interstitial spaces and by blockade of Ca^{2+} channels [35]. Overall, the low concentration of phosphate achieved during exercise and its poor vasodilator properties mitigate against phosphate playing a major role in sustained exercise vasodilation, but it may be important in short-term control in oxidative fibres.

Hyperosmolarity makes a negligible contribution

When muscle glycogen (high molecular weight) is metabolized to lactic acid (low molecular weight) in exercise, a net increase in osmolarity occurs because of the osmotic activity of the lactate produced. Mellander (see [1]) proposed that this increase in osmolarity caused vasodilation. However, the increase in osmolarity is limited to the first few minutes of exercise and topically applied hyperosmotic solutions only cause a minor vasodilation [2]. A rise in osmotic pressure is, therefore, unlikely to be a contributor to sustained exercise hyperaemia.

The role of prostaglandins, histamine and endothelial factors has yet to be determined

There has been some suggestion that prostaglandins and histamine may be involved in exercise hyperaemia in skeletal muscle [1, 17]. The arterioles of skeletal muscle release prostaglandins in response to increases in the velocity of blood flow [36]. However, indomethacin, an inhibitor of prostaglandin synthesis, can attenuate reactive but not functional hyperaemia, though it can shorten the duration of functional hyperaemia in dog skeletal muscle [17]. Thus there appears to be no strong evidence for a role of this substance in exercise hyperaemia.

There is strong evidence that vascular endothelial cells can continuously synthesize nitric oxide from the aminoacid L-arginine and that the vasodilator action of endothelium-derived relaxing factor is due to nitric oxide [37]. It has been suggested that nitric oxide plays a significant role in the control of arteriolar tone and blood pressure [38]. Acetylcholine and histamine have been reported to cause a release of an endothelium-derived hyperpolarizing factor, a K^+-channel activator, from vascular endothelial cells. This factor hyperpolarizes the smooth muscle membrane and so relaxes smooth muscles. Therefore, the endothelium-dependent relaxation of smooth muscle seems to be due to both endothelium-derived relaxing and hyperpolarizing factors, the former by an increased production of cyclic GMP and the latter by activation of K^+channels [39]. Since both factors are released when blood flow increases, it is likely that both these endogenous vasodilators are also involved in the mediation of exercise hyperaemia (fig. 12), though no experimental evidence for this is currently available.

References

1. Shepherd JH (1983) Circulation to skeletal muscle. In: Shepherd JT, Aboud FM (eds) Handbook of physiology. The cardiovascular system. Regulation of circulation to individual vascular beds. Am Physiol Soc, Bethesda MD, sect 2, vol III, pp 319-370
2. Renkin EM (1984) Control of microcirculation and blood-tissue exchange. In: Renkin EM, Michel CC (eds) Handbook of physiology. The cardiovascular system. Control of microcirculation and blood-tissue exchange. Am Physiol Soc, Bethesda MD sect 2, vol IV, pp 627-689
3. Johnson PC (1991) The myogenic response. News Physiol Sci 6: 41-42
4. Folkow B (1949) Intravascular pressure as a factor regulating the tone of the small vessels. Acta Physiol Scand 17: 289-310
5. Sparks HV (1980) Effect of local metabolic factors on vascular smooth muscle. In: Bhohr DF, Somlyo AP, Sparks HV (eds) Handbook of Physiology. The cardiovascular system. Effect of local metabolic factors on vascular smooth muscles. Am Physiol Soc, Bethesda MD, sect 2, vol II, pp 475-513
6. Johnson PC (1986) Autoregulation of blood flow. Circ Res 59: 483-594
7. Meininger GA, Mack CA, Fehr KL, Bohelen HG (1987) Myogenic vasoregulation overrides local metabolic control in resting rat skeletal muscle. Circ Res 60: 861-870
8. Cowley AW, Hinojosa-Laborde C, Barber BJ, Harder DR, Lombard JH, Green AS (1989) Short-term autoregulation of systemic blood flow and cardiac output. News Physiol Sci 4: 219-225
9. Robinson BF, Benjamin N (1988) Autoregulation in skeletal muscle. In: Vanhoutte PM (ed) Mechanisms of vasodilatation. Raven Press, New York, pp 395-399
10. Johnson PC (1980) The myogenic response. In: Handbook of physiology, the cardiovascular system, vascular smooth muscle. Am Physiol Soc, Bethesda MD, vol II, pp 409-442
11. Karim F (1987) Mechanical effects of respiration on heart function. Appl Cardiopulm Pathophysiol 1: 109-126
12. Mitchell JH (1990) Neural control of the circulation during exercise. In: Medicine and Science in sports and exercise. American College of Sports Medicine, vol 22 pp 141-154
13. Shepherd JT, Vanhoutte PM (1979) The human cardiovascular system: Facts and Concepts. Raven Press, New York, p 93

14. Rowell LB (1988) Muscle circulation in exercise: metabolic vasodilatation versus reflex vasoconstriction. In: Vanhoutte PM (ed) Mechanisms of vasodilatation. Raven Press, New York, pp 339-344
15. Nimo AJ, Whitaker EM, Karim F, Morrison JFB, Castair S (1989) Localisation of ß-adrenoceptor subtypes in the coronary arteries and gracilis arteries of the dog and pig. Proc Brit Pharmacol Soc Abst 98: 725
16. Christensen NJ, Galbo H (1983) Sympathetic nervous activity during exercise. Ann Rev Physiol 45: 139-153
17. Hudlicka O (1985) Regulation of muscle blood flow. Clinical Physiol 5: 201-229
18. Karim F, Ballard HJ, Cotterrell D (1988) Role of adenosine in exercise vasodilatation. In: Vanhoutte PM (ed) Mechanisms of vasodilatation. Raven Press, New York, pp 383-388
19. Goonewardene IP, Karim F (1991) Attenuation of exercise vasodilatation by adenosine deaminase in anaesthetized dogs. J Physiol (Lond) 442: 65-79
20. Ballard HJ, Cotterrell D, Karim F (1987) Appearance of adenosine in venous blood from the contracting gracilis muscle and its role in vasodilatation in the dog. J Physiol (Lond) 387: 401-413
21. Karim F, Goonewardene IP (1991) The effect of adenosine deaminase on pacing induced coronary functional hyperaemia in denervated hearts of anaesthetized dog. J Physiol (Lond) 438: 134P
22. Daut J, Maier-Rudolph W, Von Beckerath N, Mehrke G, Gunther K, Goedel-Meinen L (1990) Hypoxic dilatation of coronary arteries is mediated by ATP-sensitive potassium channels. Science 247: 1341-1344
23. Merkel LA, Lappe RW, Rivers LM, Cox BF, Perrone M H (1992) Demonstration of vasorelaxant activity with an A1-selective adenosine agonist in porcine coronary artery: involvement of potassium channels. J Pharmacol Exp Ther 260: 437-443
24. Lindinger MI, Sjogaard G (1991) Potassium regulation during exercise and recovery. Sports Medecine 11: 382-401
25. Medbo JI, Sejersted OM (1990) Plasma potassium changes with high intensity exercise. J Physiol (Lond) 421: 105-122
26. Standen NB (1992) Potassium channels, metabolism and muscle. Exp Physiol 77: 1-25
27. Paterson DJ, Vejlstrup N, Willford D, Hogan MC (1992). Effect of a sulphonyurea on dog skeletal muscle performance during fatiguing work. Acta Physiol Scand 144: 399-400
28. Edwards FR, Hirst GDS, Silverberg GD (1988) Inward rectificiation in rat cerebral arterioles: involvement of potassium ions in autoregulation. J Physiol (Lond) 404: 455-466
29. Kjeldsen K, Norgaard A, Han C (1990) Exercise induced hyperkalaemia can be reduced in human subjects by moderate training without change in skeletal muscle Na/K, atpase concentration. Europ J Clin Invest 20: 642-647
30. Juel C, Bangsbo J, Graham T, Saltin B (1990) Lactate and potassium fluxes from human skeletal muscle during and after intense dynamic knee extensor exercise. Acta Physiol Scand 140: 147-159
31. McKelvie RS, Lindinger MI, Heigenhauser GJF, Jones NL (1991) Contribution of erythrocytes to the control of electrolyte changes of exercise. Can J Physiol Pharmacol 69: 984-993
32. Bystrom SEG, Mathiassen SE, Franson-Hall CF (1991) Physiological effects of micropauses in isometric handgrip exercise. Eur J Appl Physiol 63: 405-411
33. Weavers RA, Joosten EMG, Margot JB, Theewes AGM, Veerkamp JH (1990). Excessive plasma K^+ increase after ischaemic exercise in myotonic muscular dystrophy. Muscle & Nerve 13: 27-32
34. Cotterrell D, Karim F (1982) Effects of adenosine and its analogues on the perfused hindlimb artery and vein of anaesthetized dogs. J Physiol (Lond) 323: 473-482
35. Hudlicka O, El-Khelly F (1988) The role of inorganic phosphate in functional hyperaemia in skeletal muscle. In: Vanhoutte PM (ed) Mechanisms of vasodilatation. Raven Press, New York, pp 377-382

36. Kotter A, Kaley G (1990) Prostaglandins mediate arteriolar dilation to increased blood flow velocity in skeletal muscle microcirculation. Circ Res 67: 529-534
37. Moncada S, Palmer R M J, Higgs E A (1991) Nitric Oxide: Physiology, Pathophysiology and Pharmacology. Pharmacol Rev 43: 109-142
38. Kuo L, Davis MJ, Chilian WM (1992) Endothelial modulation of arteriolar tone. News Physiol Sci 7: 5-9
39. Suzuki H, Chen G (1990) Endothelium-derived hyperpolarising factor: an endogenous potassium channel activator. News Physiol Sci 5: 212

CHAPTER 9
Potassium channels in the pulmonary vasculature

RZ Kozlowski and PCG Nye

The vessels of the pulmonary circulation form the only significant route through which blood can pass from the right to the left side of the heart and ultimately to the rest of the body (fig. 1). This single feature determines both the nature and the behaviour of pulmonary vessels because it makes the pulmonary circulation entirely subservient to the summed needs of all the body's tissues. This means that the lung, unlike other tissues, cannot raise or lower its overall blood flow according to the demands of the task at hand; it must be predominantly passive, accepting all the blood that is delivered to it. Consistent with this passivity, the most distinctive property of the pulmonary circulation is the low resistance and high compliance of its vessels which ensure that blood circulates at a low hydrostatic pressure regardless of the total flow. A low capillary pressure is essential because it allows the vessels to retain fluid that would otherwise flood the alveoli and prevent gas exchange.

Although they are denied the freedom to alter total pulmonary blood flow, the vessels of the lung do actively regulate the local distribution of flow, ensuring that it is reasonably even and that it avoids regions of poor ventilation. This matching of local perfusion to local ventilation reduces the contribution that poorly ventilated regions of the lung make to the composition of arterial blood and therefore helps to optimise arterial blood gas tensions.

Pulmonary blood flow responds to alveolar Po_2 and Pco_2

The primary stimulus responsible for the local regulation of pulmonary blood flow is alveolar hypoxia reinforced by alveolar hypercapnia. The gas tensions of mixed venous blood have little influence except in so far as they can affect the composition of alveolar gas. This is appropriate because in exercise, when alveolar gas tensions remain close to their resting values, the dominance of alveolar gas tensions prevents the inappropriate constriction of vessels by hypoxic venous blood. Those who want to discover the mechanisms responsible for the hypoxic constriction of pulmonary vessels should therefore be looking for a way in which modest reductions in extravascular Po_2, in the range of 60 to 100 Torr, are sensed and transduced into the excitation of vascular smooth muscle.

The small vessels of the systemic circulation are either dilated or unaffected by modest asphyxia (fig. 2). They are never constricted by it, and this contrast with the pulmonary circulation lends added interest to the study of pulmonary vascular responses to hypoxia. The excitation of pulmonary smooth muscle cells by hypoxia generates action potentials that are carried by the influx of calcium [1], and because the tone of all blood vessels is determined by the level of intracellular calcium [2] the critical question is "what links extravascular hypoxia to the production of calcium-dependent action potentials by pulmonary vascular smooth muscle?". To focus on calcium leads us directly to

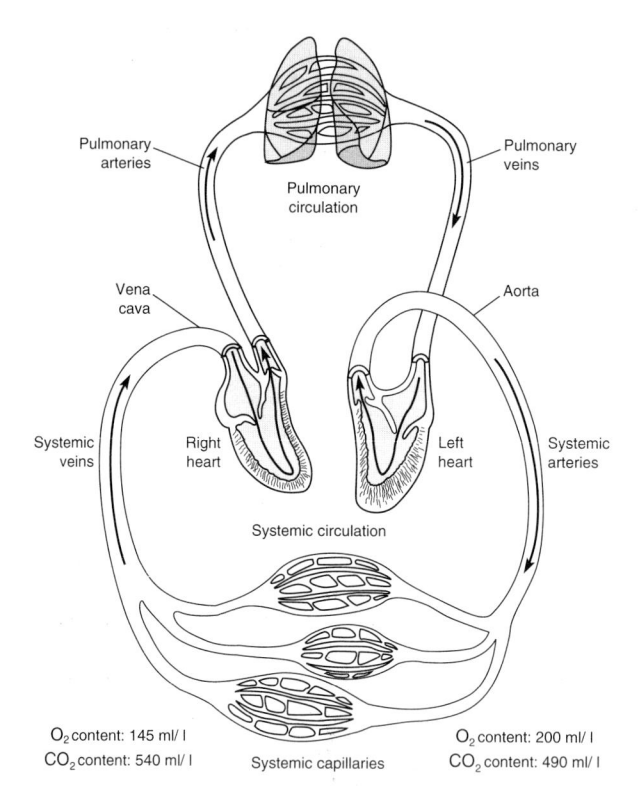

Fig. 1. The cardiovascular system drawn to emphasize the fact that the two sides of the heart are in series and connected only by the vessels of the pulmonary circulation. This means that overall pulmonary blood flow cannot be controlled independently of the systemic circulation

the likely involvement of calcium channels, the most likely candidate being the voltage-activated, L-type channel (L for Large and Long-lasting). Potassium channels may also play a role since it is their activity that primarily determines the membrane potential which regulates the activity of voltage-activated calcium channels. Activation of K^+ channels shifts the membrane potential towards the potassium equilibrium potential (about -85 mV in most mammalian cells) by allowing the outward movement of potassium ions. This stabilizes the membrane potential, inhibiting the generation of action potentials and therefore prevents calcium influx through voltage-activated calcium channels. Thus the net effect of K^+ channel activation is a reduction in muscle tone and conversely the effect of their closure is depolarization and the activation of voltage-sensitive Ca^{2+} channels. The following series of events may therefore be postulated to occur in hypoxia:

$\downarrow P_{O_2}$ → K^+ channel closure → depolarization → Ca^{2+} channel opening → ↑tension

Consistent with this scheme is the observation that Ca^{2+} channel blockers such as nifedipine, verapamil and nisoldipine [3, 4, 5] inhibit hypoxic pulmonary vasoconstriction and the calcium channel opener BAY K8644 enhances it [6]. Type I carotid body cells are also reported to be excited by the closure of potassium channels when exposed to hypoxia or acidity [7, 8]. The ability of potassium channels to modulate hypoxic pul-

Fig. 2. Contrast between the responses of pulmonary and systemic resistance vessels to hypoxia. In this example the resistance of the systemic circulation of a dog is little affected by hypoxia but that of the pulmonary circulation is nearly tripled. Many systemic vessels relax when made hypoxic. Both classes of vessel are relaxed by the potassium channel opener, cromakalim. After [28]

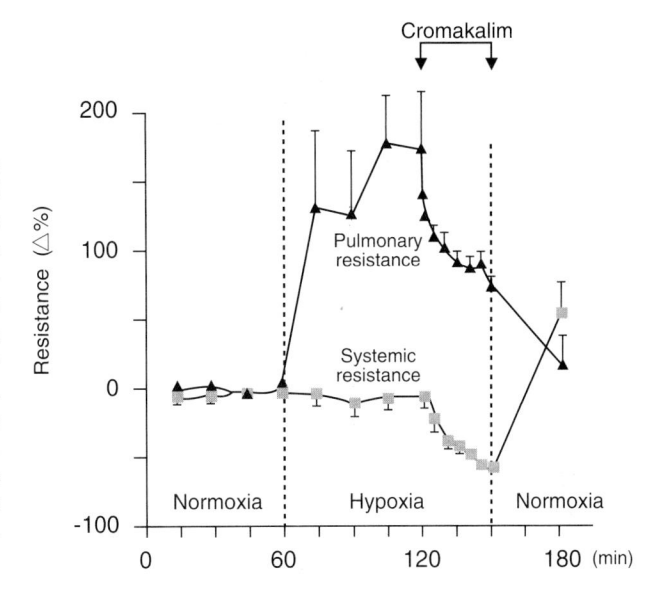

monary vasoconstriction is shown by the powerful inhibitory effects of potassium channel openers such as diazoxide, cromakalim, levcromakalim, pinacidil, RP 49356, and minoxidil [3, 9, 10, 11, 12] and the excitatory effects of potassium channel blockers such as TEA and 4-aminopyridine (4-AP) [5, 10, 12, 13, 14] discussed in detail below. However, taken alone these observations do not demonstrate direct involvement of potassium channels in the transduction of hypoxia to contraction. They only indicate that, under experimental conditions, potassium channel activity produces marked changes in pulmonary vascular tone that may, or may not, be involved in the mechanism of excitation by hypoxia. Recent support for the involvement of potassium channels in the control of pulmonary vascular tone is provided by the demonstration that calcium-activated potassium channels from the main pulmonary artery are activated by ATP [15] and may close when intracellular ATP falls below that present in normoxia (see below).

An understanding of the mechanisms which determine the resistance of pulmonary vessels and which underlie hypoxic vasoconstriction could have considerable therapeutic implications for the treatment of some clinical conditions. These include pulmonary hypertension, in which pulmonary arterial pressure rises because of excessive vasoconstriction. Hypertension is also commonly associated with pulmonary œdema, which forms in areas downstream of vessels that have failed to constrict to the same extent as the rest. Global alveolar hypoxia, which causes inappropriate global vasoconstriction, whether from hypoventilation or from the low inspired oxygen of altitude, is one cause of pulmonary hypertension that is frequently lethal. Another is the rare, unexplained and life-threatening primary pulmonary hypertension which strikes pubescent girls and causes pulmonary arterial pressure to rise above 100 mmHg. The identification of a drug that relaxes pulmonary vessels without simultaneously causing systemic hypotension could result from studies of the factors responsible for the differences between pulmonary and systemic vessels, and would be of great potential value in the treatment of pulmonary hypertension. We suspect that the hypoxic vasoconstriction that characterizes the behaviour of the pulmonary bed is mediated by the presence of specific potassium channels, and if this is so, drugs that open these channels could be useful in the treatment of pulmonary hypertension. At present the most effective treatment for this condition is transplantation of heart and lungs.

This chapter is concerned with the classes of potassium channel that are found in the lung and how the distinctive characteristics of these channels might account for the different responses of pulmonary and systemic vessels to hypoxia. The chapter ends with a section that emphasizes recent work in which recordings have been made directly from the ionic channels of pulmonary vascular smooth muscle.

Several types of potassium channel may control pulmonary vascular tone

There are three main groups of potassium channel which are likely to be involved in controlling pulmonary vascular tone and ultimately in mediating hypoxic pulmonary vasoconstriction. These include Ca^{2+}-activated, ATP-sensitive and voltage-activated K^+ channels. The major properties of these channel types are summarized below.

Ca^{2+}-activated K^+ channels

There are two major classes of calcium-activated potassium channel: a high conductance, 200 to 300 pS, maxi K^+ or BK (Big channel) and a small conductance (SK, <80 pS) channel. Both SK and BK channels are activated by calcium ions applied to their intracellular surface with maximal sensitivity usually falling over the range of 10^{-8} to 10^{-7} M Ca^{2+}. BK channels are also activated by depolarization which markedly increases their probability of opening. Two specific blockers of Ca^{2+}-activated K^+ channels have been identified: apamin (bee venom) and charybdotoxin (from the venom of the scorpion *Leiurus quinquestriatus*). Apamin is selective for SK channels, while charybdotoxin is selective for BK channels. Ca^{2+}-activated K^+ channels are involved in a variety of processes ranging from the modulation of secretion in both endocrine and exocrine tissue, to the regulation of repetitive firing and after-hyperpolarization in excitable cells. Their exact role in vascular smooth muscle has yet to be determined, although it is likely that they are involved in repolarizing the membrane during action potentials or in the control of slow wave activity. Recent evidence suggests that they may be involved in the control of myogenic tone [16]. The properties of Ca^{2+}-activated K^+ channels have been extensively reviewed elsewhere [17].

ATP-sensitive K^+ channels

ATP-sensitive K^+ channels have been found in cardiac, pancreatic, neuronal, and skeletal and vascular smooth muscle [18]. These channels have a conductance of approximately 50 to 60 pS and are inhibited by intracellular ATP with an IC_{50} of 10 to 200 μM. ATP-sensitive K^+ channels are inhibited by sulphonylureas, such as glibenclamide which binds with high affinity to the channel or to some closely associated membrane protein. In vascular smooth muscle a functional role for ATP-sensitive K^+ channels has not yet been established, although it has been suggested that when intracellular ATP falls in severe hypoxia or fatigue, the opening of ATP-sensitive K⁺ channels and the resulting vasodilation may reduce energy expenditure and raise local blood flow. Recent reports suggest that ATP-sensitive K⁺ channel activation may be largely responsible for the effects of potassium channel activators [18, 19, 20] so it is likely that ATP-sensitive K⁺ channels can determine vascular tone through their effects on membrane potential. In support of this hypothesis is the observation that arterial dilation induced by hyperpolarizing vasodilators is reversed by glibenclamide and that ATP-sensitive K⁺ channels in the rabbit mesenteric artery, activated by cromakalim or calcitonin gene related peptide, are also inhibited by glibenclamide [19, 21]. However, care must be exercised when interpreting results from these pharmacological studies since K⁺ channel activators have been reported to activate other types of potassium channel [22]. Furthermore, the specificity of glibenclamide has been questioned [23]. It can reverse cromakalim-induced activation of BK channels [24] and it can also relax the constriction of vessels caused by thromboxane [25] and prostaglandin F2 α [26].

Voltage-activated K⁺ channels

Voltage-activated K⁺ channels are widely distributed and play a multitude of roles, which include repolarization of the action potential and controlling the membrane potential between and during bursts of electrical activity. The best studied classes of voltage-gated potassium channel are the delayed rectifier and the transient outward K⁺ channels. These potassium channels are distinguished by their behaviour in voltage-clamp experiments rather than by their ionic selectivity or pharmacology, since no specific blockers have been developed for them. However, as the activity of these channels is largely governed by voltage and not by intracellular second messengers or metabolic factors they are unlikely to play a role in controlling the membrane potential, and thus vascular tone. Consequently, they are unlikely to be directly involved in hypoxic pulmonary vasoconstriction.

Pharmacological evidence suggests that K⁺ channels are involved in hypoxic pulmonary vasoconstriction

K⁺ channel activators relax pulmonary arteries and abolish hypoxic pulmonary vasoconstriction

Both in vivo and in vitro experimental approaches have been used to investigate the possible involvement of potassium channels in controlling pulmonary vascular tone and ultimately hypoxic pulmonary vasoconstriction. Evidence obtained in vivo for this supposition is derived predominantly, as mentioned in the introduction, from the observations that potassium channel activators such as cromakalim reduce or abolish hypoxic pulmonary vasoconstriction, and that constriction is restored by glibenclamide [27, 28]. This suggests that the opening of ATP-sensitive K⁺ channels is responsible for the relaxation, and that their closure then restores constriction. Furthermore, the in vivo administration of endothelin-1 (ET-1; a member of the endothelin family of peptides produced by endothelial cells which also includes ET-2 and ET-3) induces a profound vasodilation

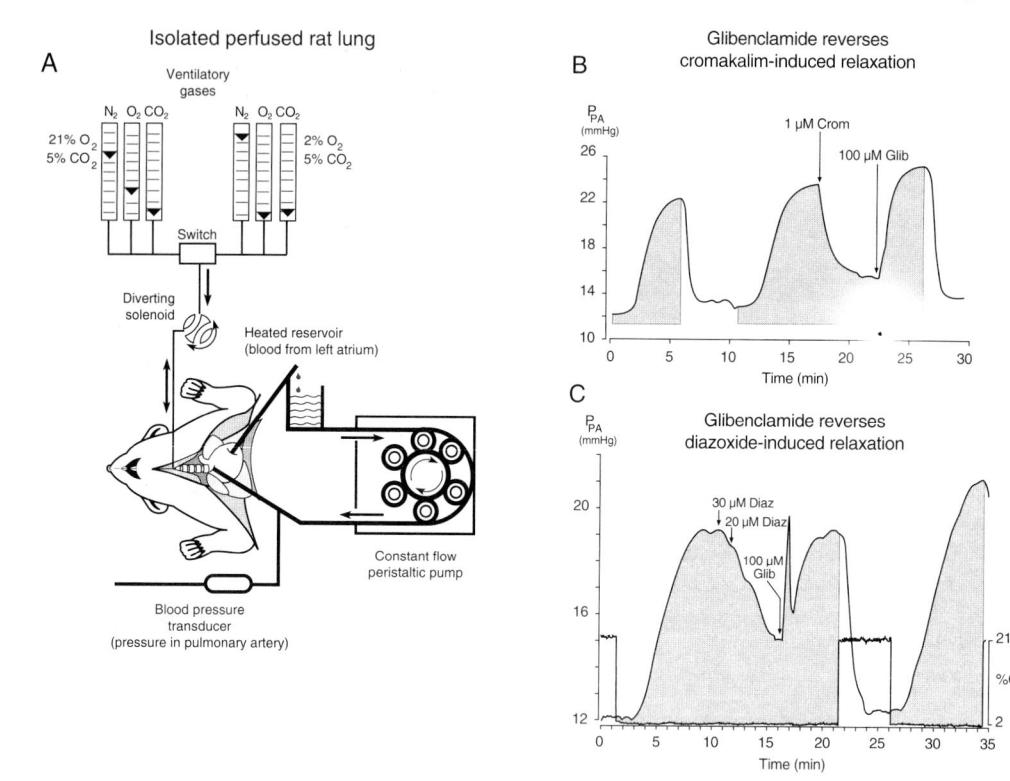

Fig. 3 A-D. The isolated perfused lung preparation and traces obtained from it. **A** The lung, cradled by the carcass of a rat, is perfused with autologous blood at a constant flow of ca. 20ml/min. Hypoxic challenges are made by switching the gas ventilating the lungs from 21% O_2, 5% CO_2 to 2% O_2, 5% CO_2. This gives maximal constriction. Pulmonary vascular resistance is recorded as pulmonary arterial pressure. **B** The response to a hypoxic challenge (shaded area) is virtually abolished by 1 µM cromakalim and then restored by 100 µM glibenclamide. **C** As for **B** but with relaxation by another potassium channel opener, diazoxide. The sharp contraction upon addition of glibenclamide is the transient response to the DMSO used to dissolve the glibenclamide. **D** Lack of effect of glibenclamide on hypoxic pulmonary vasoconstriction of the isolated, perfused lung. No concentration of glibenclamide could be made to alter hypoxic pulmonary vasoconstriction. The transient contraction caused by DMSO (see also fig. 5C) is the only observable effect, even at this very high concentration of 100 µM glibenclamide. The failure of a blocker of ATP-sensitive K^+ channels to change hypoxic pulmonary vasoconstriction suggests that these channels are not required for the response and that they are normally closed in both normoxia and hypoxia. Modified from [45]

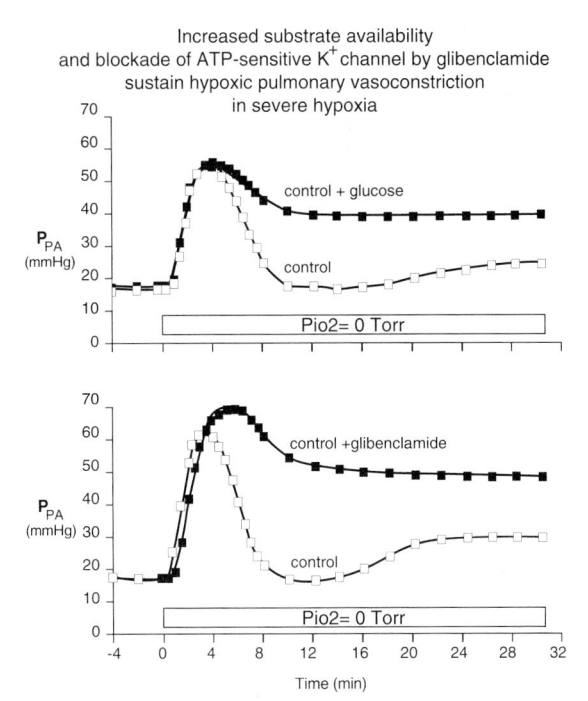

Fig. 4. The secondary relaxation of the isolated, perfused rat lung exposed to severe hypoxia is greatly reduced by the addition of either glucose or glibenclamide to the perfusate. These responses suggest that the relaxation is caused by the opening of ATP-sensitive K^+ channels when intracellular ATP falls. After [32]

followed by a potent constrictor response in both systemic and pulmonary vessels [29]. The constriction is associated with the closure of K^+ channels and membrane depolarization. Intravenous injection of ET-1, ET-2 or ET-3 under conditions of elevated pulmonary vasomotor tone at first dilates both pulmonary and systemic vessels. Application of glibenclamide inhibits endothelin-induced relaxation of the pulmonary vasculature without affecting systemic pressure, suggesting that in the lung, endothelins open glibenclamide-sensitive, ATP-sensitive K^+ channels [29, 30]. However, it is often difficult to interpret results from whole-animal studies since many physiological responses can act simultaneously to mask or complicate a given effect. Consequently, the isolated perfused lung (fig. 3), a preparation free from extra-pulmonary influences such as endocrine effects and reflex responses (other than axon reflexes), has been used as a model to study the responses of pulmonary blood vessels.

In vitro pharmacological studies on isolated, perfused lungs and isolated vessels also suggest that these contain ATP-sensitive K^+ channels. Several studies have described the relaxation of isolated, perfused lungs or rings of pulmonary vessel by cromakalim or diazoxide (see fig. 3B, C), and the reversal of this relaxation by glibenclamide [11, 31]. Other evidence for the existence of ATP-sensitive K^+ channels in the vessels of the lung comes from the patch-clamp experiments described at the end of this chapter.

Glibenclamide has no effect on the tone of pulmonary vessels in either normoxia or in hypoxia of sufficient intensity to cause maximal constriction [5] (fig. 3D). This lack of response to glibenclamide suggests that the ATP-sensitive K^+ channels of the lung are almost always closed and that they remain closed in hypoxia. Thus ATP-sensitive K^+

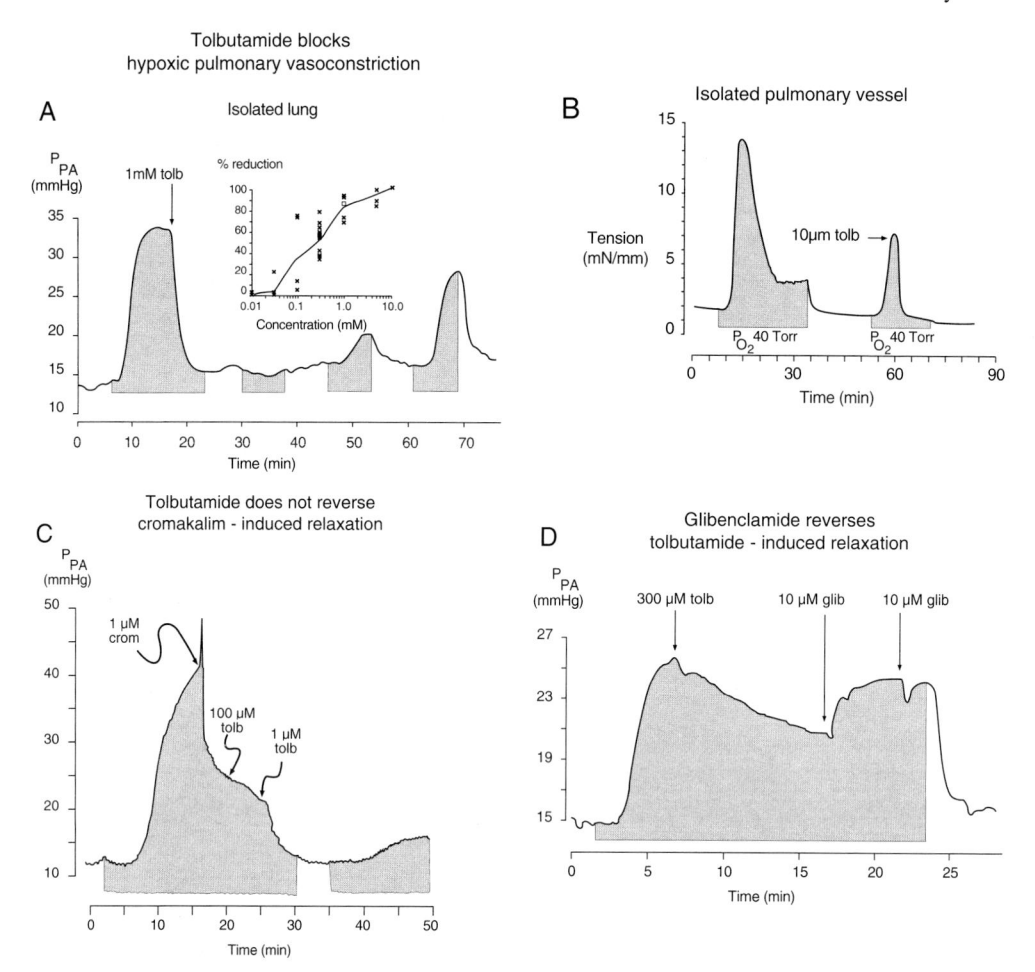

Fig. 5. A Paradoxical relaxation of isolated perfused lung by the sulphonylurea tolbutamide. The response is paradoxical because in other tissues tolbutamide acts like glibenclamide and is excitatory. The insert is a dose-response curve showing that the IC_{50} for inhibition of hypoxic pulmonary vasoconstriction by tolbutamide is around 300 μM. **B** Lower concentrations of tolbutamide block hypoxic vasoconstriction in isolated pulmonary vessels. This is presumably because the blood perfusing the lung contains proteins that bind the drug while the Ringer's solution bathing the isolated vessel does not. **C** In contrast to the effect of glibenclamide (fig. 4.) tolbutamide does not reverse cromakalim-induced relaxation. It adds to the relaxation. **D** Hypoxic constriction relaxed by tolbutamide could often be restored by glibenclamide

channels do not appear to be involved in the mechanism responsible for hypoxic pulmonary vasoconstriction, but they do appear to be responsible for the failure of constriction to be sustained when hypoxia is very severe (P_{O_2} close to 0 Torr [32, 33]). This is suggested by the observation that both glibenclamide (which mimics the effect of ATP at ATP-sensitive K^+ channels) and raised glucose levels in the perfusate (which is assumed to allow the synthesis of more ATP) restore constriction after its failure in severe hypoxia (fig. 4).

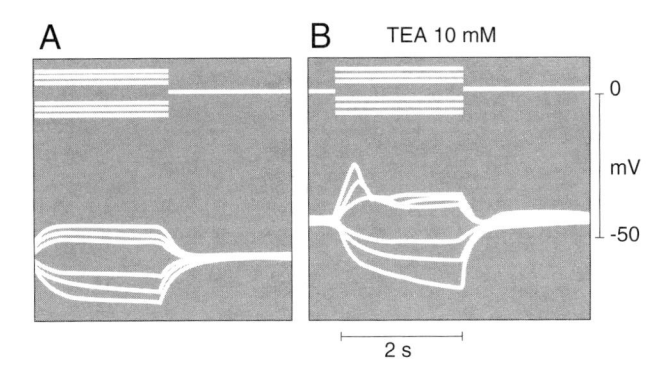

Fig. 6 A, B. Currents recorded during voltage-clamp experiments on pulmonary vascular smooth muscle **A** before and **B** after addition of 10 mM TEA to the bath. The presence of TEA excites the generation of action currents in the cells. From [13]

Tolbutamide also relaxes hypoxic pulmonary vasoconstriction

The relaxation of isolated perfused lungs or vessels by cromakalim and diazoxide and the abolition of this relaxation by glibenclamide provide pharmacological evidence for the existence of ATP-sensitive K^+ channels on the smooth muscle cells of pulmonary vessels. However, tolbutamide, a sulphonylurea which, like glibenclamide, blocks ATP-sensitive K^+ channels in other tissues, paradoxically relaxes hypoxic pulmonary vasoconstriction with an IC_{50} of around 300 μm (fig. 5A) [34]. It also relaxes isolated vessels and does so at much lower concentrations (fig. 5B). This relaxation contrasts with the glibenclamide-like behaviour of tolbutamide on systemic vessels [35] and highlights a pharmacological difference between the two classes of vessel which may be associated with the mechanism responsible for their opposite responses to hypoxia. Tolbutamide does not reduce the relaxation caused by cromakalim (it adds to it, fig. 5C) and most strikingly, the relaxation brought on by modest concentrations of tolbutamide can be reversed by glibenclamide (fig. 5D). One possible interpretation of these results is that tolbutamide opens ATP-sensitive K^+ channels in the pulmonary vasculature which are closed by glibenclamide.

4-aminopyridine and TEA constrict pulmonary arteries

The constriction of normoxic vessels by 4-aminopyridine (4-AP) and TEA suggests that a background K^+ conductance normally keeps the membrane hyperpolarized and thereby maintains the low vascular resistance of the normoxic lung. However, the identity of the channels underlying the hyperpolarization remains uncertain, although the evidence cited above rules out glibenclamide-sensitive channels because these are almost always closed.

The reported excitation of pulmonary vascular smooth muscle by 4-AP or TEA (fig. 6) [5, 13], and the 4-AP-sensitivity of the ATP-activated potassium channels described in their laboratory by Robertson [15], led the authors to compare the response of small pulmonary arteries to this drug with that to hypoxia. We found that tension often followed almost identical waveforms after both stimuli (fig. 7A, B) and that after the addition of 10 mM 4-AP, normoxia usually constricted and hypoxia relaxed the vessels (fig. 7C). It was as if the mechanism responsible for hypoxic excitation had been saturated by the drug to reveal an underlying response similar to that of systemic vessels. It is

Fig. 7 A-C. Resemblance between responses of isolated small pulmonary vessels to hypoxia and 4-AP. Traces of responses to **A** hypoxia and **B** 4-AP recorded on similar time bases. Both give strikingly similar biphasic responses. **C** Adding 4-AP to the bath just before the initial peak of a response to hypoxia in two vessels from the same bath. There is no apparent effect until the gas is switched to normoxia. This immediately constricts the vessels and now hypoxia relaxes them. The response to hypoxia is reversed by 10 mM 4-AP

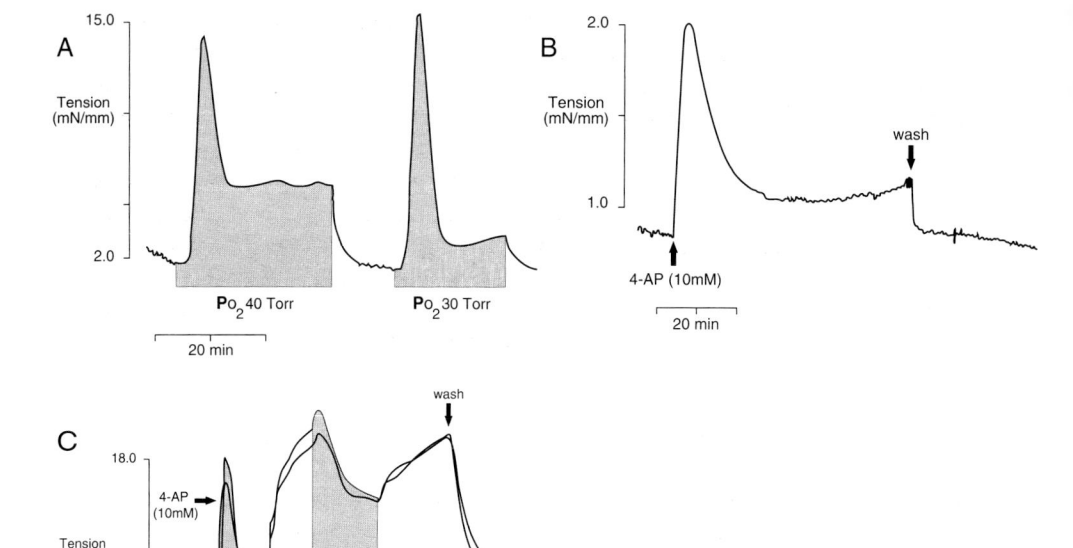

therefore tempting to speculate that closure of 4-AP-sensitive K^+ channels may be involved in the initiation of hypoxic vasoconstriction.

The role of endothelial relaxing factors is not yet clear

Recent studies have revealed that the endothelium of rat pulmonary arteries releases both endothelium-dependent relaxing factor (EDRF), and endothelium-dependent hyperpolarizing factor (EDHF) [36]. The release of EDRF, or of its primary constituent nitric oxide, is not diminished and is probably enhanced in hypoxia [36-38], so hypoxic pulmonary vasoconstriction does not depend on a mechanism that blocks the action of EDRF. Evidence implicating EDHF (which relaxes vascular smooth muscle by activating K^+ channels) in the response to hypoxia is at present speculative. It is based on the observation that TEA blocks histamine-induced vasodilation, which is thought to act via

EDHF [36]. EDHF is widely thought to act via glibenclamide-sensitive ATP-sensitive K⁺ channels and, as argued above, these are not directly involved in hypoxic pulmonary vasoconstriction. Indeed, the significance of the endothelium in the generation of pulmonary vasoconstriction by hypoxia is an issue that has not been settled. One group [39] has claimed that the endothelium is essential for any hypoxic constriction, another [40] claims that it is required for its full expression and a third group [11] claims that it acts only to reduce the magnitude of constriction because, in their experiments, mechanical removal of the endothelium enhances the response to hypoxia. These observations, taken together, suggest that at least part of hypoxic constriction occurs in the absence of the endothelium, so it seems reasonable to assume, for simplicity, that the smooth muscle cells of pulmonary vessels are the primary site responsive to hypoxia.

Electrophysiological studies of pulmonary arterial K⁺ channels

Early electrophysiological recordings [13], which used classical microelectrodes to penetrate cells in sections isolated from the main pulmonary artery revealed a membrane potential of approximately -50mV. This was reduced by the potassium channel blockers TEA or 4-AP, in the presence of which action potentials could, on occasion, be evoked by depolarizing current pulses (fig. 6). These observations are consistent with the suggestion that K⁺ ions are involved in controlling the membrane potential of pulmonary arterial smooth muscle cells.

The above microelectrode studies of cells in situ within the walls of blood vessels gave some insight into their basic electrophysiological properties, however, in recent years the development of methods for the isolation of single cells combined with the patch-clamp recording technique has allowed single ionic channels to be studied. Investigators who have used these techniques on the blood vessels of the lung have so far studied only the ionic channels of smooth muscle cells from the main pulmonary artery because these cells are larger than those from small vessels. We know of no successful recording from such cells and this is unfortunate because there are important differences between large and small pulmonary vessels [41]. However, because the pulmonary artery does constrict in response to hypoxia, it seems safe to assume that the same basic mechanism is shared by both sizes of vessel.

Initial whole-cell patch-clamp studies showed that smooth muscle cells isolated from the main pulmonary artery possess an outward current which is relatively insensitive to TEA and is blocked by external application of 10 mM 4-AP [42-44]. This outward current differs from the classical 4-AP sensitive current seen in neurones since it is not augmented by membrane hyperpolarization and is not rapidly inactivating.

Transient and slow outward K⁺ currents have been described in pulmonary arteries

More recent whole-cell studies have demonstrated that the outward current in single cells from the rabbit pulmonary artery is largely carried by the opening of both voltage-dependent and independent potassium channels [43]. Two types of voltage-activated K⁺ current have been reported; a rapidly activating and inactivating current (I_{tran}) and a slowly

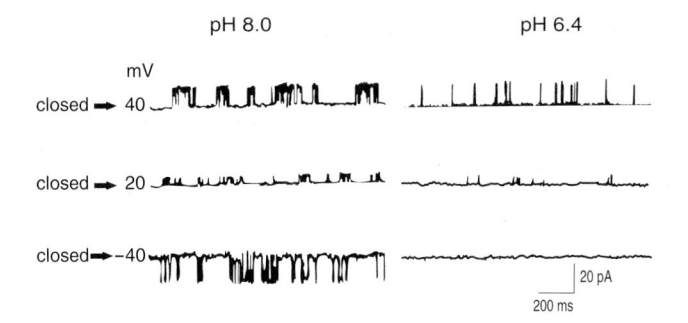

Fig. 8. Patch-clamp records from a single voltage-sensitive potassium channel in the membrane of a pulmonary vascular smooth muscle cell. It is inhibited by acidity. From [46]

activating current (IK_{so}). I_{tran} may not have been detected by others, e.g. [42], because its activity has a tendency to decrease (run-down) with time. IK_{so} is likely to result from the opening of Ca^{2+}-activated K^+ channels since it is sensitive to changes in intracellular Ca^{2+} and blocked by TEA. Furthermore, its biophysical properties closely resemble those of Ca^{2+}-activated currents observed in other tissues. However, unlike other Ca^{2+}-activated K^+ currents IK_{so} is inhibited by extracellularly applied 4-AP, which also inhibits I_{tran}. I_{tran}, in terms of its biophysical properties resembles i_{to}, it is however also sensitive to changes in intracellular Ca^{2+} concentration, a feature not generally exhibited by transient outward K^+ channels. In addition to these voltage-activated currents Clapp and Gurney [43] have reported a time- and voltage-independent outward current insensitive to intracellularly applied ATP, but blocked by 4-AP, and a spontaneous outward current which may require the presence of intracellular GTP.

Single channel studies have revealed little ATP-sensitive K^+ channel activity of the type recorded in pancreatic ß cells in either rat [15, 45] or rabbit [46] main pulmonary artery. The most abundant channel is a large Ca^{2+}-activated K^+ channel, which in rat cells is inhibited by intracellularly applied 4-AP. The characteristics of this channel closely resemble those of Ca^{2+}-activated K^+ channels from the rabbit portal vein and guinea-pig mesenteric artery. All have similar current-voltage relationships in symmetrical potassium with conductances of around 250 pS. The Ca^{2+}-activated K^+ channels of rabbit pulmonary artery have a larger conductance of 360 pS and are inhibited by lowering pH [46] (fig. 8), which shifts the relationship between internal Ca^{2+} and open probability to higher concentrations of Ca^{2+} and reduces voltage sensitivity. It has therefore been suggested [46] that Ca^{2+}-activated K^+ channel regulation by pH may be important in electrically quiescent vessels, since the voltage sensitivity of these channels observed at physiological pH would tend to result in their activation at the resting membrane potential. The inhibition of channel activity by acidity might also contribute to the enhancement of hypoxic constriction that occurs in response to hypercapnia [47].

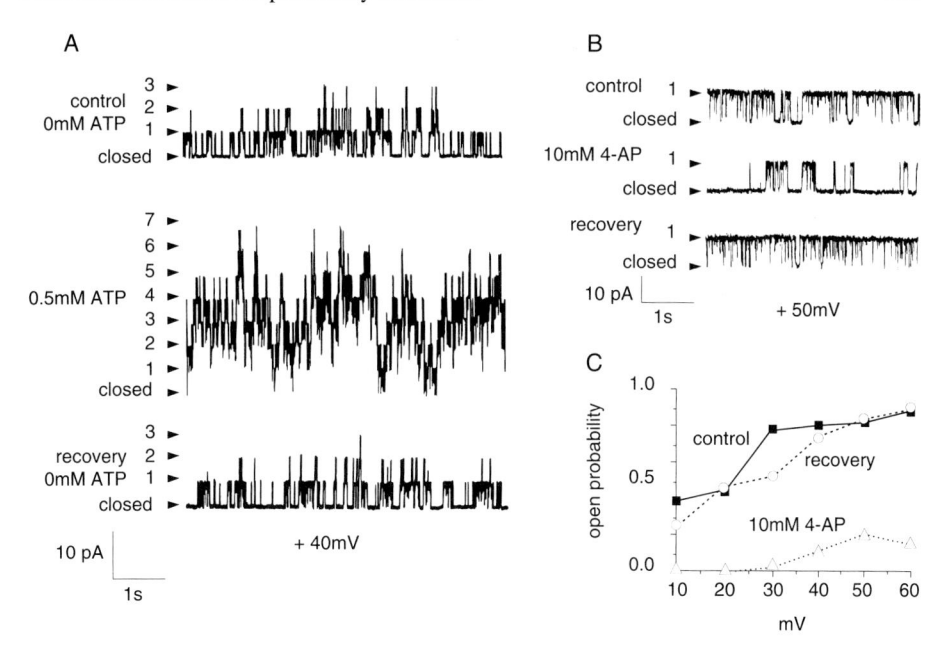

Fig. 9 A-C. Patch-clamp records of calcium-activated potassium channels from a pulmonary arterial smooth muscle cells which are opened by intracellular Mg-ATP. **A** Seven channels in a single inside-out patch. **B** A single calcium-activated potassium channel from another patch is blocked by 10 mM 4-AP. **C** Voltage sensitivity of the open probability of the channel in **B** before, during, and after application of 4-AP

A K⁺ channel activated by both Ca^{2+} and intracellular Mg-ATP may play a role in hypoxic pulmonary vasoconstriction

Recently, it has been reported that Ca^{2+}-activated K⁺ channels in the rat main pulmonary artery can be activated by intracellular ATP [15], an effect which appears to involve a recruitment of channels rather than simply a change in open probability (fig. 9). This activation appears to require the hydrolysis of ATP since the non-hydrolysable ATP analogue AMP-PNP is ineffective and Mg^{2+}, which is required for processes which involve phosphorylation, must be present. The increase in channel activity induced by ATP is readily reversible upon washout, indicating that continuous hydrolysis of the nucleotide is probably required. Thus it is possible that the channel may be autophosphorylated or closely associated with a modulatory kinase. Their sensitivity to ATP has led us to refer to Ca^{2+}-activated K⁺ channels of the pulmonary arteriolar smooth muscle cells as $K_{Ca,ATP}$ channels: a novel subclass of Ca^{2+}-activated K⁺ channel. Interestingly Ca^{2+}-activated K⁺ channels from the rabbit pulmonary artery can also be activated by membrane stretch, sensed either by the channel itself or some closely associated membrane component [48]. Stretch activation of these channels may ensure that pulmonary vessels relax to accommodate increases in cardiac output. Circumstantial evidence in support of Ca^{2+}-activated K⁺ channel involvement in hypoxic pulmonary vasoconstriction has come from parallel whole-cell and isolated-tissue studies, which have revealed that TEA, 4-AP and charybdotoxin constrict pulmonary arterial smooth muscle in vitro and

Inhibition of whole-cell K⁺ current
in pulmonary arterial smooth muscle cells
by raised intracellular ATP and by glibenclamide

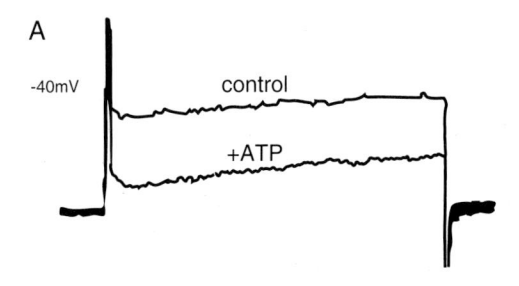

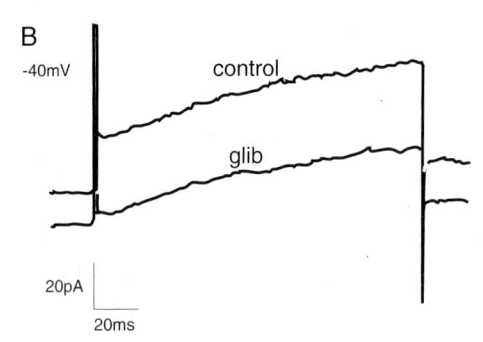

Fig. 10. A Reduction in outward potassium current after the intracellular release of caged ATP by a flash of light. **B** Similar inhibition of potassium current by glibenclamide. These observations provide direct evidence for the existence of a pulmonary vascular potassium channel that is blocked both by ATP and by glibenclamide. After [44]

in vivo [5]. These agents decrease whole-cell K^+ currents in pulmonary arterial cells, reducing a current thought to result from the opening of Ca^{2+}-activated K^+ channels, since dialysis of cells with an intracellular solution containing negligible quantities of Ca^{2+} prevents the inhibitory action of these K^+ channel blockers.

There is only one electrophysiological study to date which ascribes, with some confidence, a role for ATP-sensitive K^+ channels in maintaining the membrane potential of rat main pulmonary arterial smooth muscle cells [44]. In this study the authors used cells isolated from the pulmonary artery in the whole-cell configuration, to show that photo-released ATP inhibits an outward potassium current (fig. 10). However, the exact identity of the channels which underlie this current is uncertain, although they have a low conductance (10 to 20 pS) and are sparsely distributed. These channels are also activated by lemakalim, the active isomer of cromakalim, and inhibited by glibenclamide, agents which have hyperpolarizing and depolarizing actions respectively [44].

Summary of the possible roles of K⁺ channels in hypoxic pulmonary vasoconstriction

Pulmonary vascular smooth muscle is dilated by drugs that activate potassium channels and it is constricted by many drugs that inhibit them. It therefore appears to be modulated by the activity of potassium channels similar, or identical, to those found in systemic vessels. The responses of these vessels to cromakalim and glibenclamide suggest strongly that they possess ATP-sensitive K$^+$ channels related to the now-classical ATP-sensitive K$^+$ channels of the heart and endocrine pancreas. However, they are closed in both hyperoxia and in hypoxia of an intensity that produces maximal contraction. It seems that, as has been suggested for cardiac [49] and skeletal muscle [50], the ATP-sensitive K$^+$ channels of pulmonary vessels are activated only under extreme conditions, such as hypoxia that is so severe that it threatens the survival of the tissue. It is therefore clear that glibenclamide- and ATP-sensitive K$^+$ channels are not involved in the mechanism of hypoxic pulmonary vasoconstriction. The smooth muscle of rat main pulmonary vessels does, however, contain a different ATP-sensitive potassium channel, one that is activated, not closed, by ATP and which is also activated by calcium. This novel $K_{Ca,ATP}$ channel is the most abundant of all potassium channels in these cells [15, 45], and of all the drugs tried to date, its activity is most effectively blocked by 4-AP. When 4-AP is applied to pulmonary vascular tissue it mimics many of the characteristics of the response to hypoxia [5] (fig. 7) and high concentrations ($> = 10$ mM) reverse the responses of vessels to subsequent hypoxic challenges so that normoxia now constricts and hypoxia relaxes.

However, there is no direct evidence that $K_{Ca,ATP}$ channels are influenced by intensities of hypoxia that constrict rat main pulmonary vessels. Their activation by ATP, an appropriate response for a channel responsible for the relaxation of a muscle when Po_2 rises, is so far the only direct observation that makes them promising candidates for an important role in the process of hypoxic pulmonary vasoconstriction. Other indirect circumstantial evidence includes reports that the modulation of Ca^{2+}-activated K$^+$ and related channels is responsible for the control of smooth muscle tone in systemic vessels [16]. If $K_{Ca,ATP}$ channels of pulmonary artery smooth muscle are closed in hypoxia by a reduction in the ATP content of their immediate environment, this would depolarize the cells and constrict them by enhancing Ca^{2+} influx through voltage-gated Ca^{2+} channels. However, we must end on a note of caution because, if $K_{Ca,ATP}$ channels are indeed involved in hypoxic pulmonary vasoconstriction, their relative sensitivities to Ca^{2+} and ATP must be critical. This is because the Ca^{2+} entry, that results from their closure by reduced ATP, would tend to open them again, opposing the effect of the primary stimulus. It is therefore apparent that a description of the effects of true hypoxia on $K_{Ca,ATP}$ channels, and of the interactions between the effects of ATP and Ca^{2+} on their behaviour, will be necessary to test their significance in the control of pulmonary vascular function.

Note added in proof: Yuan and co-workers [51] have recently reported that hypoxia closes a voltage activated, Ca^{2+}-insensitive outward current in cultured smooth muscle cells from rat main pulmonary artery. These authors suggest that this current may be involved in controlling the membrane potential of those cells.

The work described in this chapter that was performed in the authors' laboratory was generously supported by the British Heart Foundation and the Wellcome Trust. The experiments were performed by Blair Robertson, Sulayma Albarwani, Niya Xia, Peter Corry, Adil Kahn and Richard Kingston, without whom the work would never have been done.

References

1. Harder DR, Madden JA, Dawson C (1985) Hypoxic induction of Ca^{2+}-dependent action potentials in small pulmonary arteries of the cat. J Appl Physiol 59: 1389-1393
2. Breemen C van, Saida K (1989) Cellular mechanisms regulating $[Ca^{2+}]_i$ smooth muscle. Ann Rev Physiol 51: 315-329
3. Palevsky HI, Fishman AP (1985) Vasodilator therapy for primary pulmonary hypertension. Ann Rev Med 36: 563-578
4. McMurtry IF, Davidson AB, Reeves JT, Grover RF (1976) Inhibition of hypoxic pulmonary vasoconstriction by calcium antagonists in isolated rat lungs. Circ Res 38: 99-104
5. Post JM, Hume JR, Archer SL, Weir EK (1992) Direct role for potassium channel inhibition in hypoxic pulmonary vasoconstriction. Am J Physiol 262: C882-C890
6. Tolins M, Weir EK, Chesler E, Nelson DP, From AHL (1986) Pulmonary vascular tone is increased by a voltage- dependent calcium channel potentiator. J Appl Physiol 60: 942-948
7. Lopez-Barneo J, Lopez-Lopez JR, Urena J, Gonzalez C (1988) Chemotransduction in the carotid body: K^+ current modulated by Po_2 in type I chemoreceptor cells. Science 241: 580-582
8. Peers C (1990) Effect of lowered extracellular pH on Ca^{2+}-dependent K^+ currents in type I cells from the neonatal rat carotid body. J Physiol (Lond) 422: 381-395
9. Wang SWS, Pohl JEF, Rowlands DJ, Wade EG (1978) Diazoxide in treatment of primary pulmonary hypertension. Br Heart J 40: 572-574
10. Eltze M (1989) Glibenclamide is a competitive antagonist of cromakalim, pinacidil and RP 49356 in guinea-pig pulmonary artery. Eur J Pharmacol 165: 231-239
11. Yuan X-J, Tod ML, Rubin LJ, Blaustein MP (1990) Contrasting effects of hypoxia on tension in rat pulmonary and mesenteric arteries. Am J Physiol 259: H281-H289
12. Hasunuma K, Rodman DM, McMurtry IF (1991) Effects of K^+ channel blockers on vascular tone in the perfused rat lung. Am Rev Respir Dis 144: 884-887
13. Casteels R, Kitamura K, Kuriyama H, Suzuki H (1977) The membrane properties of the smooth muscle cells of the rabbit main pulmonary artery. J Physiol (Lond) 271: 41-61
14. Hara Y, Kitamura K, Kuriyama H (1980) Actions of 4-aminopyridine on vascular smooth muscle tissues of the guinea-pig. Br J Pharmacol 68: 99-106
15. Robertson BE, Corry PR, Nye PCG, Kozlowski RZ (1992) Ca^{2+} and Mg-ATP activated K^+ channels from rat pulmonary artery. Pflügers Archiv 421(1): 94-96
16. Brayden JE, Nelson MT (1992) Regulation of arterial tone by activation of calcium-dependent potassium channels. Science 256: 532-535
17. Haylett DG, Jenkinson DH (1990) Calcium-activated potassium channels. In: Cook NS (ed) Potassium channels. Ellis Horwood, Chichester, pp 70-95
18. Ashcroft FM, Rorsman P (1990) Electrophysiology of the pancreatic ß-cell. Prog Biophys Mol Biol 54: 87-143
19. Nelson MT, Patlak JB, Worley JF, Standen NB (1990) Calcium channels, potassium channels, and voltage dependence of arterial smooth muscle tone. Am J Physiol 259: C3-C18
20. Brayden JE, Quayle JM, Standen NB, Nelson MT (1991) Role of potassium channels in the vascular response to endogenous and pharmacological vasodilators. Blood Vessels 28: 147-153
21. Nelson MT, Huang Y, Brayden JE, Hescheler J, Standen NB (1990) Arterial dilations in response to calcitonin gene- related peptide involve activation of K^+ channels. Nature 344: 770-773
22. Cook NS, Quast U (1990) Potassium channel pharmacology. In: Cook NS (ed) Potassium channels. Ellis Horwood, Chichester, pp 181-255
23. Caro JF (1990) Effects of glyburide on carbohydrate metabolism and insulin action in the liver. Am J Med 89: 17S-25S
24. Kajioka S, Nakashima M, Kitamura K, Kuriyama H (1991) Mechanisms of vasodilation induced by potassium-channel activators. Clin Sci 81: 129-139

25. Cocks TM, King SJ, Angus JA (1990) Glibenclamide is a competitive antagonist of the thromboxane A2 receptor in dog coronary artery in vitro. Br J Pharmacol 100: 375-378
26. Zhang H, Stockbridge N, Weir B, Krueger C, Cook D (1991) Glibenclamide relaxes vascular smooth muscle constriction produced by prostaglandin F2α. Eur J Pharmacol 195: 27-35
27. Weir EK, Chidsey CA, Weil JV, Grover RF (1976) Minoxidil reduces pulmonary vascular resistance in dogs and cattle. J Lab Clin Med 88: 885-894
28. Hicks PE, Martin D, Dumez D, Zazzi-Sudriez E, Armstrong JM (1989) Pulmonary vascular and renal effects of cromakalim in anaesthetized dogs. Pflügers Arch 414 (Suppl 1): S192-S193
29. Lippton HL, Cohen GA, McMurtry IF, Hyman AL (1991) Pulmonary vasodilatation to endothelin isopeptides in vivo is mediated by potassium channel activation. J Appl Physiol 70: 947-952
30. Hasunuma K, Rodman DM, O'Brien RF, McMurtry IF (1990) Endothelin 1 causes pulmonary vasodilation in rats. Am J Physiol 259: H48-H54
31. Robertson BE, Kozlowski RZ, Nye PCG (1992) Opposing actions of tolbutamide and glibenclamide on hypoxic pulmonary vasoconstriction. Comp Biochem Physiol 102 (3): 459-462
32. Wiener CM, Dunn A, Sylvester JT (1991) ATP-dependent K^+ channels modulate vasoconstrictor responses to severe hypoxia in isolated ferret lungs. J Clin Invest 88: 500-504
33. Wiener CM, Sylvester JT (1991) Effects of glucose on hypoxic vasoconstriction in isolated ferret lungs. J Appl Physiol 70: 439-446
34. Robertson BE, Paterson DJ, Peers C, Nye PCG (1989) Tolbutamide reverses hypoxic pulmonary vasoconstriction in isolated rat lungs. Q J Exp Physiol 74: 959-962
35. Brayden JE, Nelson MT, Quayle JM, Standen NB (1989) Blockers of ATP-sensitive potassium channels reverse vasodilatation and hyperpolarization to acetylcholine, vasoactive intestinal polypeptide, cromakalim, diazoxide and pinacidil in rabbit isolated arteries. J Physiol (Lond) 416: 47P.
36. Hasunuma K, Yamaguchi T, Rodman DM, O'Brien RF, McMurtry IF (1991) Effects of inhibitors of EDRF and EDHF on vasoreactivity of perfused rat lungs. Am J Physiol 260: L97-L104
37. Brashers VL, Peach MJ, Rose CE (1988) Augmentation of hypoxic pulmonary vasoconstriction in the isolated perfused rat lung by in vitro antagonists of endothelium-dependent relaxation. J Clin Invest 82: 1495-1502
38. Archer SL, Tolins JP, Raij L, Weir EK (1989) Hypoxic pulmonary vasoconstriction is enhanced by inhibition of the synthesis of an endothelium derived relaxing factor. Biochem Biophys Res Comms 164: 1198-1205
39. Holden WE, McCall E (1984) Hypoxia-induced contractions of porcine pulmonary artery strips depend on intact endothelium. Exp Lung Res 7: 101-112
40. Rodman DM, Yamaguchi T, O Brien RF, McMurtry IF (1989) Hypoxic contraction of isolated rat pulmonary artery. J Pharmacol Exp Ther 248: 952-959
41. Leach RM, Twort CHC, Cameron IR, Ward JPT (1992) A comparison of the pharmacological and mechanical properties in vitro of large and small pulmonary arteries of the rat. Clin Sci 82: 55-62
42. Okabe K, Kitamura K, Kuriyama H (1987) Features of 4-aminopyridine sensitive outward current observed in single smooth muscle cells from the rabbit pulmonary artery. Pflügers Arch 409: 561-568
43. Clapp LH, Gurney AM (1991) Outward currents in rabbit pulmonary artery cells dissociated with a new technique. Experimental Physiol 76: 677-693
44. Clapp LH, Gurney AM (1992) ATP-sensitive K^+ channels regulate resting potential of pulmonary arterial smooth muscle cells. Am J Physiol 262: H916-H920
45. Robertson BE (1991) Hypoxic pulmonary vasoconstriction. D. Phil, Thesis, Oxford University

46. Lee SH, Ho WK, Earm YE (1991) Effect of pH on calcium-activated potassium channels in pulmonary arterial smooth muscle cells of the rabbit. Korean J Physiol 25: 17-26
47. Barer GR, Howard P, Shaw JW (1970) Stimulus-response curves for the pulmonary vascular bed to hypoxia and hypercapnia. J Physiol (Lond) 211: 139-155
48. Kirber MT, Ordway RW, Clapp LH, Walsh JV Jr, Singer JJ (1992) Both membrane stretch and fatty acids directly activate large conductance Ca^{2+}-activated K^+ channels in vascular smooth muscle cells. FEBS Lett 297: 24-28
49. Gasser RNA, Vaughan-Jones RD (1990) Mechanism of potassium efflux and action potential shortening during ischaemia in isolated mammalian cardiac muscle. J Physiol (Lond) 431: 713-741
50. Davies NW, Standen NB, Stanfield PR (1991) ATP- dependent potassium channels of muscle cells: their properties, regulation, and possible functions. J Bioenergetics Biomembranes 23: 509-535
51. Yuan XJ, Goldman WF, Tod ML, Rubin LJ, Blaustein MP (1993) Hypoxia reduces potassium currents in cultural rat pulmonary but not mesenteric arterial myocytes. Am J Physiol 264 : L116-L123

Pharmacology

CHAPTER 10
Sulphonylureas: a receptor and a potassium channel?

MLJ Ashford

The sulphonylureas have recently become an important class of compounds in scientific research. The main reasons for this are twofold. Firstly they appear to be relatively specific inhibitors of a type of potassium channel which can contribute significantly to the resting membrane potential of certain cells. Secondly they act as antagonists to the majority of actions of the relatively novel and important class of drugs known as the potassium channel openers (see chapters 12 and 13). It is the purpose of this chapter to review the principles of action of the sulphonylureas and to indicate some areas of research (scientific and clinical) in which these agents are currently making important contributions.

Sulphonylureas are drugs of current clinical usage

During the search for new antibiotics in the 1940's a group of agents were formulated called the sulphonamides. A derivative of these, 2254 RP (p-aminobenzenesulphamido-isopropylthiodiazol) (fig. 1), when tested on typhoid fever patients was found to induce hypoglycaemia in high doses. This action was only observed in experimental animals if the pancreas was intact and it was later shown that it and a related compound (carbutamide) could elicit the release of insulin directly. The first classical sulphonylurea, tolbutamide (fig. 1), was produced in 1955 and this molecule is devoid of antibiotic activity. This was followed in 1966 by the first of the second generation drugs, glibenclamide, which is greater than 100 times more effective as an insulin secretagogue than tolbutamide [1]. These drugs and their derivatives are currently commonly used as a treatment for Type II (maturity-onset, non-insulin dependent) diabetes. Although it is likely that the principle antidiabetic action of the sulphonylureas is via stimulation of insulin secretion, other, extrapancreatic, mechanisms have also been suggested (particularly for glibenclamide) and are thought to be of greater importance in long-term treatment [2]. These include improved insulin action on peripheral tissues, decreased hepatic glucose production and enhanced peripheral glucose utilization. Further details of these actions and their likely mechanisms are beyond the scope of this review but are well discussed in reference [3].

It is not entirely clear why there is such a difference in effectiveness on insulin release between the first and second generation drugs. However, the second generation sulphonylureas (see fig. 1) contain a benzamide, or related, group in addition to the basic sulphonylurea structure as epitomized by tolbutamide. Therefore one possibility is that there are two distinct but related sites of interaction on pancreatic ß-cells for agents such as glibenclamide, one site recognising the sulphonylurea moiety (SO_2NHCO), the other the benzamide (CONH) group which is not present in tolbutamide. These two separate, but closely associated, binding sites are postulated to give rise to the greater potency of the second generation sulphonylureas by their ability to interact with both sites simultaneously.

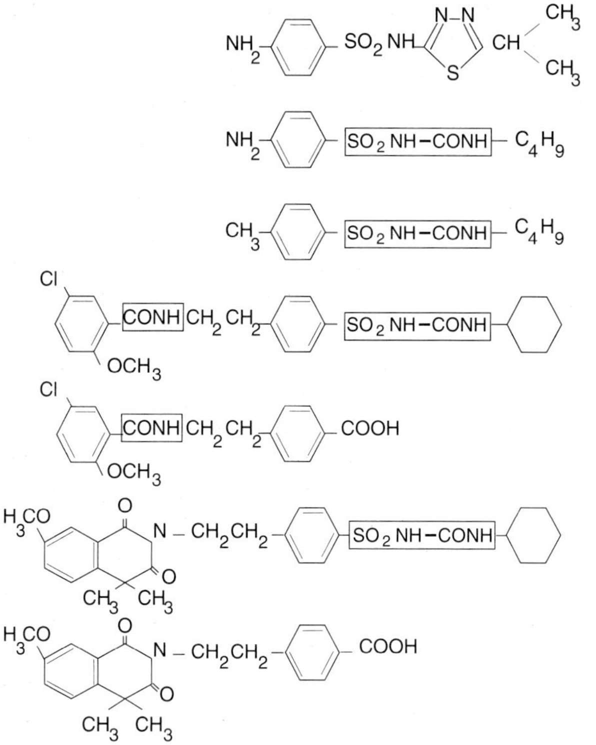

Fig. 1. Chemical structures of the hypoglycaemic agents and sulphonylureas mentioned in the text. The sulphonylurea moiety is indicated

Evidence for this contention hinges on the ability of compounds such as meglitinide (HB 699) and UL-DF9, which do not contain the sulphonylurea moiety, to elicit insulin secretion. However, the situation may be even more complex as it has been shown that both the benzoic acid and the benzamide residues of meglitinide are involved in mimicking the actions of glibenclamide on pancreatic ß-cells. Thus it has been suggested that the sulphonylurea receptor may in fact be a target for more than one chemical grouping.

Sulphonylureas bind to specific receptors

The sulphonylureas mediate their effects on cells via a membrane site of action. This has been shown clearly by the use of radiolabelled sulphonylureas, in particular tritiated (or iodinated) glibenclamide or gliquidone. These studies have shown that specific, saturable and reversible high affinity binding sites (i.e. receptors) are present on mammalian cell membranes from insulin-secreting cell lines, cardiac, skeletal and smooth muscle and central neurones. In nearly all these studies the dissociation constant (K_d; the concentration of ligand at which half the sites are occupied) for this site is in the low nanomolar range (0.05 - 2nM) the only exception reported so far being a K_d value of 10 nM for aortic smooth muscle. The average maximum number of high affinity binding sites (Bmax) reported also varies between tissues. In general the highest values are found in insulin-secreting cell lines (150 - 1,600 fmol mg^{-1} protein) and the lowest in cardiac and perhaps skeletal muscle tissue (< 100 fmol mg^{-1} protein), although it is important to note that wide variations are reported by different groups, even for the same cell type. Some of this variation may simply reflect differences in purity of the membrane preparations or in experimental protocol.

Recent use of the technique of quantitative autoradiography (whereby thin sections of tissue are cut and incubated with radioactive ligand, the tissues then washed, dried and apposed to photographic film for a period of time, the autoradiograms developed and the amount of binding determined) in conjunction with radioligand binding studies has demonstrated that the high affinity sulphonylurea receptors are heterogeneously distributed in mammalian brain. High levels (greater than 250 fmol mg^{-1} protein) of [^{3}H]-glibenclamide binding are present in areas such as the substantia nigra pars reticulata, globus pallidus, neocortex and the molecular layer of the cerebellum. Conversely, low levels of [^{3}H]-glibenclamide binding (< 100 fmol mg^{-1} protein) are present in the hypothalamic nuclei, medulla oblongata and spinal cord [4]. These differences in specific binding of radiolabelled sulphonylureas reflect variations in the density of receptors and not affinity as the K_d values are constant throughout the brain. In cardiac muscle, low to moderate numbers of binding sites are present and these are distributed homogeneously. Furthermore the binding site densities in brain, but not heart, increase post-natally and this probably reflects the development of various brain structures [4, 5].

The binding of the radiolabelled sulphonylureas to their receptor exhibits a high degree of specificity as a wide variety of agents (channel blockers or modulators, and neurotransmitter agonists or antagonists) are either virtually ineffective or produce an inhibition of binding at concentrations much greater than that required to interact at their own specific receptors. However, specific binding of radiolabelled sulphonylureas to insulin-secreting cell, cardiac and brain membrane fractions is inhibited by unlabelled sulphonylureas. Indeed the relative potencies for different sulphonylureas in displacing [^{3}H]-glibenclamide from all these tissues (fig. 2) produces an excellent correlation with their relative potency for the stimulation of insulin secretion from ß-cells and their in vivo hypoglycaemic activities. Therefore it is generally considered that the sulphonylurea receptor described above is identical, or at least very similar, in all tissues where it has so far been characterized.

Attempts to solubilize the sulphonylurea receptor, with the intention to determine its amino acid sequence and eventual cloning and receptor expression, are in progress. So far a solubilized binding site has been purified from pig brain and found to have a molecular weight of approximately 150 kDa. In addition, using photoaffinity labelling techniques (ultraviolet irradiation of radioligand inducing the formation of a covalent bond) at least two ß-cell membrane proteins have been reported. The largest of these, a 140 kDa protein is considered to be the high affinity site, and therefore the physiologic receptor, as it has a sub-nanomolar K_d and unlabelled sulphonylureas prevent binding. It has recently been shown that this protein in brain also contains ATP and ADP binding sites. The binding of radioligand to the smaller protein is less susceptible to inhibition by the presence of unlabelled sulphonylureas and may represent a separate, low affinity, site (see below). At present there is no published report of the amino acid sequence for the high affinity sulphonylurea receptor.

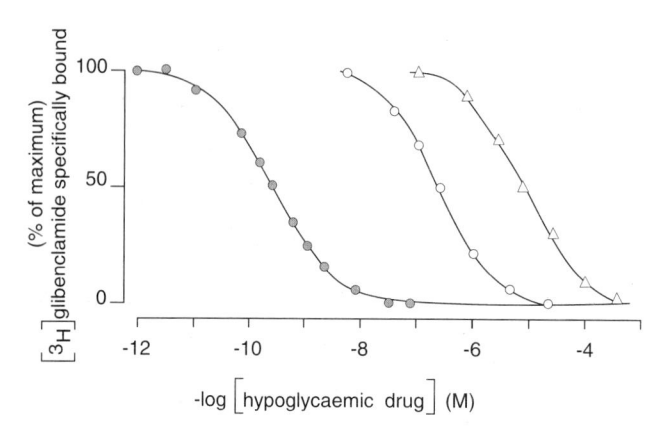

Fig. 2. Effect of increasing concentration of glibenclamide *(filled circle)*, meglitinide *(open circle)* and tolbutamide *(open triangle)* on binding of [3H]-glibenclamide to microsome membranes from insulin-secreting cells

Possibility of multiplicity of receptors and of an endogenous ligand

In studies where a much wider range of radioligand concentrations was used a second population of sulphonylurea binding sites has been identified and these have been reported to occur in membrane preparations of insulin-secreting cells, cardiac muscle, small intestine and central neurones. This second class of binding site, termed low affinity, is characterized by an approximate 100-1000 fold greater K_d and 5-20 fold greater density in comparison with the high affinity sites. Displacement of radioligand from these low affinity sites by unlabelled sulphonylureas has been reported. Furthermore there is evidence that the low affinity site is also heterogeneously distributed in the brain. The photoaffinity labelling experiments detailed above which have demonstrated the existence of more than one protein that binds glibenclamide may support the concept of receptor heterogeneity. However, it should be noted that doubts have been raised over the relevance of the low affinity site as only high affinity sites have been reported in intact ß-cells and low affinity binding is lost after solubilization of ß-cell or brain membranes. At present no physiological role has been demonstrated for this putative low affinity site.

The presence of high affinity sulphonylurea receptors (i.e. specific sites for non-naturally occurring ligands) in mammalian tissues has resulted in the quest for an endogenous ligand, in much the same way as has occurred for benzodiazepines. At present one group has reported the existence of an endogenous peptide isolated and purified from ovine brain [6]. This peptide, termed endosulphine, exists as two molecular forms, α and ß, and both recognise sulphonylurea receptors from the brain and insulin-secreting cells. ß-endosulphine was also reported to stimulate insulin secretion. This is an interesting and important development in sulphonylurea research and has profound implications for the physiological and perhaps pathological processes with which these receptors are involved (see below).

Factors and agents that alter sulphonylurea binding

Recently there have been reports that various nucleotides can displace [^{3}H]-glibenclamide from high affinity sites in brain and ß-cell membranes. In pancreatic ß-cell membranes a wide range of nucleotides including ATP, ADP, GTP, GDP and their thiophosphate derivatives have been reported [7] to inhibit the binding of [^{3}H]-gliben-

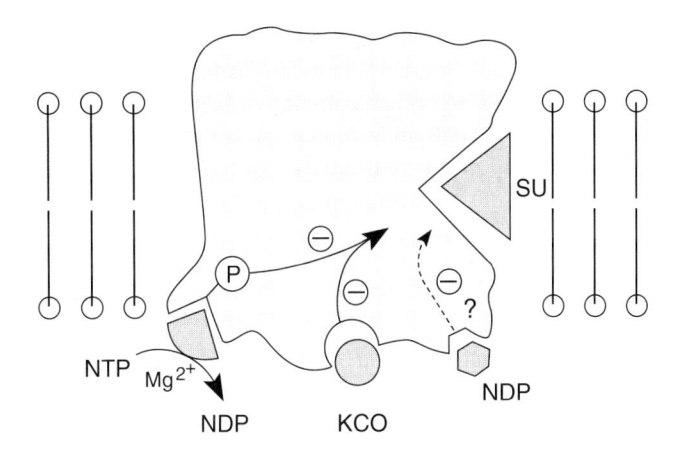

Fig. 3. Cartoon depicting possible binding sites for sulphonylureas *(SU)*, nucleotide triphosphates *(NTP)*, potassium channel openers *(KCO)* and nucleotide diphosphates *(NDP)* on the sulphonylurea receptor (shown as a transmembrane protein). The NTP site is associated with Mg^{2+}-dependent phosphorylation which results in reduced [^{3}H]-glibenclamide binding. Certain potassium channel openers will also reduce glibenclamide binding, particularly if the NTP site is occupied and phosphorylated

clamide (i.e. they increase K_d by up to sixfold). The inhibition of binding by the diphosphate nucleotides is probably due to their conversion to triphosphates by endogenous transphosphorylating enzymes (although it has also been reported that ADP in the absence of Mg^{2+} can inhibit binding). Non-hydrolysable analogues of the triphosphate nucleotides were ineffective and no inhibition of binding by nucleotides was detected in the absence of Mg^{2+} ions. These data and others have led to the suggestion that protein phosphorylation modulates the affinity (induces a decrease) by approximately three to sixfold of the sulphonylurea receptor in these cells. This lower affinity site is not equivalent to the low affinity binding site described above. Therefore, in the intact ß-cell, the relative ratios of phosphorylated (low affinity) to dephosphorylated (high affinity) receptor are likely to be controlled by the activities of endogenous protein kinases and phosphatases. It should also be noted that there is also the possibility that [^{3}H]-glibenclamide binding can be modulated by G-proteins as GTP and its analogues have been reported to inhibit binding in cardiac and ß-cell membranes.

The situation in brain membranes is similar though not identical [8]. Nucleotide triphosphates, but not monophosphates, inhibit [^{3}H]-glibenclamide binding, an action dependent upon the presence of Mg^{2+} ions. However, it was also reported that nucleotide diphosphates inhibit binding in the absence or presence of Mg^{2+} ions, and that in the absence of Mg^{2+} nucleotide triphosphates prevent the inhibition of binding by ADP. Thus it is possible that there are multiple nucleotide interaction sites with sulphonylurea receptors and that differences in these exist between tissues.

One very important class of drugs, the potassium channel openers [e.g. cromakalim, levcromakalim (BRL 38227), pinacidil, diazoxide, minoxidil sulphate and nicorandil], have their actions on various tissues, including smooth and cardiac muscle, inhibited by glibenclamide, often in a competitive manner. It has been reported that only high concentrations of diazoxide inhibit [^{3}H]-glibenclamide binding to ß-cell, brain and cardiac membranes with the other openers generally ineffective. However, recent studies on ß-cell and brain membranes have indicated that the presence of MgATP enhances the displacement of [^{3}H]-glibenclamide by diazoxide and discloses an inhibitory action for pinacidil (although high concentrations are still required) but not cromakalim or minoxidil sulphate. It is also noteworthy that pinacidil, in the presence of MgATP, acts to decrease the number of binding sites but not their affinity, which suggests that the interaction cannot be competitive. The mechanism behind these complex interactions is unclear at present (fig. 3) but may involve a separate binding site on the sulphonylurea receptor for potassium channel openers which allosterically modifies [^{3}H]-glibenclamide binding. Alternatively the actions of these openers may be mediated through protein phosphorylation as it has been reported that the MgATP enhancement of diazoxide displacement of ^{3}H-glibenclamide from ß-cell membranes is not mimicked by nonhydrolysable analogues. It is important to determine whether these or similar interactions between nucleotides, Mg^{2+}, potassium channel openers and sulphonylurea binding are general phenomena and so can also be observed with cardiac and smooth muscle tissues where the potassium channel openers are much more effective (in comparison to ß-cells).

The sulphonylurea receptor is linked to a potassium channel

The presence of a specific high affinity receptor for the sulphonylureas in various tissues does not per se confer a biological response. There must be some mechanism by which the binding of drug is linked to an effector system (signal transduction) which results in a change in cellular behaviour. The determination of the transduction target site for sulphonylureas originally came from studies on pancreatic ß-cells (or insulin-secreting cell lines). Stimulation of insulin release by sulphonylureas was shown to be associated with a reduction of the resting potassium permeability of ß-cells (leading to cell depolarization and calcium entry) thus indicating some type of functional link to a potassium channel.

The application of the patch-clamp technique for recording the ionic current flowing through a single channel has enabled the unequivocal identification of the potassium channel associated with sulphonylurea action, in at least some cell types. For example in pancreatic ß-cells from a variety of species including human, the sulphonylureas have been shown to reduce the opening of ATP-sensitive K^+ channels (fig. 4) indicating that they act to close the channel and so reduce outward current flow which results in depolarization of the pancreatic ß-cell [9,10]. This effect has been demonstrated to be fairly specific for this particular channel in ß-cells, other potassium channels (e.g. calcium or voltage-activated), voltage-activated calcium channels and non-selective cation channels (nucleotide and calcium-dependent) are unaffected. This specificity for ATP-sensitive K^+ channels over other K^+ channel types has also been noted in other tissues (e.g. heart).

The concentrations of sulphonylureas that cause half-maximal inhibition (K_i) of ATP-sensitive K^+ channel current in ß-cells compare favourably with concentrations required for stimulation of insulin release and the therapeutic plasma levels of free drug [11]. Furthermore the relative potency for ATP-sensitive K^+ channel inhibition induced by a range of these agents correlates closely to that obtained for binding affinity at the sulphonylurea receptor. Taken together these data strongly suggest that the high affinity sulphonylurea receptors present in the plasma membrane of pancreatic ß-cells are functionally linked to ATP-sensitive K^+ channels and are the sites at which these drugs

Fig. 4. Recording of single ATP-sensitive K$^+$ currents from an inside-out membrane patch isolated from an insulin-secreting cell. Under control conditions a high level of channel activity was observed (openings are shown as upward deflections) which was completely and reversibly blocked by application of 1mM ATP to the intracellular aspect of the membrane patch. Glibenclamide (100 nM) also induced closure of the ATP-sensitive K$^+$ channel (although this was not easily reversed on wash, not shown). Data were recorded in symmetrical 140mM KCl solutions at a membrane potential of +40mV

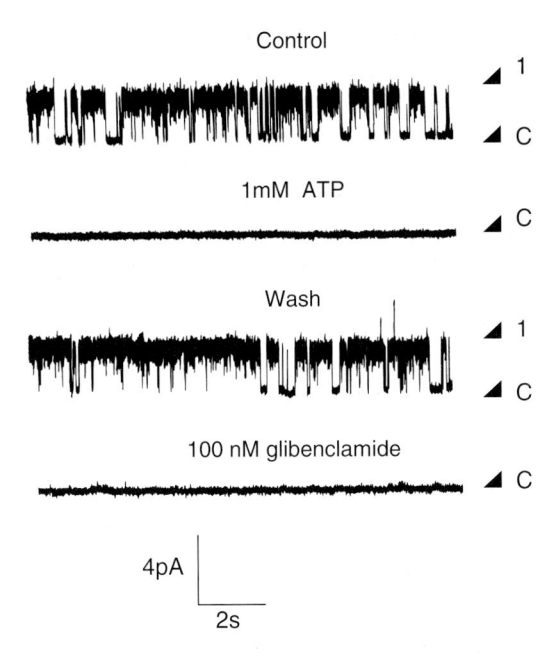

act to initiate insulin release. Investigation of the concentration-response relationships for ATP-sensitive K$^+$ channel inhibition by sulphonylureas also suggests that glibenclamide may interact with more than one site on the receptor, whereas a single site is more likely for tolbutamide. Electrophysiological studies have also shown that meglitinide, UL-DF9 and other agents which do not contain the sulphonylurea moiety are also effective inhibitors of ATP-sensitive K$^+$ channel activity (which thus explains their insulin-secretion action).

This does not exclude the possibility that these agents interfere with other K$^+$ channel types, particularly in view of recent reports that glibenclamide can inhibit a voltage-activated K$^+$ current in central neurones. However, it is not, as yet, clear to what extent this latter observation relates to the specific receptor binding sites described above.

The action of sulphonylureas is tissue specific

ATP-sensitive K$^+$ channels have been observed in many other cell types including cardiac, skeletal and smooth muscles, central and peripheral neurones and epithelia. It has also been reported that ATP-sensitive K$^+$ channels are present in the inner membrane of liver mitochondria. In general sulphonylureas have been demonstrated to inhibit ATP-sensitive K$^+$ channel activity in all cases so far described. However, it is important to realise that these channels do not all have identical biophysical, pharmacological or functional properties [12]. For example there are wide variations in the concentration of internal ATP required to inhibit channel activity ranging from K$_i$ values of 10-15 µM for ß-cells to 2-3 mM for hypothalamic neurones. The form of ATP required to inhibit the channel also varies between tissues, with the non-Mg^{2+}-bound species being the more effective in ß-cells, whereas it is MgATP that is more active in hypothalamic neurones. In skeletal and cardiac muscle both species are reported to be equally effective. Furthermore, the ATP-sensitive K^+ channel is not open (and therefore does not contribute to the resting membrane potential or action potential shape) in all cell types under normal physiological conditions. In cells that act as glucose-sensors (pancreatic ß-cells and certain hypothalamic neurones) the ATP-sensitive K^+ channels are partly open under normo-glycaemic conditions and therefore do contribute to the resting membrane potential. In contrast, ATP-sensitive K^+ channels are closed under normal conditions in heart, skeletal and smooth muscle and in some central neurones but are activated under conditions of metabolic stress (see below) which results in lowering of intracellular levels of ATP. Therefore these data and others make it likely that a multiplicity of channels exists under the general class of ATP-sensitive K^+ channels.

This heterogeneity is also reflected in the sensitivity of ATP-sensitive K^+ channel activity to the sulphonylureas. It has been reported that tolbutamide is 50 times less potent in heart cells ($K_i = 380$ µM) than pancreatic ß-cells ($K_i = 7$ µM). A similar shift in sensitivity has also been noted for glibenclamide, with high concentrations (1-20 µM) being required for block of ATP-sensitive K^+ channels in heart cells in comparison with 4-50 nM for pancreatic ß-cells. Although it should be noted that recently it has been suggested that glibenclamide has a very slow onset time for channel inhibition in cardiac cells and that consequently nM concentrations can induce significant block. High concentrations (5-20 µM) of glibenclamide are required for block of ATP-sensitive K^+ channels in skeletal and smooth muscle and the inner membrane of liver mitochondria. In general, high (0.5-100 µM) concentrations of glibenclamide reverse the effects of anoxia or potassium channel openers (purported to act via an increase in ATP-sensitive K^+ channel activity) in functional studies on substantia nigra or hippocampal neurones, although it has been reported for substantia nigra neurones that other second generation sulphonylureas (e.g. gliquidone) are much more effective than glibenclamide. The few published studies of sulphonylureas on ATP-sensitive K^+ channel activity in brain neurones indicate that their actions are not straightforward as it has been shown that although tolbutamide excites hypothalamic glucose-sensitive cells by closure of ATP-sensitive K^+ channels, glibenclamide (10-500 nM) is ineffective per se but does inhibit the action of tolbutamide. In contrast, glibenclamide (100 nM) has been demonstrated to block ATP-sensitive K^+ channel activity in vertebrate axons.

These differences in sulphonylurea potency (efficacy) for inhibition of ATP-sensitive K^+ channel activity between tissues are perhaps surprising in view of the similarities of their radioligand binding characteristics, particularly with respect to heart and brain versus ß-cells. The electrophysiological techniques (mostly patch-clamp recordings) used to determine ATP-sensitive K^+ channel function are invasive to the cell and disturb its normal environment (this is particularly true of isolated membrane patches) and so may be considered as possible contributory factors to this variation. However, almost identical conditions are employed by most investigators regardless of cell type. Thus perhaps the most likely explanations are that these differences somehow arise either because of intrinsic ATP-sensitive K^+ channel variations or from heterogeneity between the sulphonylurea receptors and their coupling to the channels.

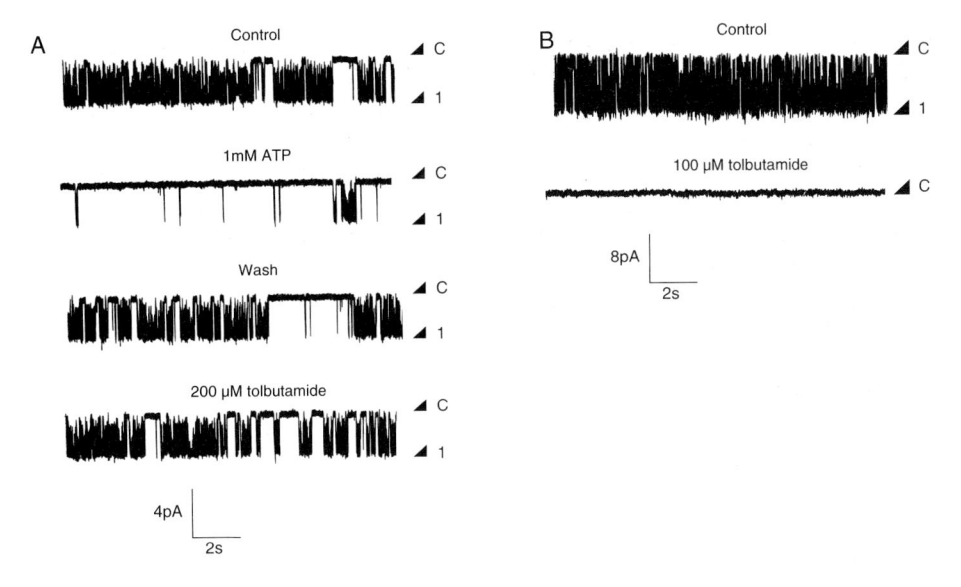

Fig. 5 A, B. Single ATP-sensitive K$^+$ channels recorded from isolated hypothalamic (ventromedial nucleus) neurones. **A** Recordings from an excised inside-out membrane patch showing that application of 1mM ATP to the intracellular aspect of the patch induced inhibition of channel activity, an effect reversible on wash. Application of the sulphonylurea tolbutamide had no effect on channel activity. Data were recorded in symmetrical 140mM KCl solutions at a membrane potential of -20mV. **B** Cell-attached recording of single ATP-sensitive K$^+$ channel activity (the cell had been exposed to glucose-free conditions for 20 min prior to recording in order to achieve a high level of channel activity). In contrast to the isolated patch recording, addition of tolbutamide (100 µM) inhibited ATP-sensitive K$^+$ channel activity completely. This effect was reversible on wash (not shown). Data were recorded with 140mM KCl solution in the electrode and normal NaCl saline in the bath at 0mV pipette voltage (i.e. the driving force for the potassium movement was from the resting membrane potential of the cell). Currents flowing when the channel was open are denoted by downward deflections in **A** and **B**

Are the ATP-sensitive K$^+$ channel and the sulphonylurea receptor a single entity or associated proteins?

At present there is no direct evidence to indicate whether the ATP-sensitive K$^+$ channel and the sulphonylurea receptor are a single entity (i.e. part of a single macromolecular receptor-channel complex) or whether they are closely linked, but separate, proteins. The evidence to date does not suggest any requirement for a freely diffusible intracellular factor (e.g. ATP, calcium or cyclic nucleotide) as a coupling molecule between the channel and receptor responsible for channel closure. Indeed it has been demonstrated using single channel recordings from patches of membrane isolated from pancreatic ß-cells that sulphonylureas are effective inhibitors of ATP-sensitive K$^+$ channel activity when applied to either the extracellular or intracellular face of the plasma membrane. Hence the sulphonylureas are currently thought to access their site of action via the lipid phase of the plasma membrane in the undissociated (i.e. uncharged) form.

It may be inferred from these observations that the sulphonylurea receptor is intima-

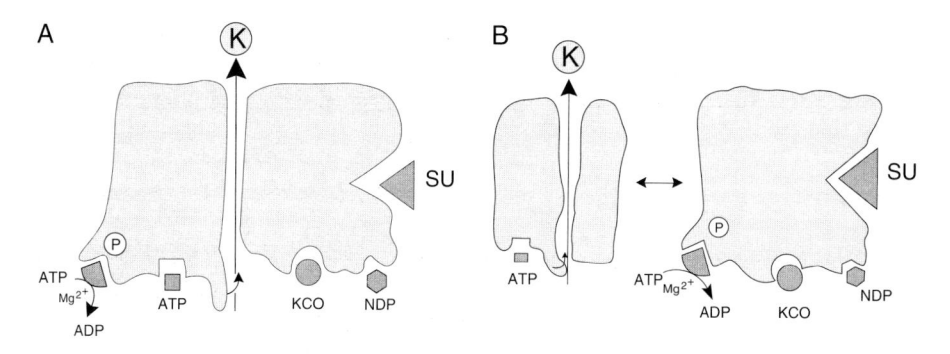

Fig. 6 A, B. Cartoons depicting the possible relationships between the ATP-sensitive K$^+$ channel and the sulphonylurea receptor. **A** The sulphonylurea receptor and the ATP-sensitive K$^+$ channel are a single entity and hence an additional site is present (compare with fig. 3) for ATP to bind (hydrolysis not required) and cause channel closure. Therefore channel gating is directly affected by conformational changes occurring on ligand binding (e.g. SU, KCO, ATP). **B** The sulphonylurea receptor and ATP-sensitive K$^+$ channel are separate but closely associated entities in the membrane and some intracellular factor enables coupling to occur. Same abbreviations as in fig. 3

tely associated with the ATP-sensitive K$^+$ channel and coupling between the bound receptor and the channel is likely to be direct via some conformational change of the receptor/channel protein complex. However, not all observations on tissues other than ß-cells fit this simple hypothesis. A particularly striking example is the lack of effect of tolbutamide on ATP-sensitive K$^+$ channel activity recorded from isolated inside-out membrane patches from hypothalamic neurones which contrasts markedly with the inhibition of these channels in intact cells (fig. 5). It has also been reported that in approximately 30% of inside-out patches isolated from heart cells tolbutamide is ineffective at inhibiting ATP-sensitive K$^+$ channel activity. Thus it may be just as feasible (fig. 6) to suppose that some intrinsic coupling factor exists (which is responsible for transduction of the coupling signal) and is lost or inactivated on formation of the excised patch in some cell types (i.e. the factor is perhaps more labile in hypothalamic neurones, less so in heart and not at all in ß-cells). Although the identity of this putative factor or coupling mechanism in these cells is at present unknown, the following observations made on ß-cell and cardiac ATP-sensitive K$^+$ channels may prove to be pertinent.

Tolbutamide is a more effective inhibitor of ß-cell ATP-sensitive K$^+$ channel activity when applied to the intact cell in comparison to the cytoplasmic aspect of inside-out patches. However, channel sensitivity to tolbutamide in inside-out membrane patches is strongly enhanced by the presence of the Mg^{2+} complex of ADP at the cytoplasmic side. This action of MgADP is not mimicked by ADP in the absence of Mg^{2+}, ATP or non-hydrolysable analogues of ATP, but these agents are reported to increase the effectiveness of ADP or its analogues. A similar action has also been reported for meglitinide inhibition of ß-cell ATP-sensitive K$^+$ channel activity. Consequently it has been suggested that the presence of these nucleotide binding sites are in some way important for the ATP-sensitive K$^+$ channel-sulphonylurea receptor interaction, and that occupation of these sites enhances sulphonylurea sensitivity. In complete contrast to this finding, it has been reported that ADP in cardiac cells reduces the sulphonylurea sensitivity of ATP-sensitive K⁺ channel currents. It will certainly be interesting to determine whether similar interactions occur in other cell types particularly with respect to the lack of effect of tolbutamide in isolated patches from hypothalamic neurones.

Sulphonylureas antagonize the effects of potassium channel openers

The potassium channel openers act to inhibit cell function mainly by hyperpolarization of the plasma membrane and are effective in many tissues including cardiac, skeletal and smooth muscles, neurones and secretory cells [13, 14]. Their inhibitory actions in all these tissues are antagonized by glibenclamide. In addition they have been demonstrated to activate ATP-sensitive K⁺ channels in heart cells (see chapter 13) and ß-cells (albeit at rather high concentrations for all except diazoxide), and this activation is also reversed by glibenclamide. Therefore from these and other observations it has generally been considered that their main site of action is on ATP-sensitive K⁺ channels. However, in the case of smooth muscle cells this is not certain. Although the potassium channel openers exhibit the greatest potency on smooth muscle the identity of the K⁺ channel involved is, at present, not completely clear. Several channel types, of large and small conductance exhibiting ATP and/or calcium sensitivity, have been reported to be activated by these agents in vascular smooth muscle (chapter 12).

The potency of the potassium channel openers in tissues containing ATP-sensitive K⁺ channels also appears to be the converse of their sensitivity to glibenclamide. Pancreatic ß-cell ATP-sensitive K⁺ channels exhibit the highest sensitivity to glibenclamide yet show the weakest response (i.e. high concentrations required) to the majority of the openers, whereas in tissues more responsive to the openers (particularly smooth muscle) much higher concentrations of glibenclamide are required to inhibit their actions. Consequently it is possible, if not probable, that there is heterogeneity of the target site for these agents, both the potassium channel openers and the sulphonylureas. This may not be particularly surprising given the diverse chemical structures of the openers in conjunction with the possibility that the sulphonylurea receptor may have multiple binding sites.

Sulphonylurea receptors influence cellular functions in many tissues

The actions and clinical relevance of the sulphonylureas have already been dealt with for pancreatic ß-cells. As outlined above ATP-sensitive K⁺ channels and sulphonylurea receptors are not confined to ß-cells but are also present in many other tissues. Therefore how do the sulphonylureas influence their cellular functions and what are the possible clinically relevant consequences? It is not feasible in this review to describe all these in detail [14] and most of the relevant information with respect to their effects on vascular smooth muscle are included in other chapters of this book. However, some examples are presented below.

Central neurones

Glucose-receptive neurones, present in the ventromedial hypothalamic nucleus, like pancreatic ß-cells have ATP-sensitive K^+ channels open at rest (i.e. at normal fasting glucose concentrations). The activity of these channels is inhibited by tolbutamide, which results in cell excitation, mimicking the action of a rise in the extracellular glucose concentration. Sub-micromolar concentrations of glibenclamide do not excite these cells but tolbutamide induced excitation is blocked by low concentrations (i.e. 20 nM) of glibenclamide [15]. These results are also indicative of multiple sites of interaction for these agents at the functional level. The ventromedial in conjunction with other hypothalamic and non-hypothalamic nuclei play key roles in regulation of food intake. Consequently sulphonylureas may potentially disrupt (either directly, in the manner of tolbutamide, or indirectly, as with glibenclamide, by interference with an endogenous molecule) the integration of the satiety and feeding signals in the brain. Actions at these sites may also result in alterations of CNS autonomic output.

Sulphonylureas have also been shown to modulate neurotransmitter release in the brain but only under conditions of anoxia or hypoglycaemia (i.e. the system must be "stressed" first). This suggests that the ATP-sensitive K^+ channels underlying these actions are present in pre-synaptic terminals, are closed at rest, and must first be activated by some metabolic crisis. One particularly interesting and potentially very important example of this is in the hippocampus and the modulation of the release of the excitatory neurotransmitter, glutamate [16]. A brief episode of anoxia results in an increased synaptic release of glutamate in this brain area resulting in depolarization of post-synaptic neurones. Application of glibenclamide potentiates this depolarization and prevents the inhibition of this depolarization by the opener, diazoxide. As abnormal glutamate action on brain cells is currently considered to be important in the aetiology of stroke and epilepsy an understanding of the means by which sulphonylureas interfere with this system is obviously crucial.

A similar role has been proposed for ATP-sensitive K^+ channel function in GABA-containing nerve terminals in the substantia nigra that control the activity of dopamine-containing neurones. Under anoxic or hypoglycaemic conditions there is a decreased release of GABA (through activation of ATP-sensitive K^+ current pre-synaptically) and it has been postulated that this contributes to the increased likelihood of generalized seizures associated with pathological events such as brain ischaemia and severe hypoglycaemia.

Cardiac muscle

Under conditions of resting oxygen tension and glucose levels (i.e. at normal physiological concentrations of intracellular ATP) cardiac ATP-sensitive K^+ channels are generally considered to be closed. Following acute myocardial ischaemia there is a decline of intracellular ATP levels and as a consequence activation of ATP-sensitive K^+ channel activity. During an ischaemic episode there is also an increase in intracellular H^+ concentration and this too will act to increase ATP-sensitive K^+ channel activity. This results in a marked shortening of cardiac action potential duration and the elevation of extracellular potassium concentration. Such an action, if not unchecked, will result in localized tissue depolarization and ultimately give rise to conduction inhomogeneities and reentrant arrhythmias. It is therefore argued that blockade of ATP-sensitive K^+ channel currents by sulphonylureas will be antiarrhythmic by preventing the ischaemia induced action potential shortening and reduction in refractory period (i.e. act similarly to class III antiarrhythmic agents which block delayed rectifier K^+ current and prolong the myocardial action potential).

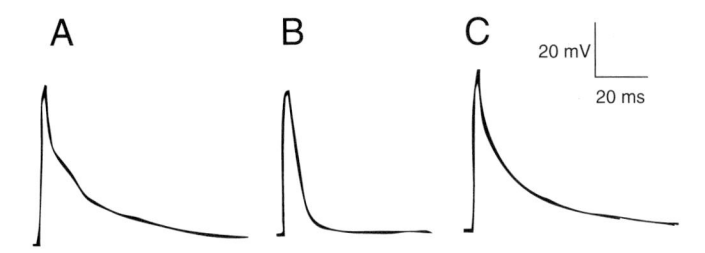

Fig. 7 A-C. Ventricular action potential recorded **A** under control, **B** under ischaemic conditions and **C** under ischaemic conditions in the presence of 10μM glibenclamide

In vitro studies using isolated heart preparations have demonstrated that glibenclamide and tolbutamide can reduce K^+ loss and inhibit the shortening of action potential duration in ischaemic hearts (fig. 7). They have also been shown to reduce significantly or abolish the incidence of ventricular fibrillation in response to ischaemia. However generally rather high concentrations of glibenclamide (10-100 μM) are required to produce these actions and such concentrations have been shown not to completely suppress ischaemia-induced K^+ loss. Yet this concentration range of glibenclamide is certainly sufficient to produce substantial if not complete block of cardiac ATP-sensitive K^+ channel currents (see above). A likely explanation for this discrepancy is that during ischaemia, not only ATP concentrations fall and H^+ levels increase inducing ATP-sensitive K^+ channel activation but also ADP levels will rise sufficiently to limit the effectiveness of sulphonylureas during ischaemic insult.

However, in vivo studies have not confirmed this therapeutic potential for sulphonylureas. Although there is a report that glibenclamide reduces post-infarct ventricular fibrillation, other studies indicate that glibenclamide-treated animals display increased likelihood of ventricular fibrillation during, or after, an ischaemic insult. The explanation for this finding is probably associated with the idea that activation of cardiac ATP-sensitive K^+ channels during ischaemia is a protective mechanism (by acting to reduce calcium entry and contractile activity) which decreases the metabolic requirements of compromised cells. Consequently prevention of this mechanism with sulphonylureas may result in some proarrhythmic action.

Therefore although there is considerable therapeutic potential [17] for modulation of cardiac ATP-sensitive K^+ channels (the potassium channel openers have also been shown to produce pro- and antiarrhythmic actions, see chapter 13) it would appear that a more detailed understanding of ATP-sensitive K^+ channel function and modulation is required before new-generation antiarrhythmics can be developed.

In recent years the sulphonylureas have become an invaluable tool of pharmacologists and clinical scientists for investigating the possible contribution of ATP-sensitive K^+ channels in tissue function. However, some note of caution may be required as their selectivity for this type of K^+ channel may not be absolute, particularly at the high (μM) concentrations required in many studies. Furthermore there is likely heterogeneity of ATP-sensitive K^+ channels and possibly also of sulphonylurea receptors. Therefore one essential future aim is that a full and detailed understanding be obtained of the targets and mechanisms by which the sulphonylureas modify potassium channel function. It is particularly important that their actions are investigated both under normal resting conditions and those of metabolic stress, as substantial changes in cellular function can occur associated with K$^+$ channel activation.

References

1. Loubatières A (1977) Effects of sulphonylureas on the pancreas. In: Volk B W, Wellmann K E (eds) The diabetic pancreas. Balliere Tindall, London, p 489
2. Lebovitz HE (1985) Oral hypoglycaemic agents. In: Alberti KGMM, Krall LP (eds) The diabetes annual. Elsevier, p 93
3. Symposium on glyburide (1990) Glyburide: new insights into its effects on the ß cell and beyond. Am J Med 89 (suppl 2A): 1S-53S
4. Mourre C, Widmann C, Lazdunski M (1990) Sulphonylurea binding sites associated with ATP-regulated K$^+$ channels in the central nervous system: autoradiographic analysis of their distribution and ontogenesis, and of their localization in mutant mice cerebellum. Brain Research 519: 29-43
5. Miller JA, Velayo NL, Dage RC, Rampe D (1991) High affinity [3H] glibenclamide binding sites in rat neuronal and cardiac tissue: localization and developmental characteristics. J Pharm Exp Ther 256: 358-364
6. Virsolvy-Vergine A, Leray H, Kuroki S, Lupo B, Dufour M, Bataille D (1992) Endosulphine, an endogenous peptidic ligand for the sulphonylurea receptor: purification and partial characterization from ovine brain. Proc Natl Acad Sci 89: 6629-6633
7. Schwanstecher M, Löser S, Brandt Ch, Scheffer K, Rosenberger F, Panten U (1992) Adenine nucleotide-induced inhibition of binding of sulphonylureas to their receptor in pancreatic islets. Br J Pharmacol 105: 531-534
8. Gopalakrishnan M, Johnson DE, Janis RA, Triggle DJ (1991) Characterization of binding of the ATP-sensitive potassium channel ligand, [3H] glyburide, to neuronal and muscle preparations. J Pharm Exp Ther 257: 1162-1171
9. Ashcroft FM, Rorsman P (1989) Electrophysiology of the pancreatic ß-cell. Prog Biophys Mol Biol 54: 87-143
10. Ashford MLJ (1990) Potassium channels and modulation of insulin secretion. In: Cook NS (ed) Potassium channels: structure, classification, function and therapeutic potential. Ellis Horwood, Chichester, pp 300-325
11. Panten U, Burgfeld J, Goerke F, Rennicke M, Schwanstecher M, Wallasch A, Zünkler BJ, Lenzen S (1989) Control of insulin secretion by sulphonylureas, meglitinide and diazoxide in relation to their binding to the sulphonylurea receptor in pancreatic islets. Biochem Pharmacol 38: 1217-1229
12. Ashcroft SJH, Ashcroft FM (1990) Properties and functions of ATP sensitive K$^+$ channels. Cellular Signalling 2: 197-214
13. Duty S, Weston, AH (1990) Potassium channel openers: pharmacological effects and future uses. Drugs 40: 785-791
14. Robertson DW, Steinberg MI (1990) Potassium channel modulators: scientific applications and therapeutic promise. J Med Chem 33: 1529-1541
15. Ashford MLJ, Boden PR, Treherne JM (1990) Tolbutamide excites rat glucoreceptive ventromedial hypothalamic neurones by indirect inhibition of ATP-K$^+$ channels. Br J Pharmacol 101: 531-540
16. Ben-Ari Y (1989) Effect of glibenclamide, a selective blocker of an ATP-K$^+$ channel, in the anoxic response of hippocampal neurones. Pflügers Arch 414 (Suppl 1): S111-S114
17. Lynch JJ Jr, Sanguinetti MC, Kimura S, Bassett AL (1992) Therapeutic potential of modulating potassium currents in the diseased myocardium FASEB J 6: 2952-2960

CHAPTER 11
Potassium channel blockers in the heart

I Baró and D Escande

Owing to the recent development of the patch-clamp technique, numerous drugs have been identified as cardiac K⁺ channel blockers. Specific K⁺ channel blockers are valuable laboratory tools used to explore the role of the different K⁺ channel categories in the heart muscle both under physiological and pathological circumstances. From a therapeutic perspective, blocking K⁺ channels prolongs the action potential duration (QT prolongation in the ECG) and lengthens the refractory period which is a classical antiarrhythmic mechanism [1]. Several pure cardiac K⁺ channel blockers are being evaluated experimentally and clinically as antiarrhythmic agents. However, K⁺ channel block in the heart is not a specific attribute of antiarrhythmics but is also demonstrated by compounds not primarily aimed at modifying cardiac electrophysiology (e.g. neuroleptics). Undesirably prolonged repolarization creates a risk of life-threatening ventricular arrhythmias [2]. This issue has recently been highlighted by the discovery that antihistamine drugs such as astemizole or terfenadine [3] are responsible in a limited number of cases for *torsades de pointe* arrhythmias. Although the incidence of such side effects is extremely low, the number of patients treated with these agents is very high and the risk/benefit ratio unacceptable for those few patients presenting delayed repolarization. Finally, it must be kept in mind that QT prolongation is not always the consequence of reduced K⁺ channel activity and that other cellular mechanisms may be involved in delayed repolarization.

Classification and nomenclature

Only those agents which block K⁺ channels in cardiac muscle are discussed in this chapter.

"Classical" K⁺ channel blockers are laboratory tools

Although exhibiting a low affinity > 1 mM, classical K⁺ blockers have some value as laboratory tools because of their specificity for K⁺ channels. Quaternary ammonium ions such as tetraethylammonium chloride (TEA) have been widely employed and could be considered as the ancestors of K⁺ channel blockers. Other quaternary ammonium ions include tetramethylammonium (TMA), tetrabutylammonium (TBA), and tetrahexylammonium (THA). TEA, whose potency usually differs according to whether it is applied internally or externally, shows poor selectivity against the various K⁺ channel categories in cardiac cells.

By contrast, 4-aminopyridine (4-AP), another classical K⁺ channel blocker, is more selective for the transient outward current, i_{to}. Although aminopyridines such as 2,3 and 3,4-diaminopyridines and others are usually more potent than 4-AP, they exhibit a comparable profile. Because of its K⁺ channel blocking properties, 4-AP facilitates the

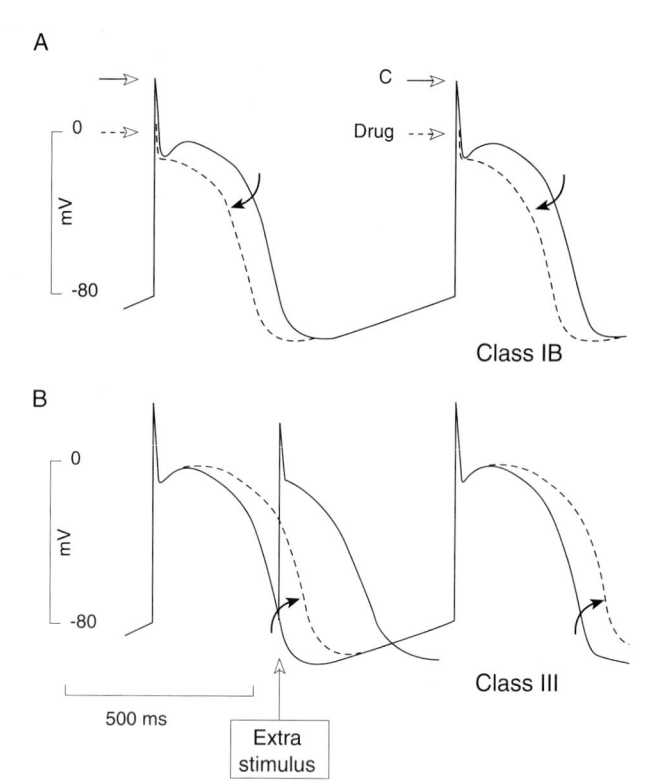

Fig. 1 A, B. Schematic representation of the effects of **A** class IB and **B** class III compounds on the normal action potential of Purkinje ventricular cells. In **B**, an extra stimulus applied during the functional refractory period triggers an abbreviated action potential. In the presence of the class III agent, the same stimulus fails to trigger a regenerative cellular response

release of transmitter at the neuro-muscular junction and has been tentatively exploited clinically as an anticurare agent by Bulgarian anaesthetists. Furthermore, it may be beneficial in patients with Alzheimer's disease [4]. Tacrine (9-amino-1,2,3,4-tetrahydroacridine, THA), an inhibitor of cholinesterase activity and a K^+ channel blocker with a structural resemblence to 4-AP, was also reported to have some beneficial effects in Alzheimer's disease.

Inorganic cations such as Ba^{2+} and Cs^+ exhibit poor selectivity against the different K^+ channel subcategories. External Ba^{2+} is frequently used in in vitro cardiac preparations to create abnormal automaticity secondary to enhanced diastolic depolarization. Intravenous Cs^+ has been employed experimentally in dogs to produce *torsades de pointe-like* polymorphic ventricular tachycardias associated with early after-depolarization potentials.

Numerous natural toxins extracted from the venom of snakes, scorpion or honey bees (dendrotoxin, noxiustoxin, charybdotoxin, apamin...) are useful for pharmacological dissection of the various K^+ channel categories expressed in several tissues (for a review see [5]). Surprisingly, these toxins are of much less interest in cardiac cells because of their poor efficacy and specificity.

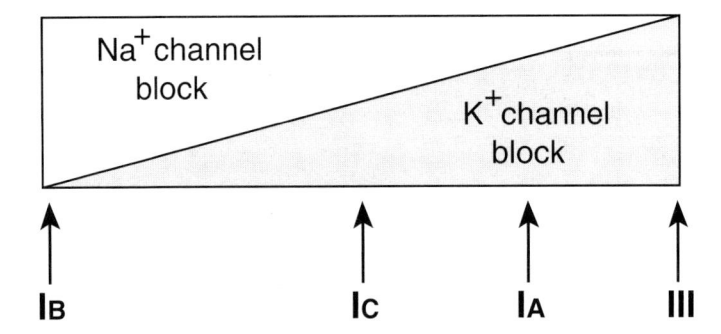

Fig. 2. Relative potency of class I and III antiarrhythmic compounds in blocking Na⁺ and K⁺ channels. Redrawn from [Colatsky, Follmer CH, (1990) Drug Dev Res 19: 129-140]

K⁺ channel blockers have therapeutic potential

Antiarrhythmic drugs

In the Vaughan-Williams classification based on the effects of drugs on the morphology of the cardiac action potential [6], class III antiarrhythmic agents are those which prolong the plateau duration and thus prolong refractoriness (Table 1). The simplest way to achieve this goal is to block myocardial K⁺ channels because less net outward current will be available to repolarize the cells. It has been proposed that antiarrhythmic drugs from both classes I and III be reclassified according to their relative potency in blocking myocardial Na⁺ and K⁺ channels [7]. Class I agents are those which depress the velocity of the depolarizing phase of the action potential as a result of predominant Na⁺ channel block and which thereby slow conduction. At the cellular level, exclusive Na⁺ channel block leads not only to decreased depolarization rate in the upstroke but also to the shortening of the action potential, because a Na⁺ current which does not inactivate operates to maintain a depolarizing level during the plateau. Class IB drugs such as lidocaine are potent Na⁺ channel blockers and exert no effects on K⁺ channels at clinically relevant concentrations: consequently, these drugs decrease both upstroke velocity and action potential duration (fig. 1A). By contrast, pure class III drugs are potent K⁺ channel blockers but exert virtually no effects on Na⁺ channels: these agents exclusively prolong the action potential duration (fig. 1B). Class IA compounds such as quinidine are both Na⁺ and K⁺ channel blockers with a predominant effect on K⁺ channels: they decrease the upstroke velocity but also increase the action potential duration. Class IC drugs such as flecainide are supposedly equipotent Na⁺ and K⁺ channel blockers: they decrease the upstroke velocity but do not significantly alter the action potential duration. Thus, as far as the relative potency of antiarrhythmics to block Na⁺ and K⁺ channels is concerned, a continuum exist between the various drugs available with Class IB and pure class III compounds at opposite ends (fig. 2).

Blockers of the ATP-sensitive K⁺ channels

Sulphonylureas such as glibenclamide (= glyburide), tolbutamide, glypizide, gliquidone..., are orally-active antidiabetic drugs acting on pancreatic ß-cells by blocking

Table 1. Vaughan-Williams classification of antiarrhythmic drugs

		Prototype	Upstroke velocity	Action potential duration
Class IA	Na⁺ channel blockers	Quinidine	Decreased	Prolonged
Class IB	Na⁺ channel blockers	Lidocaine	Decreased	Shortened
Class IC	Na⁺ channel blockers	Flecainidine	Decreased	Unchanged
Class II	ß blockers	Propanolol	Unchanged	Unchanged
Class III	K⁺ channel blockers	Sotalol	Unchanged	Prolonged
Class IV	Ca²⁺ channel blockers	Verapamil	Unchanged	Shortened

ATP-sensitive K⁺ channels (see chapter 10). Correctly speaking, they are not drugs of the cardio-vascular system but they are widely used as experimental tools to explore the role of ATP-sensitive K⁺ channels in cardiac cells. Indeed, glibenclamide and other sulphonylureas directly and specifically block ATP-sensitive K⁺ channels in muscular cells. The affinity of glibenclamide for cardiac channels is around the nanomolar range, although this value may not represent the true affinity of the drug for its receptor because the sulphonylurea receptor site may be situated in the lipid phase of the sarcolemma [8]. There is an highly specific binding site for radio-labelled glibenclamide in myocytes [9] with a K_D of around $5 \cdot 10^{-11}$ M. Similar binding sites are also found in the cortex and in the pancreas. Displacement of [³H]-glibenclamide binding to cardiac membranes by various sulphonylureas occurs with a K_i that agrees well with their respective potencies: (i) to induce hypoglycemia in vivo; (ii) to inhibit K⁺ efflux from ß-cells; and (iii) to displace the binding of [³H]-glibenclamide from both ß-cells and brain membranes. This indicates that the binding site for sulphonylureas in cardiac cells is identical with those found in the brain and in pancreatic ß-cells and that it defines the channel itself or an associated protein. Whether the block of cardiac ATP-sensitive K⁺ channels by sulphonylureas (the therapeutic plasma concentration of glibenclamide is 0.1 to 1.5 µM) leads to some deleterious or beneficial effects in diabetic patients treated with such drugs is speculative. Nevertheless, the fact remains that sulphonylureas are particularly useful tools to explore experimentally, both in vitro and in vivo, the role of ATP-sensitive K⁺ channels under physiological and pathological situations. Other substances have been reported to block these channels in pancreatic ß-cells. These include quinine and quinidine, the non-sulphonylurea antidiabetic drug linoglimide, the α-2-adrenergic agonist clonidine, the alkaloid sparteine, the tricyclic compound amantadine, the Chinese remedy component ligustrazine, and the barbiturates thiopentone and pentobarbitone. The effectiveness of these agents as blockers of ATP-sensitive K⁺ channels in the heart as well as their specificity need further investigation. Of great interest is the identification of 5-hydroxydecanoate, a lipid component of human milk, which was recently suggested to block cardiac ATP-sensitive K⁺ channels exclusively under ischaemic conditions [10].

K⁺ channel block is a cause of adverse effects

Numerous drugs delay cardiac repolarization, producing an adverse effect (for a review see [11]). Prolonged cardiac repolarization is a potent antiarrhythmic mechanism in many circumstances (see below). However, it may also lead per se to life-threatening ventricular arrhythmias, particularly when hypokalaemia, hypomagnesaemia and/or bradycardia are associated. Drugs that adversely impair repolarization include antibiotics, neuroleptics, Ca^{2+} channel blockers, antihistamine and antiserotonin compounds. In selected cases, the mechanism by which the QT interval is prolonged has been demonstrated to be K^+ channel block.

Phenothiazines may influence the normal surface ECG by prolonging the QT interval. Although several members of this class of drugs, including chlorpromazine, thioridazine, fluophenazine or trifluoperazine effectively block K^+ channels in non-cardiac preparations, the link between K^+ channel block and QT prolongation needs to be formally established.

Verapamil, diltiazem and nicardipine block the delayed rectifier K^+ current in cardiac myocytes at concentrations several hundred times greater than those required to block Ca^{2+} channels. By contrast, the methoxy-derivative of verapamil, D600, and bepridil affect different K^+ channel categories in cardiac cells at concentrations that block Ca^{2+} channels.

There is evidence that the serotonin antagonist ketanserin which is aimed at the treatment of hypertension acting at 5-HT2 receptors, prolongs the cardiac action potential and the QT interval [12]. ICS 205-930, another serotonin antagonist acting at the 5-HT3 receptor subtype, also prolongs the action potential by blocking cardiac K^+ channels. Histamine antagonists such as terfenadine and astemizole acting at the H1 receptor may induce QT prolongation and occasionally *torsades de pointe* arrhythmias. A possible relationship between serotonin or histamine inhibition and K^+ channel block needs further investigation.

The possible correlation between the sigma nonopioid receptor and K^+ channel block (reviewed in [13]) is also of great interest. Phencyclidine (PCP; also called "angel dust") is a potent fast-acting general anaesthetic which was withdrawn from medical use and became a major drug of abuse because of its euphorigenic effects. PCP binds with high affinity to two receptors (fig. 3): (i) the PCP receptor supposedly associated with the NMDA subtype of glutamate receptor-coupled ion channel and; (ii) the sigma receptor whose precise function is still unknown. The sigma receptor is highly sensitive to the butyrophenone neuroleptic haloperidol and to various phenothiazines, and is selectively labelled by $[^3H](+)3$-(3-hydroxyphenyl)-N-(1-propyl)piperidine; $[^3H](+)3$-PPP. Several ligands of the $[^3H](+)3$-PPP binding site including PCP are potent K^+ channel blockers in the brain. Their relative potency to bind to the sigma receptor parallels their ability to block brain K^+ channels and also to elicit behavioral effects in vivo. The mechanism responsible for the behavioural effects of PCP is not known with precision; however, behaviourally inactive PCP analogues fail to show the high affinity inhibition of K^+ channels; furthermore, intoxication with the K^+ channel blocker 4-AP causes a psychotic syndrome in humans that resembles the one produced by PCP. It has also been shown that PCP and haloperidol are potent blockers of K^+ channels in cardiac cells. Melperone, another butyrophenone, prolongs the QT interval and exerts some class III antiarrhythmic activities. A tentative interpretation of these data is that the sigma receptor is a K^+ channel itself or is an associated regulatory protein present both in the brain and in the heart and that inhibition of this K^+ channel has dual consequences, behavioural and cardiac. Should this hypothesis be validated, then $[^3H](+)3$-PPP binding assays may be useful to screen K^+ channel blockers with class III antiarrhythmic potential, provided that these drugs do not cross the blood-brain barrier, thus avoiding PCP-like adverse effects. In addition, the test could also be of value to identify psychotropes lacking cardiac side effects linked to the block of myocardial K^+ channels.

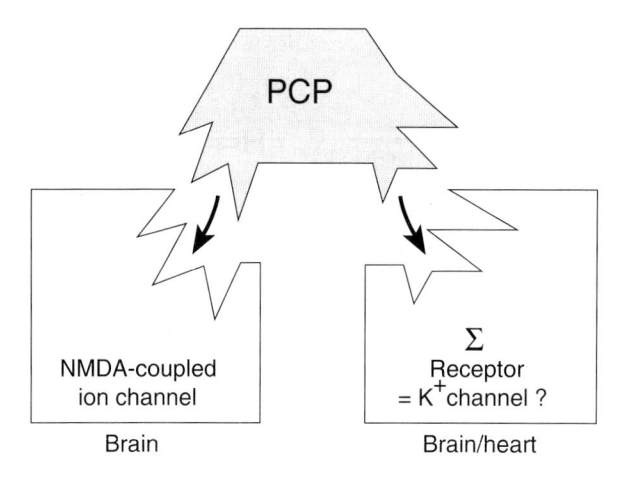

Fig. 3. Phencyclidine binds to two receptors: the excitatory amino-acid coupled ion channel and the sigma nonopioid receptor

Antiarrhythmic drugs

K^+ channel block is part of the pharmacological profile of class IA and III drugs

In cardiac myocytes, quinidine, the prototype of class IA agents, blocks the transient outward K^+ current [14], the delayed rectifier and the inward rectifier K^+ currents. However, in spite of its K^+ channel blocking properties, the effects of quinidine on the cardiac action potential duration are variable depending on the concentration, the species studied and the rate of stimulation: in certain instances, it clearly prolongs the action potential duration with minimum effects on the upstroke, whereas in other cases a substantial action potential shortening due to predominant Na^+ channel block is observed. Thus, depending on the experimental circumstances, quinidine could be classified either as a class III agent with additive class I properties or as a class I agent which additionally prolongs the cardiac action potential. In addition, the drug possesses class IV activities because it also blocks Ca^{2+} channels. For drugs with multiple mechanisms of action such as quinidine, it is almost impossible to determine which mechanism is mainly responsible for their antiarrhythmic activity. This is characteristic as far as the first generation of class III antiarrhythmics is concerned. Amiodarone, which was first approved in France in 1966, is still regarded in 1993 by numerous investigators as a gold-standard for its undeniable clinical efficacy. Because the most obvious action of chronically administered amiodarone is prolongation of repolarization, during the early 80's it became the prototype of the newly individualized class III. However, ten years later, one should realize that classifying amiodarone as a class III drug may be confusing because it is such an excessively complex and unique product exhibiting class I, class III and class IV [15] properties and also interfering with ß- and α-adrenoreceptors as well as with the metabolism of thyroid hormones. It is conjectural as to which property is crucial to its remarkable potency as an antiarrhythmic agent. Indeed, amiodarone should either be in a class by itself or should better not be classified at all. The other classical member of class III, sotalol, is a potent ß blocker (class II activity) and also a Ca^{2+} channel blocker [16], a combination that is likely to confer a broad spectrum of antiarrhythmic effects on the drug.

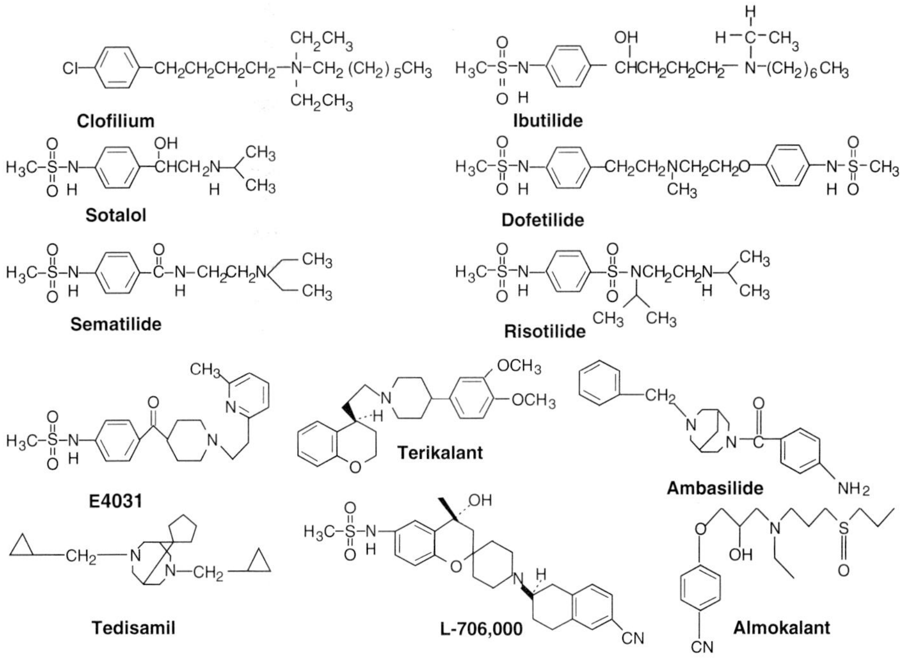

Fig. 4. Chemical structure of the main representatives of pure class III antiarrhythmic drugs

Thus drugs that exclusively prolong the action potential duration were needed to validate the theoretical concept of class III antiarrhythmics. "Second generation" class III drugs are currently being developed by several pharmaceutical groups: D-sotalol, the dextro-isomer of sotalol, lacks the ß blocking properties of its parent compound; clofilium is an analog of TEA. Furthermore, as illustrated in fig. 4, the (methylsulphonyl)amino group of sotalol has further been exploited as the chemical root of a series of pure class III drugs including dofetilide (UK 68,798), sematilide (CK-1752), risotilide (Wy-48,986), E4031, ibutilide (U70,226E), and L.706,000. Other drugs not structurally related to sotalol, such as almokalant (H 234/09), terikalant (RP 62719), tedisamil (KC-8857), and ambasilide (LU-47710), have also been discovered.

Second generation class III drugs are diverse

With the striking exception of ibutilide which acts to prolong action potential duration through an increase in a slow inactivating Na$^+$ current [17], virtually all second generation class III drugs retain voltage-dependent K$^+$ channel block as their mechanism of action [18-24] (fig. 5). In most cases the current involved seems to be the delayed rectifier K$^+$ current or more precisely its rapid component termed i_{Kr} [21], as is the case with D-sotalol, dofetilide, sematilide, risotilide, E4031, and almokalant. In selected cases however, block of the transient outward current (i_{to}; e.g. tedisamil and clofilium) or of the inward rectifier (i_{K1}; e.g. terikalant) accounts for prolongation of the action potential. This distinction is of importance because: (i) the various K$^+$ currents do not participate

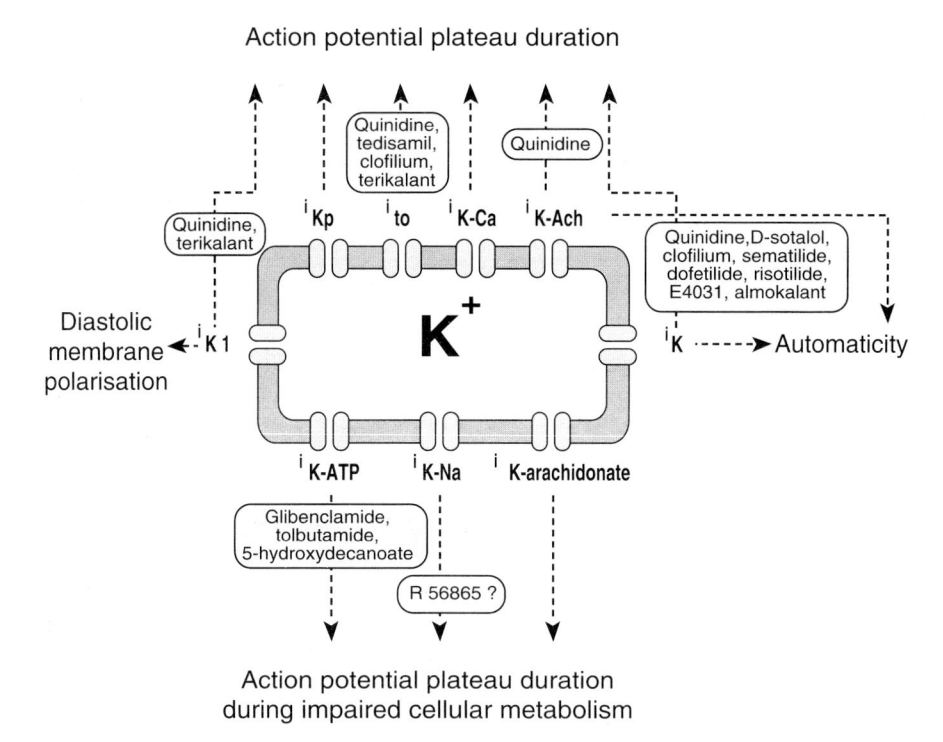

Fig. 5. The various K$^+$ channel blockers affect different cardiac K$^+$ channel categories involved in several cell functions

equally in repolarizing the cells; (ii) the distribution of K$^+$ channels markedly varies among species and also varies within the same species among the different cardiac tissues. The transient outward current mainly operates during the initial repolarization phase; in contrast, the inward rectifier operates during final repolarization while the delayed rectifier K$^+$ current operates over the entire voltage-range of the action potential. Consequently, a blocker of the inward rectifier (e.g. terikalant) or a blocker of the delayed rectifier (e.g. E4031) modifies the action potential shape in different ways in as much as the former does not prolong the plateau at voltages above 0 mV whereas the latter does [25] (fig. 6). It is well known that the cardiac action potential shortens with increasing stimulation rate (fig. 7). Although the cellular mechanisms responsible for the rate-dependent plateau abbreviation have been incompletely elucidated, drugs with different targets are expected to show different frequency-dependent profiles (see below) because the participation of the different K$^+$ currents in repolarization varies markedly in relation to the rate of depolarization. For example, the relative importance of the transient outward current decreases at high rates as a consequence of incomplete reactivation. As far as cardiac K$^+$ currents are concerned, marked species differences exist. In the guinea-pig ventricle the delayed rectifier and the inward rectifier K$^+$ currents are the major repolarizing currents whereas the transient outward current is virtually absent. In contrast, in the rat ventricle the transient outward current is the major repolarizing current whereas outward current through the inward rectifier and the delayed rectifier are small. In the dog heart,

Fig. 6. Schematic drawing of the effects of a blocker of the inward rectifier, i_{K1}, and of the delayed rectifier, i_{Kr}, on the normal action potential shape in guinea-pig ventricular cells. Adapted from [5]

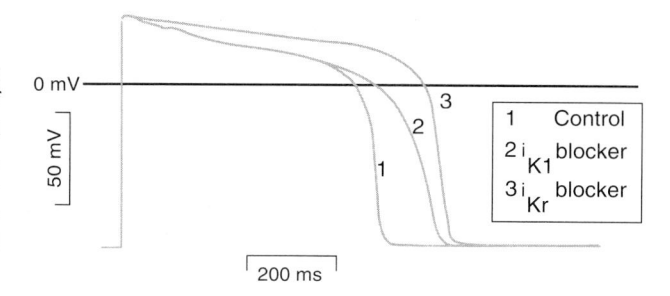

marked differences have also been observed in the amplitude of the transient outward current between endocardial and epicardial regions of the ventricle, as well as between conductive and contractile ventricular tissues. Thus a selective K^+ channel blocker may be effective in one species but not in another. The obvious consequence is that precise knowledge of the distribution of the various K^+ currents within the different regions of the human heart is highly desirable in order to predict the best compound for clinical testing. Unfortunately, available information is limited to the atrial level where the situation strongly resembles that of the rat ventricle with a predominant transient outward current [26]. A major goal of future research will be to determine precisely the picture in other areas of the human heart and most importantly in the ventricle.

Since many second generation class III agents are methylsulphonamide derivatives, one could postulate that their pharmacological profile will be very similar among themselves and also to that of D-sotalol. The effectiveness of methylsulphonamide compounds has been progressively improved and the most recently discovered drugs (e.g. dofetilide) are at least a thousand times more potent than D-sotalol. Furthermore, structure-activity investigations in the methylsulphonamide series and also in quaternary nitrogen compounds (e.g. clofilium, see [27]) or benzopyran derivatives (e.g. terikalant), have clearly shown that minor changes in the molecular structure may have dramatic consequences not only with regard to their potency but also with regard to the category of K^+ current blocked.

Second generation class III drugs share common pharmacological characteristics

Although of diverse chemical structure, pure class III drugs have at least four pharmacological properties in common (fig. 8): (i) bradycardia; (ii) slight positive inotropic effects; (iii) an unfavorable frequency-dependence profile; (iv) antiarrhythmic properties in various animal models. In addition, they all introduce an unpredictable risk of *torsades de pointe*.

Bradycardia is a common feature of class III agents. This behaviour is preferentially related to the prolongation of the action potential in sinus node cells rather than to a decrease in the spontaneous diastolic depolarization (fig. 9; see also chapter 4). For example, dofetilide, a potent and selective blocker of the i_{Kr} current in guinea-pig myocytes, achieves a maximum 20% reduction in rate in isolated right atrium at 100 nM, a concentration that produces a 42% prolongation of the action potential in ventricular tissues [28] but which does not affect the rate of phase 4 depolarization. This striking finding suggests that i_{Kr} is distinct from the K^+ current participating in spontaneous diastolic depolarization in nodal cells. Negative chronotropy is also observed with tedisamil, a

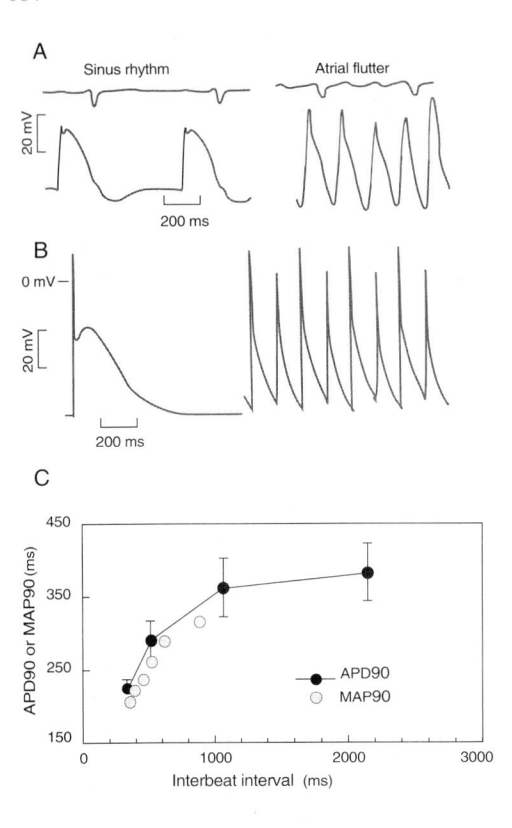

Fig. 7 A-C. Action potential duration shortening with increasing stimulation rates in human atrial myocardium. **A** Monophasic action potentials (lower traces) recorded simultaneously with ECG lead V1 (upper traces) in a patient during sinus rythm and during atrial flutter. **B** Transmembrane action potentials recorded in vitro from an human atrial muscle strip sampled during open heart surgery, and stimulated at 60 (left) and 375 (right) beats.min^{-1}. **C** In vitro transmembrane action potential duration at 90% repolarization (APD90) and in vivo monophasic action potential duration (MAP90) plotted as a function of the interbeat interval. **A** and **B** reproduced from [Lauribe et al. (1989) Cardiovasc Res 23: 159-168] with permission

blocker of i_{to}, and with terikalant, a drug selective for i_{K1} in guinea-pig cells but which also blocks i_{to} in rat myocytes. Thus far bradycardia appears to be independent of the K$^+$ channel category blocked, although an extensive comparative study of the various class III agents is needed before such a conclusion can be reached.

In stark contrast to class I drugs, pure class III agents do not exhibit negative inotropic properties but on the contrary moderately increase contraction [29] (see chapter 5). In in vitro studies, D-sotalol, dofetilide, E4031, terikalant, and almokalant increase peak developed force by less than 50%. Although in vivo the haemodynamic consequence of this positive inotropic action is weak, it remains that pure class III agents do not impair cardiac contractility and that this particular property may prove useful, especially in patients with impaired left ventricular function who run a higher risk of cardiac arrhythmia. The mechanism of the positive inotropic effects of class III drugs is likely to be the prolongation of the action potential. Similar behaviour is also observed with DPI 201-106, a drug that prolongs repolarization by delaying Na$^+$ channel inactivation. Furthermore, the dose-effect relationship of pure class III drugs in increasing peak developed force paralleled their ability to prolong action potential duration: in the case of drugs with a bell-shape dose-effect relationship on action potential prolongation such as terikalant, the dose-response curve for positive inotropy is also bell shaped [29]. An increased action potential augments the time during which the cell remains in the depolarized state thus allowing more calcium to flow into the myocytes through opened voltage-activated calcium channels. As a consequence, cytoplasmic calcium slightly

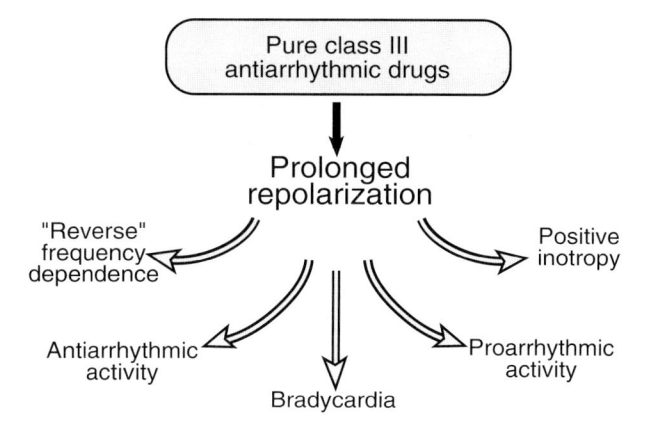

Fig. 8. Pure class III antiarrhythmic drugs share common pharmacological characteristics

increases on a beat-to-beat basis. Another possibility is that K$^+$ channel blockers interfere with sarcoplasmic reticulum K$^+$ channels leading to an increased calcium release from intracellular stocks during systole [30]. Under the action of K$^+$ channel blockers, myocardial relaxation has been reported to be either accelerated (terikalant, dofetilide) or slowed (E4031, almokalant).

Since the 1950s, it has been recognized that class I agents typically exhibit greater pharmacological activity (i.e. more conduction slowing) as the heart rate is increased. This profile is desirable since these drugs are supposedly more effective when the heart rate is elevated, i.e. during tachycardia, than when the heart rate is slow, i.e. during baseline sinus rhythm. The rate-dependent profile of Na$^+$-channel blockers is usually explained on the basis of a periodical systolic binding and diastolic unbinding of these molecules on their receptor (modulated-receptor theory). In opposition to class I antiarrhythmic drugs, class III agents have an unfavorable frequency-dependent profile (so-called "reverse frequency-dependence") since they are more active at slow than at fast rates [31]. Theoretically, this may limit their activity during tachycardia and may also produce excessive prolongation of the action potential when the cardiac rate is slow, leading to *torsades de pointe* arrhythmias. The mechanism of reverse frequency-dependence is unknown. It has been hypothesized that the block produced by class III drugs may be relieved during depolarization and enhanced by hyperpolarization, which is the reverse of what occurs with class I substances. However, in patch-clamp experiments, the effects of almokalant, E4031 and dofetilide on the delayed rectifier, as well as the effects of clofilium on the transient outward current, actually increase rather than decrease with augmented stimulation rate. Furthermore, the effects of terikalant on the inward rectifier are essentially rate-independent (fig. 10). All these drugs exhibit an unfavourable frequency-dependence profile, thus demonstrating that the above hypothesis is invalid. It is intriguing to note that reverse frequency-dependence is observed irrespective of the category of K$^+$ channel (i_K, i_{K1} or i_{to}) blocked although the different K$^+$ current categories significantly vary in amplitude with the stimulation rate. The simplest explanation is that the relative importance of K$^+$ currents in determining repolarization decreases as the rate increases and that other cellular mechanisms, such as incomplete repriming of Ca^{2+} channels during a shortened diastole, may be the primary mechanism responsible for action potential and QT abbreviation during tachycardia. Whatever the

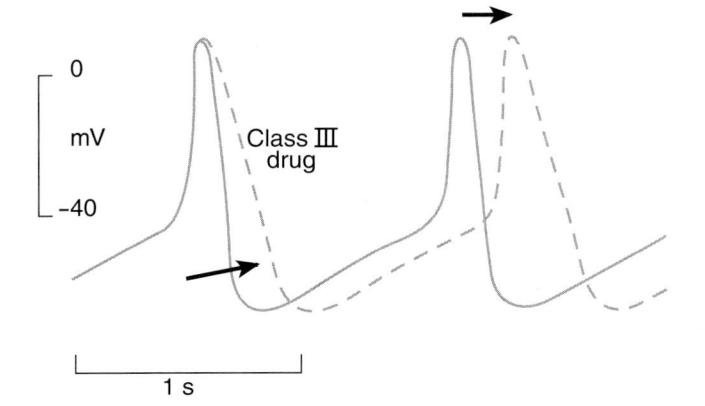

Fig. 9. Schematic drawing of the effects of a pure class III antiarrhythmic drug on the sinus node spontaneous electrical activity

case may be, substances that prolong repolarization are extremely effective during tachyarrhythmia; thus their unfavorable frequency-dependence profile does not greatly diminish their antiarrhythmic efficacy compared with class I drugs.

In experimental and in clinical models, the antiarrhythmic profile of pure class III agents differs from that of class I drugs [32]. For example, in the clinical setting, the former show poor efficacy in preventing ventricular premature beats, a classical criterion for evaluating a class I drug. Likewise, class III antiarrhythmic drugs are inactive against arrhythmias that occur in dogs 24 hours after a two-stage occlusion of the LAD (Harris model) or against arrhythmias induced by ouabain in rats or in dogs. By contrast, QT-prolonging drugs effectively prevent fatal arrhythmias induced by acute occlusion-reperfusion of the LAD in dogs or in pigs. In the same experimental model, class IC drugs are not only inactive but actually increase the incidence of fatal arrhythmias. Similarly, class III agents are more active than class I in preventing arrhythmias induced by programmed electrical stimulation. Finally, drugs such as terikalant have been shown to restore sinus rhythm during acute atrial fibrillation.

Proarrhythmia, i.e. the ability of a drug to originate or to aggravate arrhythmias, is a common theoretical drawback of any QT-prolonging drug, although delayed repolarization does not necessarily imply a high risk of *torsades de pointe*. For example, it is rarely observed during hypothyroidism or during profound hypocalcaemia, which both lengthen the QT and the QTc intervals. Furthermore, the rarity of amiodarone-related *torsades de pointe* is striking when one considers the amount of QT prolongation that this agent produces. Amiodarone interferes with most of the ionic currents of cardiac cells (i.e. sodium, calcium and potassium currents) and it is not known at present which of its numerous properties contributes most to the low incidence of *torsades de pointe*. At the cellular level, *torsades de pointe* is thought to be the consequence of early after-depolarizations arising during the repolarization phase of the action potential (fig. 11). To create an early after-depolarization, inward currents have to exceed temporarily outward currents during phase III of the action potential. This can be achieved when repolarizing K^+ currents are reduced (as during hypokalaemia, a classical cause of *torsades de pointe*) and/or when an additional inward current is created. The precise nature of the inward current responsible for early after-depolarizations is still a matter of discussion, although the long-lasting Ca^{2+} current is likely to be involved. Should this mechanism be elucidated,

Fig. 10 A, B. Frequency-dependent profile of RP 58866, a pure K$^+$ channel blocker acting on the inward rectifier K$^+$ current. **A** Action potential duration as a function of the stimulation period determined in an isolated guinea-pig ventricular myocyte in the absence and presence of 1 µM RP 58866. **B** The percentage of block of i_{K1} at −40 mV produced by 1 µM RP 58866 under voltage-clamp conditions. Although the block of i_{K1} is largely rate-independent, the action of RP 58866 on APD exhibits "reverse" frequency-dependence in guinea-pig ventricular cells

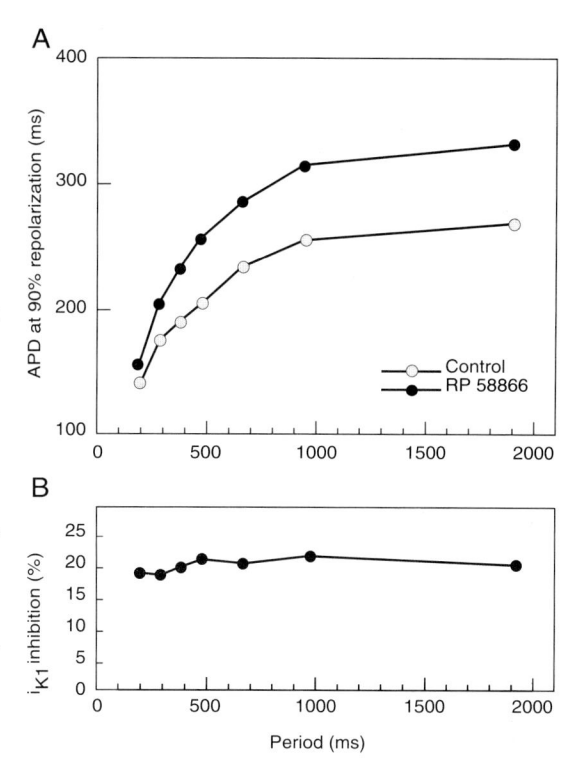

it might then be possible to identify K$^+$ channel blockers that also interfere with the inward current generating early after-depolarizations and so decrease the risk of long QT-related arrhythmias. Further basic research programs are needed, in particular to gain a better understanding of the consequence of hypothyroidism on the kinetics of slow inward current reactivation. Finally, early after-depolarizations usually arise in the Purkinje network and eventually spread to the neighbouring myocardium; thus drugs that preferentially prolong repolarization in the specialized conducting tissues are more prone than others to producing *torsades de pointe* arrhythmias. In this respect, class III drugs, which also block the slowly-inactivating component of the Na$^+$ current in Purkinje cells, may have a more favorable proarrhythmic profile.

Quinidine blocks K$^+$ channels opened by vagal stimulation

At the atrial level, vagal stimulation leads to the opening of K$^+$ channels, resulting in an additional outward current. The acetylcholine-induced K$^+$ current is responsible for: (i) decreased refractory period in atrial muscle as a result of an abbreviated action potential plateau; (ii) decreased automaticity in sinus node cells because less net inward current is available for diastolic depolarization; (iii) decreased conduction velocity in the atrio-ventricular node secondary to a decreased rate of depolarization in this tissue. Abbreviated refractory periods facilitate episodes of atrial fibrillation in diseased tissues. It has been shown that quinidine dose-dependently depresses the vagally-induced K$^+$ current because of a direct effect on muscarinic receptor-coupled K$^+$ channels [33], a

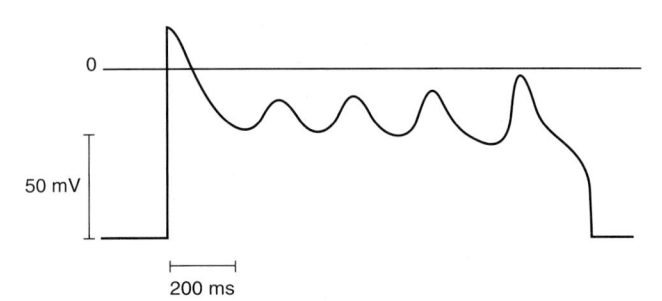

Fig. 11. Early after depolarization potentials (EADs) recorded under the influence of 1 µM dofetilide in a guinea-pig papillary muscle. Reproduced from [Tande et al. (1990) J Cardiovasc Pharmacol 16: 401-410] with permission

mechanism responsible for its anticholinergic effects. Direct modulation of muscarinic K$^+$ channels may also be considered as an antiarrhythmic mechanism.

Do drugs that block ATP-sensitive K$^+$ channels exhibit antiarrhythmic activity?

When the metabolism of cardiac cells is impaired, e.g. during ischaemia, ATP-sensitive K$^+$ channels open because intracellular ATP and pH decrease and ADP increases. Outward current through opened channels shortens the action potential duration in the ischaemic zone and therefore creates an inhomogeneity in refractoriness between injured and normally oxygenated tissues (fig. 12). Thus it might be expected that the block of ATP-sensitive K$^+$ channels would limit dispersion in refractory periods and thus would be antiarrhythmic. However, it should be realized that the activation of these channels in the injured cells is a natural protective mechanism which markedly increases their chance of surviving throughout the ischaemic period (see chapter 13). Indeed, an abbreviated action potential plateau limits the amount of Ca^{2+} entering the cells and ultimately suppresses both electrical and mechanical activity in the ischaemic cells: local akinesia saves the amount of energy necessary for cell survival whereas arrest of the electrical activity is highly desirable to prevent the emergence of foci of abnormal automaticity and impaired conduction in partially depolarized ischaemic tissues. Thus, from a theoretical standpoint, inhibition of ATP-sensitive K$^+$ channels can be considered as antiarrhythmic and also as proarrhythmic. Unfortunately it is still difficult to get a clear view as regards the pro/antiarrhythmic effects of ATP-sensitive K$^+$ channel block due to the lack of convincing experimental data with blockers or with activators of ATP-sensitive K$^+$ channels (see chapter 13).

Conclusion and perspectives

K$^+$ channel block in the heart may be one of the most efficient antiarrhythmic mechanisms identified so far. The problem is that any drug that prolongs repolarization has intrinsically associated risk of *torsades de pointe* arrhythmias. However, prolonged repolarization does not necessarily imply that *torsades de pointe* will occur, as exemplified by hypothyroidism. This makes the identification of a "third generation" of non-proarrhythmic class III agents conceivable, although a strong impediment to this approach is that no validated model exists to assess experimentally the risk of *torsades de pointe* for a given substance.

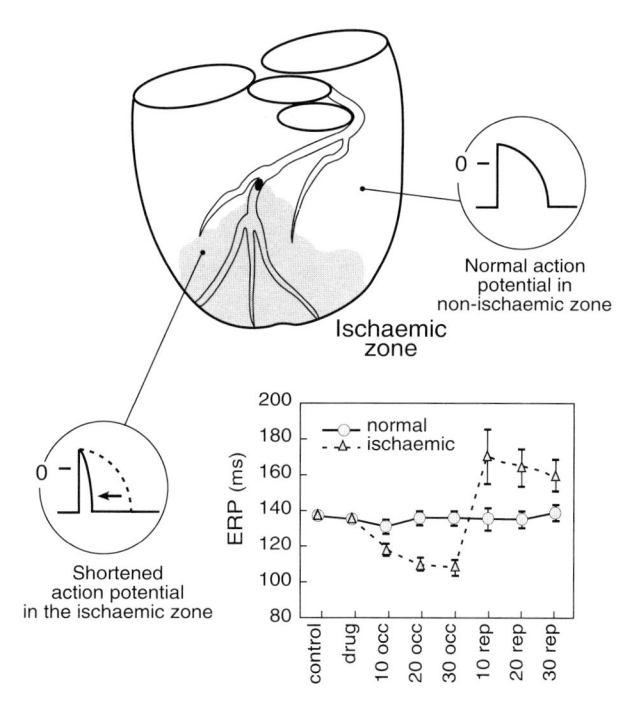

Fig. 12. Inhomogeneity in ventricular effective refractory period *(ERP)* between ischaemic and normally oxygenated tissues during acute coronary artery occlusion *(occ; 10, 20, 30 occ:* 10, 20, 30 min after occlusion) and reperfusion *(rep)* in anaesthetized dogs. Reproduced from [32] with permission

Furthermore, since the K⁺ channels responsible for repolarization actually differ between the atrium and the ventricle, it may be possible to identify K⁺ channel blockers that will be active against supraventricular arrhythmias but that will not prolong the QT interval, and thus will not be proarrhythmic. Information on the various K⁺ currents responsible for repolarization in the human ventricle as well as on the cellular mechanisms leading to early after depolarisation is urgently needed to select safe class III agents.

References

1. Singh B, Nademanee K (1985) Control of cardiac arrhythmias by selective lengthening of repolarization: theoretic considerations and clinical observations. Am Heart J 109: 421-30
2. Lathers CM, Lipka LJ (1987) Cardiac arrhythmia, sudden death, and psychoactive drugs. J Clin Pharmacol 27: 1-14
3. Warin RP (1991) Torsades de Pointes complicating treatment with terfenadine. Br Med J 303: 6793
4. Wesseling H, Agoston S, Van Dam GBP, Pasma J, De Wit DJ, Havinga H (1984) Effects of 4-aminopyridine in elderly patients with Alzheimer's disease. N Eng J Med 310: 988-989
5. Castle NA, Haylett DG, Jenkinson DH (1989) Toxins in the characterization of potassium channels. Trends Neurosci 12: 59-65
6. Vaughan-Williams EM (1984) A classification of antiarrhythmic actions reassessed after a decade of new drugs. J Clin Pharmacol 24: 129-147
7. Colatsky TJ, Follmer CH (1989) K⁺ channel blockers and activators in cardiac arrhythmias. Cardiovasc Drug Rev 7: 199-209

8. Findlay I (1992) Inhibition of ATP-sensitive K$^+$ channels in cardiac muscle by the sulphonylurea drug glibenclamide. J Pharmacol Exp Ther 261: 540-545
9. Gopalakrishnan M, Johnson DE, Janis RA, Triggle DJ (1991) Characterization of binding of the ATP-sensitive potassium channel ligand, [3H]glyburide, to neuronal and muscle preparations. J Pharmacol Exp Ther 257: 1162-1171
10. McCullough JR, Normandin DE, Conder ML, Sleph PG, Dzwonczyk S, Grover GJ (1991) Specific block of the anti-ischaemic actions of cromakalim by sodium 5-hydroxydecanoate. Circ Res 69: 949-958
11. Zehender M, Hohnloser S, Just H (1991) QT interval prolonging drugs: mechanisms and clinical relevance of their arrhythmogenic hazards. Cardiovasc Drugs Ther 5: 515-530
12. Van Zwieten PA, Blauw GJ, Van Brummelen P (1990) Pharmacological profile of antihypertensive drugs with serotonin receptor and α-adrenoreceptor activity. Drugs 40: 1-8
13. Sonders MS, Keana JFW, Weber E (1988) Phencyclidine and psychotomimetic sigma opiates: recent insights into their biochemical and physiological sites of action. Trends Neurosci 11: 37-40
14. Imaizumi Y, Giles WR (1987) Quinidine-induced inhibition of transient outward current in cardiac muscle. Am J Physiol 253: H704-H708
15. Nishimura M, Follmer CH, Singer DH (1989) Amiodarone blocks calcium current in single guinea-pig ventricular myocytes. J Pharmacol Exp Ther 251: 650-659
16. Singh BN, Nademanee K (1987) Sotalol: a ß blocker with unique antiarrhythmic properties. Am Heart J 114: 121-139
17. Lee KS (1992) Ibutilide, a new compound with potent class III antiarrhythmic activity, activates a slow inward Na$^+$ current in guinea-pig ventricular cells. J Pharmacol Exp Ther 262: 99-108
18. Castle NA (1991) Selective inhibition of potassium currents in rat ventricle by clofilium and its tertiary homolog. J Pharmacol Exp Ther 257: 342-350
19. Carmeliet E (1985) Electrophysiologic and voltage-clamp analysis of the effects of sotalol on isolated cardiac muscle and Purkinje fibres. J Pharmacol Exp Ther 232: 817-825
20. Dukes ID, Morad M (1989) Tedisamil inactivates transient outward K$^+$ current in rat ventricular myocytes. Am J Physiol 257: H1746-H1749
21. Sanguinetti MC, Jurkiewicz NK (1990) Two components of cardiac delayed rectifier K$^+$ current: differential sensitivity to block by class III antiarrhythmic agents. J Gen Physiol 96: 194-214
22. Carlsson L, Abrahamsson C, Almgren O, Lundberg C, Duker G (1991) Prolonged action potential duration and positive inotropy induced by the novel class III antiarrhythmic agent H234/09 (almokalant) in isolated human ventricular muscle. J Cardiovasc Pharmacol 18: 882-887
23. Gwilt M, Arrowsmith JE, Blackburn KJ, Burges RA, Cross PE, Dalrymple HW, Higgins AJ (1991) UK-68798: a novel, potent and highly selective class III antiarrhythmic agent which blocks potassium channels in cardiac cells. J Pharmacol Exp Ther 256: 318-324
24. Escande D, Mestre M, Cavero I, Brugada J, Kirchhof C (1992) RP 58866 and its active enantiomer RP 62719 (Terikalant): blockers of the inward rectifier K$^+$ current acting as pure class III antiarrhythmic agents. J Cardiovasc Pharmacol 20: 106-113
25. Martin CL, Chinn K (1992) Contribution of delayed rectifier and inward rectifier to repolarization of the action potential: pharmacologic separation. J Cardiovasc Pharmacol 19: 830-837
26. Escande D, Coulombe A, Faivre JF, Deroubaix E, Coraboeuf E (1987) Two types of transient outward currents in adult human atrial cells. Am J Physiol 252: H142-H148
27. Arena JP, Kass RS (1988) Block of heart potassium channels by clofilium and its tertiary analogs: relationship between drug structure and type of channel blocked. Mol Pharmacol 34: 60-66
28. Rasmussen HS, Allen MJ, Blackburn KJ, Butrous GS, Dalrymple HW (1992) Dofetilide, a novel class III antiarrhythmic agent. J Cardiovasc Pharmacol 20 (2): S96-S105

29. Beregi JP, Escande D, Coudray N, Mery P, Chemla D, Mestre M, Lecarpentier Y (1993) Positive inotropic and lusitropic effects of a pure class III antiarrhythmic agent. J Pharmacol Exp Ther (in press)
30. Ishida Y, Honda H, Watanabe TX (1992) Ca^{2+} release from sarcoplasmic reticulum of guinea-pig psoas muscle induced by K^+ channel blockers. Br J Pharmacol 103: 764-765
31. Colatsky TJ, Follmer CH, Starmer CF (1990) Channel specificity in antiarrhythmic drug action: mechanism of channel block and its role in suppressing and aggravating cardiac arrhythmias. Circulation 82: 2235-2242
32. Gibson JK, Kersten JA (1990) In vivo assessment of class III agents and their antiarrhythmic activity. Drug Dev Res 19: 173-185
33. Nakajima T, Kurachi Y, Ito H, Takikawa R, Sugimoto T (1989) Anti-cholinergic effects of quinidine, disopyramide, and procainamide in isolated atrial myocytes: mediation by different molecular mechanisms. Circ Res 64: 297-303

CHAPTER 12
Pharmacological profile of potassium channel openers in the vasculature

I Cavero and JM Guillon

During the resting state, in excitable cells such as smooth muscle myocytes, the concentration of K^+ outside the membrane ($[K^+]_o \sim$ 3-5 mM) is at least 25-fold lower than the K^+ concentration in the intracellular fluid ($[K^+]_i \sim$ 130-160 mM). Therefore, the opening of K^+ channels and the resulting efflux of positive charges carried by K^+ generate an outward current [1]. The efflux of K^+ hyperpolarizes the membrane until the equilibrium potential for K^+ (E_K), which is described by the Nernst equation, is reached (-80 to -90 mV). E_K is the value of the membrane potential at which the chemical driving force for K^+ efflux is in equilibrium with the electrical field of the membrane, which retains K^+ intracellularly (see chapter 1). The resting membrane potential in vascular myocytes is mainly determined by the resting K^+ conductance. In the vascular smooth muscle cell at rest, the concentration of Ca^{2+} is approximately 2 mM in the extracellular space and much lower (0.01 to 1 mM) in the intracellular milieu. The Nernst equilibrium potential for Ca^{2+}, E_{Ca}, is over +150 mV. Thus, the opening of L-type Ca^{2+} channels is followed by the entry of Ca^{2+} into the cell. Ca^{2+} entry (as well as Na^+ entry or Cl^- efflux) reduces the natural polarization of the plasmalemma. The respective values for E_{Na} and E_{Cl} are +50 mV and -20 mV ($[Na^+]_i$ = 10-20 mM, $[Na^+]_o$ = 150 mM, $[Cl^-]_i$ = 40-70 mM and $[Cl^-]_o$ = 140 mM) [1]. The opening of K^+ channels is a physiological means for counteracting, restricting or preventing the depolarizing activity caused by inward currents associated mainly with the entry of Ca^{2+} and Na^+ and the efflux of Cl^-. Its electrophysiological and functional consequence in vascular smooth muscle cells is either relaxation or enhanced cellular resistance to excitatory stimuli.

    This review is not meant to be all-inclusive, since several publications have recently appeared on this topic [2-7], but its primary aim is to deal with the preclinical pharmacological profiling of agents which open K^+ channels in the vascular smooth muscle. Firstly, the best known chemical entities which have so far been shown to possess K^+ channel opening or blocking properties will be briefly presented. We will then describe a cascade of in vitro and in vivo pharmacological tests by which a given compound can be assigned to the K^+ channel opener class. Finally, the clinical potential of K^+ channel opening drugs for treating cardiovascular diseases will be outlined. Before beginning, it is necessary to propose a simple definition of a K^+ channel opener. For the purpose of this review, K^+ channel openers will be considered as compounds which increase membrane K^+ conductance and consequently produce vasorelaxation. Their functional, biochemical and electrophysiological effects on the vasculature can be antagonized by glibenclamide or similar sulphonylureas. This general definition is not strictly accurate because it does not specify the type (voltage- or ligand-gated) of K^+ channel that is opened by this class of drugs. However, this is also the case with other established terms, such as "calcium antagonists" or "calcium entry blockers", which are used for drugs blocking only one (L-type) of the four classes (L, N, T, P) of voltage-gated calcium channels.

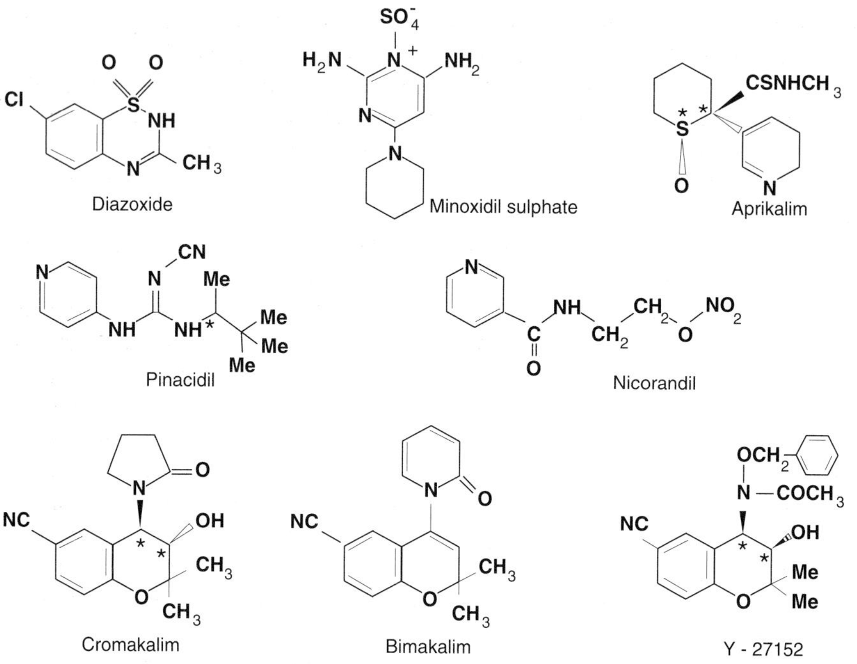

Fig. 1. Chemical structures of some K⁺ channel openers. An asterisk indicates a chiral center

Established and novel K⁺ channel openers

Numerous compounds of entirely different chemical structures possess K^+ channel opening properties (fig. 1). A few clinically studied antihypertensive drugs (diazoxide, minoxidil and pinacidil), classified originally as direct vasodilator agents, have subsequently been demonstrated to be openers of K^+ channels, although they are not as specific as more recently described agents, such as aprikalim or the benzopyran derivatives of which cromakalim is the prototype [8].

Diazoxide. A benzothiadiazine derivative which is chemically related to the thiazide diuretic chlorothiazide without sharing its pharmacological properties, diazoxide has been used for the i.v. treatment of hypertensive emergencies. Its administration is accompanied by serious and undesirable effects of cardiovascular reflex nature. In addition, diazoxide treatment can increase plasma glucose concentration since it restrains the release of insulin from pancreatic ß-cells by opening their ATP-sensitive K^+ channels. This non-cardiovascular action of diazoxide is sometimes exploited to treat pathological forms of hypoglycaemia, particularly in children [9]. Thus, diazoxide can open K^+ channels either in vascular smooth muscle cells to produce relaxation [10] or in the pancreas to reduce insulin release [11]. Of the well known K^+ channel openers presented here, only diazoxide has a significant tropism for pancreatic ATP-sensitive K^+ channels (see section on Effects on insulin secretion).

Minoxidil. Minoxidil, a pyrimidine derivative, is a prodrug which, when given by oral route, is metabolized by hepatic sulphotransferases to minoxidil sulphate, which has recently been found to open K^+ channels [12]. Like diazoxide, minoxidil is clinically used to treat hypertension. However, its prescription is confined to severely hypertensive patients refractory to other combination therapies, since minoxidil intake is followed by tachycardia and fluid retention due to reflex autoregulation. In addition, minoxidil (and also to a lesser extent, diazoxide) produces hypertrichosis, an effect which may be due to increased blood perfusion of the hair follicles. For this reason, this compound has been subsequently developed for preventing hair loss and stimulating hair growth [13].

Pinacidil. A pyridylcyanoguanidine derivative, pinacidil contains a chiral carbon in its trimethylpropyl moiety. The biological activity of this racemate is associated with the (-)-(R)-enantiomer. Pinacidil produces active vasodilation, an effect recently attributed, to a large extent, but not exclusively, to activation of vascular K^+ channels [14]. Pinacidil has been developed as an antihypertensive drug but its prescription is very limited due to the high incidence of adverse effects (headache, oedema, palpitations, tachycardia) resulting from its potent peripheral vasodilator action. P1075 is an analogue of pinacidil without a chiral center since the moiety conferring asymmetry [$CH - CH_3 -C(CH_3)_3$] (fig. 1) is replaced by [$(CH_3)_2 - CH_2-CH_3$] [4]. This compound is 30-fold more potent than pinacidil and is used as a radioligand for labeling K^+ channels in blood vessels (see section on Radioligand binding assay).

Nicorandil. A nicotinamide derivative possessing a nitrate moiety, this compound was the first pharmacological agent to be shown to open K^+ channels [16]. In addition, it activates soluble guanylate cyclase in a manner similar to that of nitrovasodilators [17]. Under in vivo conditions, the arterial vasodilator effects of nicorandil appear to be mostly mediated by K^+ channel opening [18]. Nicorandil is clinically used for the treatment of angina, and there is no tolerance to its beneficial effects. A novel compound resembling nicorandil is KRN2391 [4].

Benzopyran derivatives (for ample details on this series see [19, 20]). The standard reference for K^+ channel openers is the benzopyran derivative cromakalim (BRL 34915). This compound is a trans-racemate containing two enantiomers (resulting from the chiral carbon number 3 on the benzopyran moiety) designated BRL 38226 and BRL 38227, which is the (-)-3S,4R-enantiomer possessing biological activity. The initially proposed international non-proprietary name of BRL 38227 was lemakalim; however, this designation has now been replaced by levcromakalim. Levcromakalim is under clinical investigation as an antihypertensive agent. In the large series of benzopyran derivatives, bimakalim (code names: SR 44866 and EMD 52692), celikalim [6] and Y-27,152 should also be mentioned. The last of these compounds is a prodrug, which after oral administration is converted by the liver cytochrome P450 enzyme complex to the desbenzyl derivative Y-26763, which is the biologically active form [21]. Bimakalim is under clinical investigation as an antianginal drug, whereas Y-27,152 is presently being evaluated in humans as an antihypertensive and antianginal agent. The clinical development of celikalim has been discontinued.

Aprikalim. Aprikalim is the enantiomeric active form [(-)-(1R, 2R)] of RP 49356, which is a trans-diastereoisomer, carbothiamide derivative. The inactive form of RP 49356 is designated RP 61499 [22, 23]. Aprikalim is undergoing clinical trials for use in the treatment of coronary artery disease.

Drugs that antagonise the vascular effects of K⁺ channel openers

Antidiabetic sulphonylureas. These compounds reduce plasma glucose content by increasing insulin secretion from pancreatic ß-cells where they selectively block K⁺ channels which are regulated by ATP ([24]; see also chapter 10). The best known members of this series are glibenclamide (glyburide) (fig. 2), glipizide, glibornuride, glisoxepide and tolbutamide. Presently, glibenclamide is used as a standard, selective, rather specific and apparently competitive antagonist of the vasorelaxant effects of K⁺ channel openers.

Tedisamil. This sparteine derivative is not only a bradycardic agent but also a class III antiarrhythmic compound which blocks the transient outward K⁺ current (fig. 2; see also chapter 11). Recently, it was reported to be also a weak inhibitor of the vasorelaxant effects of cromakalim and minoxidil sulphate [25].

Ciclazindol. A pyrimidoindol derivative clinically used as an antidepressant agent, ciclazindol (fig. 2) blocks noradrenaline uptake. Ciclazindol was recently found to inhibit the vasorelaxant effects of BRL 38227 [26] and aprikalim (personal unpublished results).

Imidazoline derivatives. The bradycardic agent alinidine and the non-selective α-adrenoceptor antagonist phentolamine have been shown to antagonize the vascular smooth muscle effects of K⁺ channel openers in vitro. However, their potency as in vivo blockers of K⁺ channel opener effects is very low [27].

U-37,883A. This is a relatively new potent antagonist of the vascular effects of K⁺ channels (fig. 2) [28].

Sodium 5-hydroxydecanoate. This compound (fig. 2) does not inhibit the vascular effects of cromakalim [29] or aprikalim (personal observation) and thus it should not be included in this list. However, it merits mention because it inhibits the cardioprotective effects of cromakalim [29]. This finding indicates that vascular K⁺ channels may be somewhat different from the cardiac myocyte K⁺ channels (see also chapter 13).

Many preclinical pharmacological tests exist for characterizing K⁺ channel openers

In vitro tests can be used to detect and demonstrate K⁺ channel opening activity

The sequence of events produced by K⁺ channel openers in vascular smooth muscle cells is given in figure 3. These compounds open cell membrane K⁺ channels. This leads to membrane repolarization and/or hyperpolarization which is followed by a reduction in cytosolic free Ca^{2+} and/or inhibition of mechanisms producing increases in cytosolic free Ca^{2+}. Thus, K⁺ channel openers can: (i) reduce open probability of voltage-dependent L-and T-type Ca^{2+} channels; (ii) decrease agonist-induced Ca^{2+} entry via receptor-activated Ca^{2+} channels or restrict agonist-induced Ca^{2+} release from intracellular sources through impaired inositol triphosphate synthesis [30] and possibly; (iii) accelerate the clearance of intracellular free Ca^{2+} via plasmalemmal Na/Ca exchange, Ca/Mg-ATPase

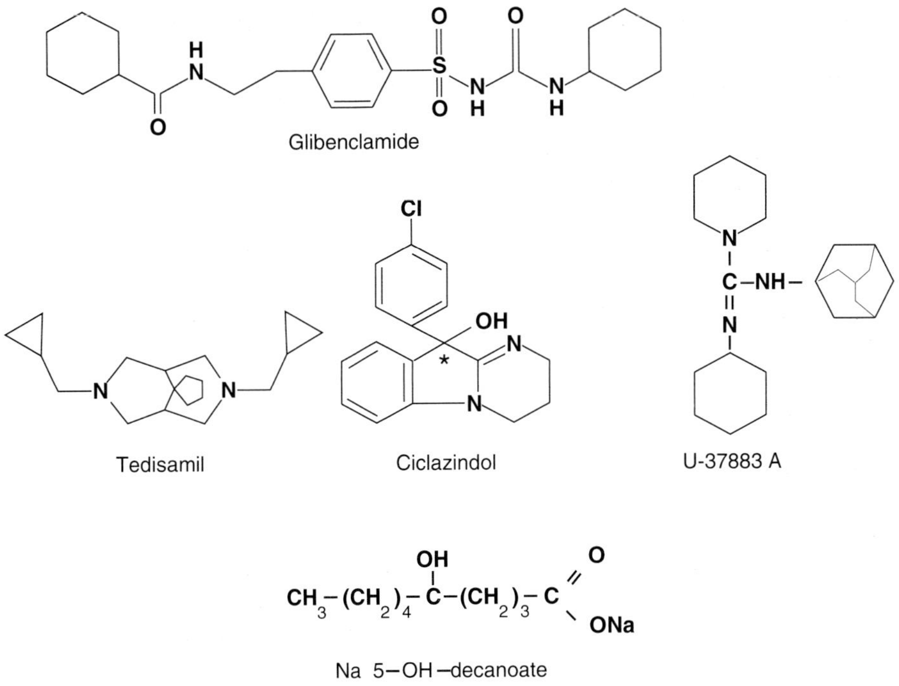

Fig. 2. Chemical structures of a few antagonists of the vascular effects of K⁺ channel openers. For sodium 5-hydroxydecanoate see text

pump and/or the Ca^{2+}-activated pump of the sarcoplasmic reticulum. The functional outcome of all these effects is vasorelaxation and/or reduced membrane excitability resulting in a greater cellular resistance to activation by excitatory agonists. K⁺ channel openers may also exert direct intracellular effects which can inhibit agonist-induced elevation in intracellular Ca^{2+} mobilization. However, this mechanism is not yet fully clarified. Theoretically, a screening test should be simple, quick, specific for the desired mechanism, and, whenever possible, inexpensive. Nowadays, the most widely applied techniques fulfilling these criteria are radioligand binding assays. However, radioligands labelling vascular K⁺ channels (activated by K⁺ channel openers) in a membrane preparation are still lacking, although efforts to find them are being made and some interesting results have recently become available as detailed in a later section.

Vascular preparations contracted with a low concentration of KCl

Rapid, simple, widely used tests to uncover the K⁺ channel opening effects of a compound include in vitro vascular preparations, such as the rat [31-33], rabbit [34] and guinea-pig [28] aorta and the guinea-pig pulmonary artery [35], precontracted with a low concentration of KCl, or norepinephrine, to elevate the low initial tone, as well as the spontaneously contracting rat portal vein [31, 36]. In our laboratory we use the rat aortic rings cleared of the endothelium to eliminate possible effects of vasoactive endothelial factors that might

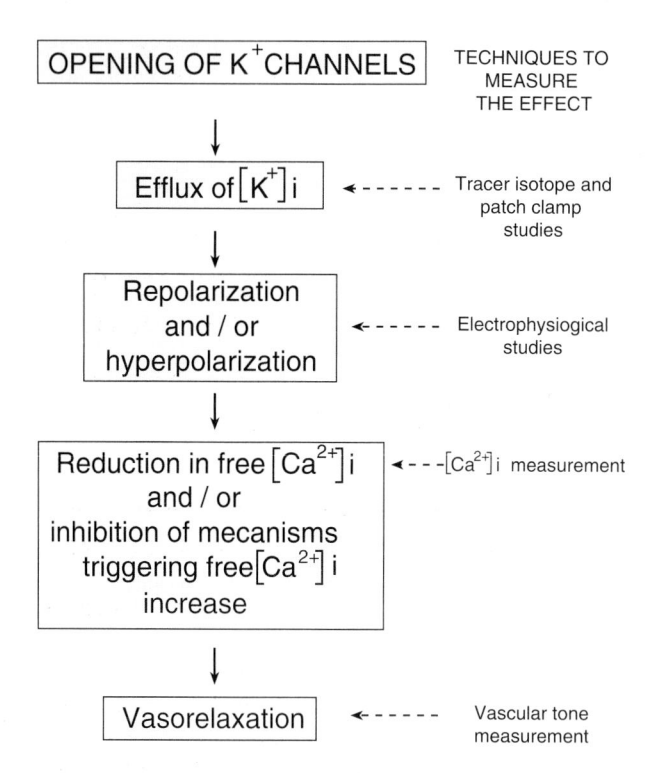

```
OPENING OF K⁺ CHANNELS        TECHNIQUES TO
         │                       MEASURE
         ▼                      THE EFFECT
    Efflux of [K⁺]i    ◄------ Tracer isotope and
         │                       patch clamp
         ▼                         studies
    Repolarization
     and / or          ◄------ Electrophysiogical
    hyperpolarization             studies
         │
         ▼
Reduction in free [Ca²⁺]i  ◄---- [Ca²⁺]i measurement
    and / or
inhibition of mecanisms
triggering free[Ca²⁺]i
    increase
         │
         ▼
    Vasorelaxation     ◄------ Vascular tone
                                measurement
```

Fig. 3. Effects produced in vascular smooth muscle by openers of K^+ channels and some experimental approaches for their study

be released by the compound under study [32, 33]. However, it should be mentioned that the presence of endothelium does not substantially modify the vasorelaxant potency of K^+ specific channel openers [17]. The rings are contracted with a low concentration of KCl (20-25 mM). When a steady-state response is attained, increasing cumulative concentrations of the compound under study are added until no further significant relaxant effect is obtained. A K^+ channel opener will almost fully antagonize the contractile effects of a low concentration of KCl. The decreases in tension are expressed as percentages of the maximal response to KCl, and plotted against the cumulative concentrations of the compound tested. The sigmoid curve obtained (fig. 4) is analyzed to calculate the effective inhibitory concentration of the compound producing 50 % relaxation (IC_{50}) [32]. From IC_{50}, a pIC_{50}, which is the negative log of IC_{50}, can be easily calculated. For example, cromakalim ($IC_{50} = 0.24 \pm 0.01$ μM; n = 6 ; $pIC_{50} = 6.6$) is as potent as aprikalim ($IC_{50} = 0.26 \pm 0.01$ μM; n = 6 ; $pIC_{50} = 6.6$).

However, this screening test lacks specificity for K^+ channel openers since calcium antagonists (fig. 4), nitrovasodilators, and direct vasodilators, such as papaverine, can also fully relax aortic rings contracted with a low concentration of KCl. Thus, to verify whether the observed vasorelaxation is caused by K^+ channel opening, it is suggested that the effects of the relaxant agent are studied in a similar manner as described above after the tissue has been incubated for 30 min in a bathing solution containing glibenclamide (0.3 to 3 μM), which is a selective and highly specific antagonist of the vasorelaxant effects of K^+ channel openers [2, 32]. Glibenclamide produces a rightward shift of the control concentration-vasorelaxant response curve to aprikalim (fig. 5), cromakalim [32, 33] and other K^+ channel openers [9, 13, 17, 21, 25]. Glibenclamide antagonism appears to be of a

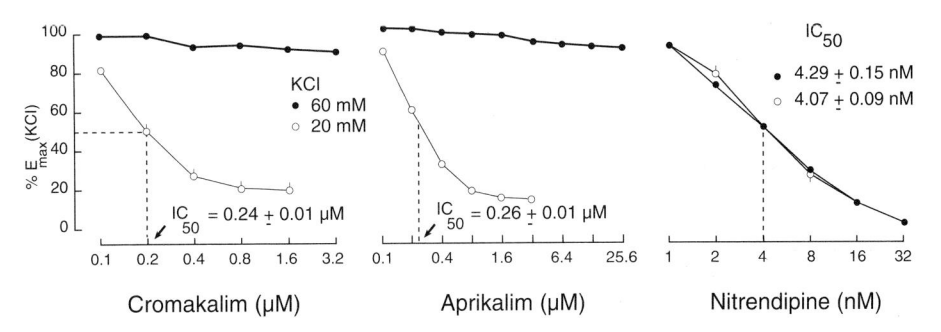

Fig. 4. Effects of cromakalim, aprikalim and nitrendipine in rat aortic rings precontracted with 20 and 60 mM KCl. The IC$_{50}$ (concentration of compound producing 50 % relaxation) is also indicated. The increase in tension produced by 20 and 60 mM KCl is 1.5 to 2.5 g

competitive nature, at least within the concentration range (0.1-3 µM) studied in the rat aortic ring preparation, since the slope of the Schild plot is close to unity (fig. 5). The pA$_2$ of glibenclamide is approximately 7 when determined against cromakalim [32] or aprikalim [22]. It should be remembered that the pA$_2$ is the negative logarithm of the concentration of the antagonist (glibenclamide) requiring twice the concentration of the agonist (in our case, the K$^+$ channel opener) to obtain the same vasorelaxant response as in control (vehicle-pretreated) preparations. By comparison, the pA$_2$ of the sulphonylurea derivatives glipizide, glisoxepide and glibornuride against aprikalim are 5.6, 5.0 and 4.4, respectively [22], and thus they are less potent antagonists than glibenclamide. Glibenclamide does not interfere with the relaxant activity of calcium channel blockers (e.g. nitrendipine and diltiazem) [32], nitrovasodilators [33] or papaverine. However, glibenclamide also has an additional vascular property; it behaves as a competitive antagonist of the contractile effects induced by the thromboxane mimetic U-46619 in the rat aorta (pA$_2$ = 6.7, personal unpublished results) and dog coronary artery rings [37].

Vascular preparations contracted with a high concentration of KCl

These preparations allow pure K$^+$ channel openers to be distinguished from vasodilators possessing other mechanisms of action. K$^+$ channel openers producing relaxation in low KCl contracted preparations do not relax or relax weakly rat aortic rings contracted with high concentrations of KCl (> 40 mM) (fig. 4) [9, 11, 31-33]. The gradual loss of vasodilating activity with the elevation of extracellular K$^+$ can be easily explained on the basis of the Nernst K$^+$ equilibrium potential (E$_K$, see chapter 1). Fig. 6 illustrates the calculated values of E$_K$ and the experimental values of membrane potentials (V$_m$) as a function of extracellular potassium concentrations [38, 39]. As the extracellular concentration of K$^+$ is increased, E$_K$ becomes less negative. When [K$^+$]$_o$ is greater than 40 mM, E$_K$ and V$_m$ have similar values. As already stated, the vasodilator effect of K$^+$ channel openers is the direct consequence of membrane repolarization or hyperpolarization mediated by the efflux of intracellular K$^+$ (fig. 3). When E$_K$ approaches or matches V$_m$, the opening of K$^+$ channels is no longer followed by an outward K$^+$ current since the electrochemical gradient for [K$^+$]$_i$ across the membrane is zero. Therefore, if, as the [K$^+$]$_o$ is raised from 15 to 50 mM, the vasorelaxant potency of a compound in in vitro artery preparations gradually disappears,

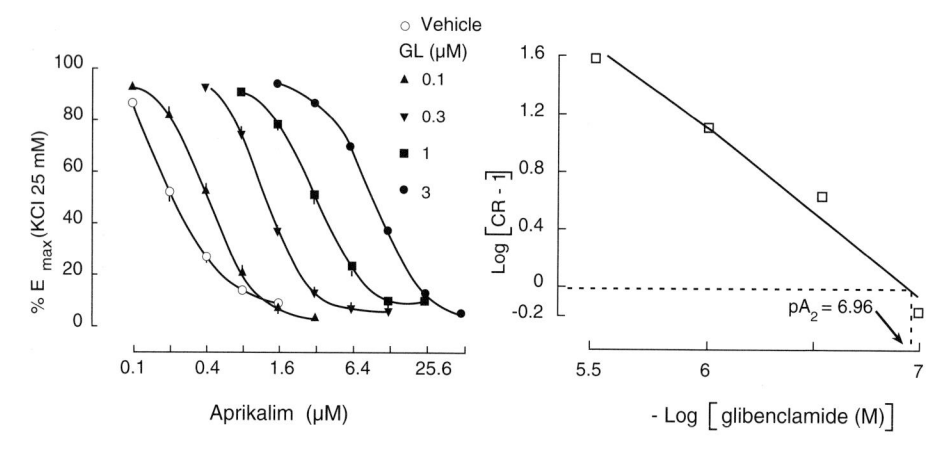

Fig. 5. Effects of various concentrations of glibenclamide (GL) after 30 min incubation time on the vasorelaxant effects of aprikalim in rat aortic rings precontracted with KCl (25 mM). The calculation of pA2 with a Schild plot (slope = -1.14) is also represented. CR are the ratios of the IC_{50} of aprikalim after each concentration of glibenclamide and the IC_{50} determined in control (vehicle) preparations

then the compound is likely to produce vasorelaxation exclusively via the opening of K^+ channels. Increasing $[K^+]_o$ depolarizes the membrane, an event which produces contraction by opening voltage-gated L-type Ca^{2+} channels. A K^+ channel opener can overwhelm this response by repolarizing the cell. This can occur only if E_K lies below the activation threshold of L-type calcium channels, which is around -45 mV in vascular smooth muscle cells. By contrast, calcium entry blockers can relax vascular preparations contracted with either a low or a high concentration of KCl (fig. 4) [32] since they inhibit the entry of Ca^{2+} through voltage-gated calcium channels which are opened by KCl depolarization.

Determination of the antivasoconstrictor spectrum

The antivasoconstrictor spectrum of K^+ channel openers can be determined by assessing the antagonist effects of these compounds against vasoconstrictors as noradrenaline, phenylephrine, angiotensin II, 5-hydroxytryptamine (serotonin), vasopressin, endothelin-1, histamine, U-46619, etc. It is important to note that specific K^+ channel openers do not antagonize the vasoconstrictor effects of agonists by occupying agonist receptor sites. Rather, they decrease the responsiveness of the cell to the agonist due to repolarization and/or hyperpolarization which drives the cell to rest or puts it in a state of increased resistance (reduction of excitability) to excitatory stimuli. Therefore, K^+ channel openers, by impairing the efficiency of the intracellular coupling processes (since they decrease the availability of intracellular Ca^{2+} for the contractile cascade), should be more appropriately considered functional rather than non-competitive antagonists of excitatory agonists. The most complete report published to date compares the effects of cromakalim and the dihydropyridine calcium antagonist isradipine on the contractile responses of rabbit aortic rings to angiotensin II, 5-hydroxytryptamine and noradrenaline [34]. The 5-hydroxytryptamine and angiotensin II concentration-response curves were antagonized non-competitively by cromakalim (0.1-10 μM : 40-55 % depression of the maximum). In a study performed in

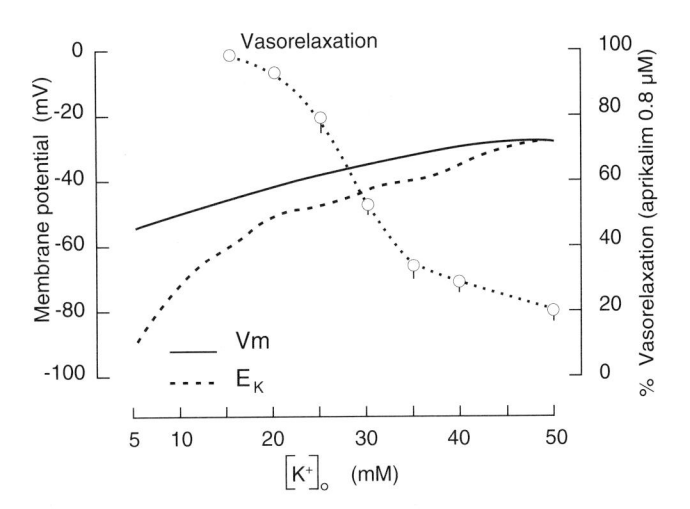

Fig. 6. Relationship between extracellular K$^+$ concentration and resting membrane potentials (V$_m$) and the equilibrium potentials for K$^+$ (E$_K$) in rat aortic rings. Adapted from [39]. The maximal relaxant effects produced by a supramaximal concentration of aprikalim in rat aortic rings contracted with various concentration of KCl are also reported

our laboratory, aprikalim inhibited in a similar manner the vasoconstrictor effects of KCl, endothelin-1, phenylephrine and, to a lesser extent, those of U-46619 (pIC$_{50}$ = 6.4, 6.3, 6.3 and 5.8, respectively). By contrast, nitrendipine was very potent only against KCl-induced contractions (pIC$_{50}$ = 7.6). The pIC$_{50}$ for nitrendipine against phenylephrine and endothelin-1 were 6.5 and 6.1, respectively. However, nitrendipine was unable to antagonize completely the contraction produced by these agonists and was virtually without effect against U-46619 effects. In a recent study [40], benzopyran-derived K$^+$ channel openers (EMD 52892, Ro 31-6930, BRL 38227, SDZ PCO 400), RP 49356 and pinacidil evoked concentration-related relaxant effects in rat aortic rings contracted by either KCl (20 mM; pIC$_{50}$ values: 8.3, 7.8, 7.6, 7.2, 6.5 and 6.6, respectively) or endothelin-1 (10 nM; pIC$_{50}$ values: 7.2, 6.2, 6.2, 6.1, 6.3 and 6.5, respectively). The benzopyran derivatives were 10- to 40-fold more potent in relaxing rat aorta contracted by KCl than by endothelin-1. Furthermore, they did not fully antagonize the endothelin-1 responses. By contrast, RP 49356 and pinacidil relaxed fully and with the same potency the preparations contracted with either KCl or endothelin-1. In the rabbit mesenteric artery, 5-hydroxytryptamine concentration-response curves were depressed by cromakalim, but were only slightly affected by nimodipine. However, in the basilar artery, only the first component of the 5-HT concentration-response curve was inhibited by cromakalim; nimodipine, in contrast, depressed both components of the concentration-response curve [41]. Isolated pressurized rat mesenteric arteries (diameter 134 ± 6 µm) dilated when perfused with a solution containing cromakalim or pinacidil. However, cerebral arteries of similar diameter were not relaxed by these K$^+$ channel openers [42]. Cromakalim and pinacidil, unlike the calcium entry blocker dazodipine, and the guanylate cyclase stimulant, sodium nitroprusside inhibit the contractile effects which result from caffeine-induced Ca^{2+} release in rabbit isolated renal arteries. The two K$^+$ channel openers also reduced the tonic (due to Ca^{2+} entry) but not the phasic (due to intracellular Ca^{2+} release) component of noradrenaline response. However, cromakalim impaired both components of histamine-evoked contractile response. Glibenclamide pretreament antagonizes these effects of cromakalim [43].

Overall, the results of studies described in this section indicate that under in vitro conditions, the antivasoconstrictor spectrum of K^+ channel openers at the level of large (conductance) arteries (which are of major pathological importance for the human coronary vasospasm) is generally much broader than that of calcium antagonists. Furthermore, the vasorelaxant responsiveness of blood vessels to K^+ channel openers and, in general, to direct vasorelaxant agents, appears to be a function of several variables, such as the diameter of the vessel, the vascular region, the animal species, the excitatory agonist used to precontract the vessel and its concentration.

Membrane potential measurements in vascular preparations

K^+ channel openers can hyperpolarize the membrane up to values matching E_K. Most electrophysiological studies on specific K^+ channel openers have used cromakalim. Cromakalim and other K^+ channel openers increased the membrane resting potential in rat [43-45] and rabbit [46] aorta, guinea-pig mesenteric artery and vein [47] and rat portal vein [36] as did pinacidil [48] and minoxidil sulphate [49]. Nicorandil was the first pharmacological agent reported to increase the resting membrane potential in porcine and guinea-pig coronary arteries [50]. In the rat portal vein a low concentration (0.1 µM) of cromakalim initially reduced the duration, the amplitude and the frequency of the spontaneous multispike discharges, then completely abolished electrical activity of the preparation. A 50-fold higher concentration of cromakalim produced, in addition to these effects, a persisting hyperpolarization by moving the membrane potential close to the theoretical E_K which is -88 mV [36]. In the portal vein, the concentrations of cromakalim producing relaxation are lower than those that cause a hyperpolarizing effect. This finding may imply that the mechanical and electrophysiological effects of cromakalim are entirely dissociated. However, it is possible that minor changes in the resting potential are sufficient to reveal a significant vasorelaxant effect with a K^+ channel opener, although that may be difficult to measure. In the rat aorta the resting membrane potential was found to be -62 ± 2 mV. This value increased to -82 ± 3 mV after incubation with 10 µM cromakalim but was not modified by the 1 µM concentration. KCl (20 mM), noradrenaline (0.3 µM) and 5-hydroxytryptamine (1 µM) depolarized the membrane by 18, 13 and 23 mV, respectively. By contrast, cromakalim at 1 and 10 µM repolarized KCl-treated preparations by 16 and 23 mV, respectively. Similar effects were found when cromakalim was added to the aorta that had been depolarized with noradrenaline and 5-hydroxytryptamine (fig. 7). These results clearly indicate that a low concentration of cromakalim, devoid of effects on basal resting potential, can repolarize spasmogen-depolarized aorta preparations. Higher concentrations of cromakalim are necessary to hyperpolarize the aortic myocyte under basal conditions [45].

Isotope flux techniques in vascular preparations

Tracer experiments are aimed at assessing the unidirectional movements of specific ions. The efflux of K^+ can be measured by using ^{42}K. The K^+ tracer ^{86}Rb, however, is often pre-

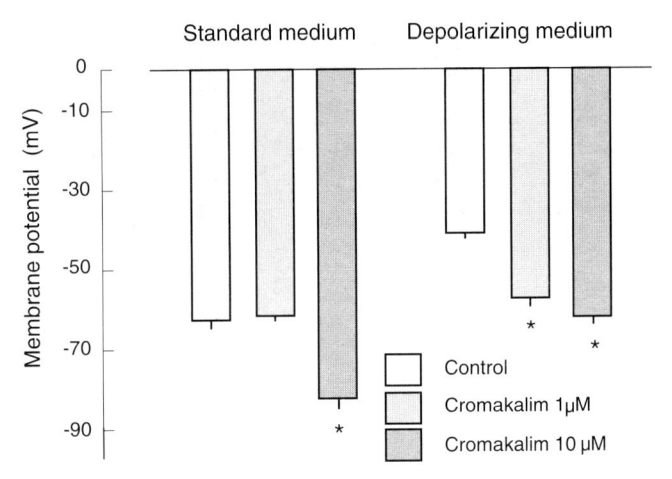

Fig. 7. Effects of vehicle and cromakalim (1 and 10 μM) on the membrane potential measured in cells from the rat aorta under resting conditions and after KCl (20 mM) depolarization. An asterisk indicates a significant effect. Drawn from [45]

ferred to ^{42}K, since it is easier to use, due to its longer half life (16 days instead of 12 h for ^{42}K), and safer due to its ionization power which is smaller than that of ^{42}K. The permeability to ^{86}Rb is slightly lower than that to ^{42}K (for vascular smooth muscle the ratio ^{86}Rb/^{42}K is approximately 0.76) [51]. Interpretation of results obtained with ^{86}Rb should take into account the permeability coefficient, as well as possible minor differences among similar types of K$^+$ channels present in different vascular regions. Thus, results obtained with ^{86}Rb can differ from those determined with ^{42}K, as recently reported for minoxidil sulphate [49]. The technique of Quast and Baumlin [51] for measuring ^{42}K or ^{86}Rb effluxes, allows the effects of a compound on vascular tone to be assessed concurrently. Rat or rabbit aorta rings are incubated for 90 min in warm, oxygenated physiological salt solution to which ^{42}K or ^{86}Rb have been added to load the cytosol with these isotopes. At the end of this period, the rings are mounted in a superfusion chamber which allows contractile force to be measured. Cromakalim [44] (fig. 8), its active enantiomer BRL 38227 [52], celikalim [53], diazoxide [9], pinacidil [48], nicorandil [54], and minoxidil sulphate [11] increase the spontaneous efflux of ^{42}K or ^{86}Rb. An excellent correlation was demonstrated between the vasorelaxant effects and the tracer flux increases for most of the K$^+$ channel openers studied. However, minoxidil sulphate and nicorandil were evident exceptions. For nicorandil, the explanation might be that the relaxation it produces in the rat aorta can be almost exclusively accounted for by the stimulation of guanylate cyclase [17]. For minoxidil sulphate, the atypical result may be due to its lack of specificity for K$^+$ channels and possibly also its partial agonism at vascular K$^+$ channels [55]. A consistent finding in these studies is that the concentrations of K$^+$ channel openers required to produce vasorelaxation are lower than the concentrations enhancing tracer effluxes (fig. 8). This may indicate that opening of a few K$^+$ channels produces a small repolarization current that is not accompanied by a large K$^+$ efflux but which may be sufficient to activate those mechanisms leading to vasorelaxation. However, for the K$^+$ efflux to become measurable, the sustained recruitment of a large population of K$^+$ channels may be required. Glibenclamide inhibits the tracer effluxes induced by K$^+$ channel openers [56]. Thus, experiments with this antagonist can confirm whether the increase in K$^+$ efflux produced by a given compound results from K$^+$ channel opening activity.

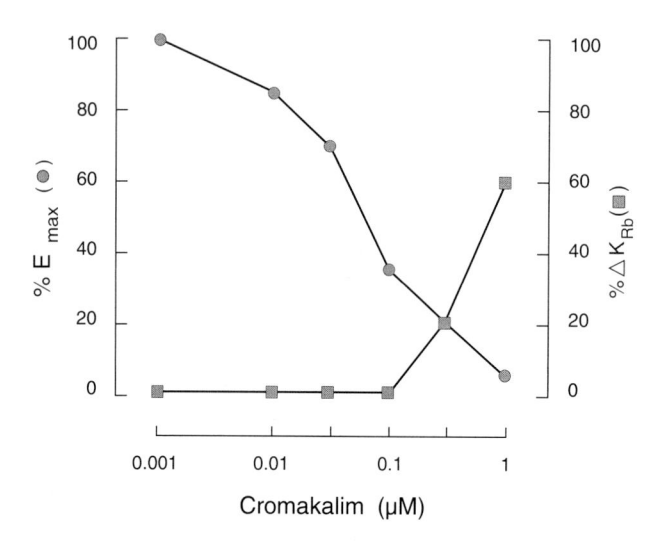

Fig. 8. Effects of cromakalim on developed tension (% E_{max}) and ^{86}Rb efflux in rat aortic segments contracted with KCl (23 mM). Redrawn from [51]

Radioligand binding assay for K^+ channel openers

A recent report from Bray and Quast [57] describes for the first time, to our knowledge, a radioligand binding assay for K^+ channel openers although this is carried out in rat aortic strips and not in membrane preparations. The radioactive ligand used is [^{3}H]P1075, a K^+ channel opener analogue of pinacidil. The K_D value for [^{3}H]P1075 was found to be 6 ± 1 nM and the B_{max} (maximal binding capacity) 21 ± 3 fmol/mg. This high affinity corresponds well to the potency of P1075 as a vasorelaxant agent. The specific binding was inhibited completely by several K^+ channel openers of unrelated chemical structure, such as unlabelled P1075, RP 49356, RP 61674, cromakalim and its active enantiomer BRL 38227 (fig. 9), SDZ PCO 400, minoxidil sulphate, nicorandil and diazoxide. The Hill coefficient of the inhibition curves was generally close to unity, indicating that the agents studied interacted with a single binding site. The pharmacological relevance of these results is strongly suggested by the excellent correlation (r = 0.93) between the potencies of the K^+ channel openers in displacing the radioligand from its binding sites (Ki values) and in increasing $^{86}Rb^+$ efflux from the rat aorta. An exception to this correlation was minoxidil sulphate, which is probably a partial agonist at the K^+ channel [57]. The antidiabetic sulphonylureas glibenclamide and glipizide also completely inhibited [^{3}H]P1075 binding by causing a 4-fold increase in the dissociation rate of the [^{3}H]P1075-receptor complex. In contrast, the K^+ channel openers did not produce this effect and behaved like competitive antagonists of the radioligand. To explain these results, Bray and Quast [57] advanced the hypothesis that the sulphonylurea binding site is possibly distinct from the binding site of the openers, but allosterically coupled to it in a negative fashion. It should be recalled that [^{3}H]glibenclamide has very high affinity constants (< 1 nM) in membranes prepared from the guinea-pig longitudinal ileum muscle [58], the heart [51], the pancreas [60] and the brain [60, 61]. K^+ channel openers do not displace [^{3}H]glibenclamide from these sites, suggesting that the high affinity sulphonylurea receptor present in these cell types is entirely distinct from the low affinity sulphonylurea binding sites found in the aorta.

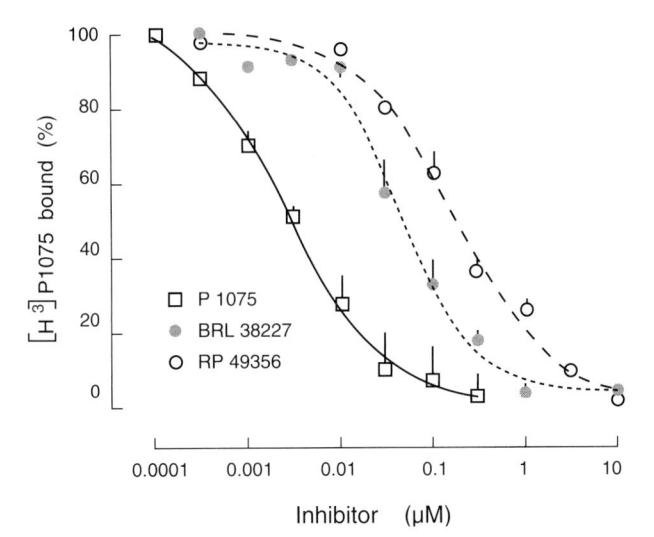

Fig. 9. Displacement of [H³]P1075 binding to endothelium-denuded rings of rat aorta by unlabeled P1075, BRL 38227 and RP 49356. The values of pKi for these compounds were 8.5 ± 0.03, 7.3 ± 0.08 and 6.8 ± 0.16, respectively [55]. Some of these results were kindly provided by Dr U Quast

Patch-clamp studies

The type of channel activated by K⁺ channel openers to produce hyperpolarization of vascular smooth muscle cells has not yet been unequivocally identified. However, if electrophysiological studies on K⁺ channel openers carried out in pancreatic ß-cells and in cardiac and smooth muscle cells are considered together, the most likely target for these compounds appears to be the ATP-gated and glibenclamide-sensitive channel [62] although the properties of the channels (e.g. susceptibility to activators and inhibitors, unitary conductance, modulation, etc.) are substantially different in these different tissues. Furthermore, vascular myocytes from different blood vessels may contain distinct populations of channels sensitive to K⁺ channel openers. In rat portal vein myocytes voltage-clamped at -60 mV, cromakalim, pinacidil and nicorandil generated an outward K⁺ current which increased in amplitude when the membrane was depolarized [62]. In more recent experiments, BRL 38227 was also found to induce a non-inactivating K⁺ current which activated slowly to its full size [63]. Under voltage-clamp conditions the current induced by BRL 38227 ran above the control current-voltage relationship, crossed it at -80 mV, which is the actual E_K value, and exhibited outward rectification, a property not seen in the control current-voltage curve. The unitary conductance of the channel which underlies the macroscopic current induced by BRL 38227 was calculated to be close to 17 pS. The effects of BRL 38227 were concentration-dependently inhibited by glibenclamide [63] and ciclazindol [26]. However in the same preparation, glibenclamide had no effect on a transient outward K⁺ current which was inhibited by BRL 38227 [63]. In patches excised from rat and rabbit mesenteric artery myocytes, cromakalim (1 μM) increased the open-state probability of a K⁺ channel which was maintained closed by a low concentration (1 μM) of ATP. This effect was blocked by glibenclamide. The conductance of this channel (135 pS with 60 mM [K⁺]$_o$ and 120 mM [K⁺]$_i$) is very large and 2 to 3-fold higher than the conductance of ATP-sensitive K⁺ channels present in the heart and the pancreatic ß-cells. The gating of this ATP-sensitive channel is insensitive to internal Ca^{2+} concentrations [64]. In rabbit mesenteric artery myocytes, BRL

38227 (levcromakalim) and metabolic inhibition with iodoacetate and dinitrophenol (to prevent production of ATP) activate a current which shared several properties with the non-smooth muscle ATP-sensitive K^+ current. In particular, this current has little voltage sensitivity and is completely inhibited by glibenclamide. These findings were proposed to support the view that hypoxia-induced vasodilation arises from the activation of ATP-sensitive K^+ channels [65]. Finally, smooth muscle cells dissociated from rabbit main pulmonary artery exhibited a mean resting potential of -55 mV when recorded with pipettes containing 1 mM ATP. This value was -70 mV when ATP was omitted. BRL 38227 hyperpolarized these action potentials and increased actively a basal, glibenclamide-sensitive, time-independent K^+ current. Glibenclamide depolarized myocytes which were studied with pipettes containing 0 or 1 mM ATP. By contrast, glibenclamide had little or no effect on the membrane potential when the ATP concentration in the pipette was raised to 3 mM. This implies that K^+ channels underlying the resting potential can be effectively inhibited by both ATP and glibenclamide [66]. Since the intracellular ATP concentration in the vascular smooth muscle under normal conditions is rather low (~ 1.5 mM) [67], it is possible that an ATP-sensitive K^+ current contributes to the normal maintenance of resting membrane potential and basal vascular tone. Cromakalim may open Ca^{2+}-dependent K^+ channels in vascular smooth muscle myocytes [64]. It should be recalled that, on the basis of ionic conductance, Ca^{2+}-dependent K^+ channels have been classified into three distinct subtypes. The large (BK_{Ca}) and intermediate (IK_{Ca}) conductance channels are selectively blocked by charybdotoxin, a scorpion venom toxin whereas small conductance K^+ channels (SK_{Ca}) are blocked by the bee venom toxin apamin. However, these toxins do not modify cromakalim-activated K^+ currents or K^+ channel opener-induced vasorelaxation [68].

In vivo cardiovascular experiments can unveil K^+ channel opening activity

The main purpose of in vivo cardiovascular studies is to see if results obtained under in vitro conditions can be replicated in an integrated system which is controlled by physiological homeodynamic mechanisms. These experiments also allow the selectivity of action to be determined and can provide clues for possible clinical development of the drug being investigated. It is essential to keep in mind that in vitro techniques to study vasorelaxant agents generally involve the use of large arteries, (e.g. the aorta), which are conductance vessels that do not contribute to peripheral vascular resistance. By contrast, through in vivo studies the effects of a compound on resistance vessels can be determined although effects on conductance vessels (e.g. coronary artery) of large experimental animals can also be studied.

Effects on blood pressure and heart rate in pentobarbitone-anaesthetized rat

This preparation is very appropriate for disclosing the possible K^+ channel opening activity of a compound. Briefly, rats are anaesthetized with pentobarbitone and artificially ventilated with room air. A carotid artery is cannulated for the measurement of blood pressure. Intravenous injections are made via the femoral veins. As a pretreatment, receptor antagonists are administered slowly (for 5 min) 10 min before the i.v. infusion

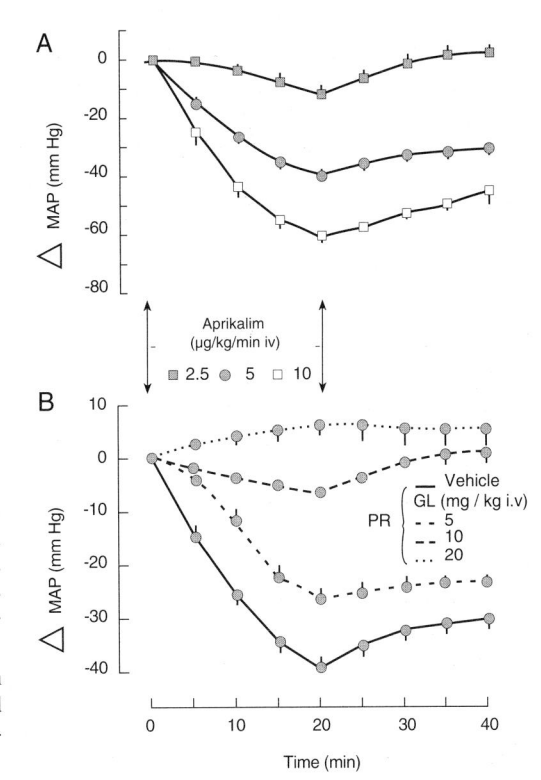

Fig. 10. **A** Dose-related decreases in mean carotid artery blood pressure (ΔMAP) produced by aprikalim in pentobarbitone anaesthetized rats (n=5-7/group). **B** Dose-related inhibition of aprikalim hypotensive effects by glibenclamide *(GL)* pretreatment *(PR)*. The baseline MAP before aprikalim administration was 125 ± 2 mmHg and 134 ± 2 mmHg in vehicle and glibenclamide-pretreated animals

(20 min) of a K^+ channel opener [32]. In this preparation, aprikalim [23], cromakalim [32], nicorandil [33], diazoxide [69], and, in general, K^+ channel openers [2] produce marked, dose-related decreases in blood pressure. During the infusion of aprikalim and cromakalim, the onset of this effect was gradual and a maximum was reached between 15 and 20 min after starting the administration. With cromakalim approximately 50% of the maximal response disappeared within 10 to 15 min after ending the infusion and the remaining effect attained a novel apparent steady-state [32]. However, with aprikalim there was only a small decrease in the maximal effect during this period (fig. 10). The full duration of the hypotension was not studied since it is difficult to maintain a rat at a constant level of surgical anaesthesia with pentobarbitone for a prolonged period. K^+ channel openers generally do not modify heart rate in pentobarbitone anaesthetized rats since this anaesthetic agent inhibits vagal drive and stimulates sympathetic tone. The latter effect strongly elevates baseline heart rate. Therefore, possible reflex increases in heart rate, which are the physiological response to decreases in aortic blood pressure, produced by peripheral vasodilators, should be studied in conscious animals.

Determination of the pharmacological mechanism of blood pressure decreases

Numerous mechanisms can decrease blood pressure in the anaesthetized rat. In order to clarify whether this effect is due to the opening of K^+ channels, we use glibenclamide which antagonizes in a rather specific manner the blood pressure lowering activity of K^+

channel openers [2, 32]. The inhibition is dose-related, and at the high dose of 20 mg/kg i.v. glibenclamide blocks entirely a decrease in blood pressure of approximately 40 mmHg produced by aprikalim (fig. 10). Similar results were obtained with other K$^+$ channel openers [2, 71]. The specificity of glibenclamide as an antagonist of K$^+$ channel opener effects in vivo in the anaesthetized rat is supported by the finding that this sulphonylurea does not modify the hypotensive effects of calcium antagonists, nitrodilators, papaverine, dihydralazine, acetylcholine, adenosine, calcitonin gene-related peptide, platelet activating factor (PAF) or salbutamol [2]. These results imply that the decrease in blood pressure produced by these agonists in the rat does not require the opening of channels which are the site of action of K$^+$ channel openers. However, glibenclamide can inhibit responses to certain endogenous vasodilator substances, such as adenosine, in the guinea-pig perfused heart [70]. Thus, the receptors stimulated by these vasodilators in some vascular beds of certain species can be coupled directly or indirectly to K$^+$ channels susceptible to glibenclamide blockade. The hypotensive action of aprikalim [22] and cromakalim [32] appears to be exclusively due to activation of vascular K$^+$ channels since these effects are not modified by many standard agents which block the receptors or mechanisms mediating decreases in blood pressure (fig. 11) [2]. These antagonists are idazoxan (central α_2-adrenoceptors), methylatropine (muscarinic receptors), methysergide (5-HT2-receptors), dexchlorpheniramine (histamine receptors), propranolol (ß$_2$-adrenoceptors), RP 57229 (PAF receptors) phentolamine (peripheral α_1 and α_2-adrenoceptors), SCH 23390 (D1 dopamine receptors), S-sulpiride (D2 dopamine receptors), SKF 100273 (vasopressin receptors). Diclofenac is an inhibitor of cyclooxygenase, which can prevent the formation of vasodilator prostanoids. Since this compound did not prevent aprikalim from inducing hypotension, the latter effect is not due to endogenous substances produced by the cycloxygenase pathway during aprikalim administration. Furthermore, since bivagotomy and ligation of carotid arteries failed to modify the effects of aprikalim on blood pressure, afferent cardiopulmonary nervous pathways do not participate in this effect. A direct (independent of the central nervous system) effect of cromakalim [32] or aprikalim (fig. 11) on the vasculature is clearly indicated by the ability of these compounds to lower blood pressure in pithed rats in which the low level of the baseline blood pressure (due to destruction of cardiovascular centers by pithing) is raised by an infusion of vasopressin. This conclusion is supported by the finding that the intracerebral (cisterna magna) administration of aprikalim was not accompanied by hypotensive effects in pentobarbital-anaesthetized rats [22]. In intact as well as in pithed vasopressin-supported rats pretreated with enalapril (an inhibitor of angiotensin I converting enzyme) or losartan (an antagonist of angiotensin II receptors), the hypotensive effects of aprikalim and cromakalim [32] were enhanced. This indicates that, in intact rats, the full decrease in blood pressure produced by K$^+$ channel openers is partly opposed by angiotensin II (see section on Interaction with the renin-angiotensin system).

Effects on blood pressure and heart rate in conscious animals

In conscious normotensive or spontaneously hypertensive rats, oral administration of aprikalim [22] or cromakalim [32] produced dose-related, long-lasting decreases in blood pressure accompanied by heart rate increases. Both effects were inhibited by pre-

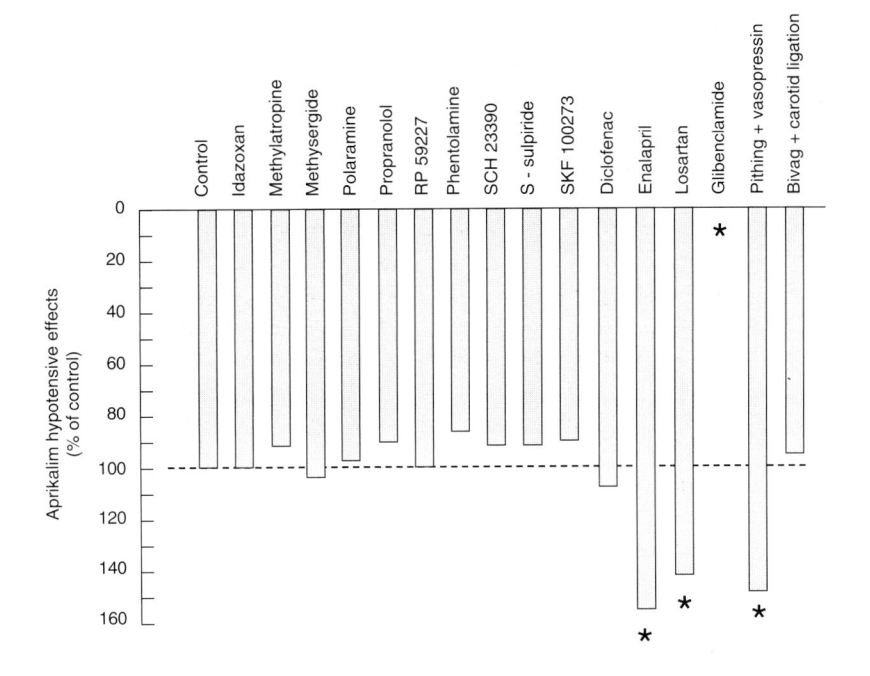

Fig. 11. Changes in mean carotid artery blood pressure (ΔMAP) produced at the end of a 20-min i.v. infusion of 5 µg/kg/min aprikalim in pentobarbitone anaesthetized rats pretreated with saline (control group) or various vascular receptor antagonists (see text) or subjected to pithing or bilateral vagotomy *(bivag)* plus ligation of carotid arteries. An asterisk indicates a significant difference from control response

treating the animals with glibenclamide by the i.v. (fig. 12) but not by the oral route [32]. Furthermore, glibenclamide antagonized the established antihypertensive and tachycardic effects of cromakalim [32]. In spontaneously hypertensive rats, the tachycardic effect of aprikalim disappeared faster than the decrease in blood pressure [22]. Furthermore, in spontaneously hypertensive rats the increase in heart rate was of a very small magnitude when blood pressure was decreased progressively by administering aprikalim as a slow i.v. infusion [22]. The strict dependence of the heart rate increase on the rate of decrease in blood pressure following the administration of a K^+ channel opener is clearly supported by results obtained in spontaneously hypertensive rats with Y-27,152, a prodrug K^+ channel opener of the benzopyran series. Following the administration of 1 mg/kg p.o. of this compound, the onset of the blood pressure decrease was gradual with a maximum being attained 7 h after dosing. The effect lasted over 24 hours. Despite a substantial fall in blood pressure, there was only a small initial increase in heart rate. By contrast, orally administered BRL 38227 and nifedipine rapidly lowered blood pressure and this effect was accompanied by a pronounced tachycardia (fig. 13). Interestingly, an i.v. bolus of Y-26763, the active metabolite of Y-27,152, decreased blood pressure quickly and increased markedly heart rate, but, the latter effect did not occur when the decrease of blood pressure was achieved gradually by a slow i.v. infusion of Y-26763 [21]. In conscious unrestrained renal hypertensive dogs, Y-27,152 at doses of 0.1, 0.3 and 1.0 mg/kg p.o. lowered blood pressure with a slow onset and a long duration of action and

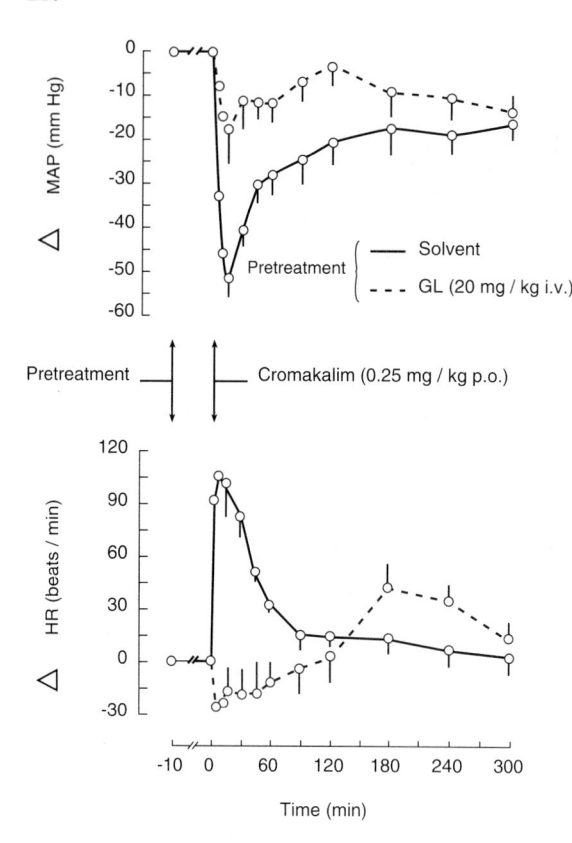

Fig. 12. Decreases in mean carotid artery blood pressure *(ΔMAP)* and increases in heart rate *(ΔHR)* produced by orally administered cromakalim in spontaneously hypertensive rats. Pretreatment of rats with glibenclamide (GL) inhibits both effects of cromakalim

had only a minimal effect on heart rate, whereas the antihypertensive action of both BRL 38227 and nifedipine were followed by a striking tachycardia [21]. In conscious normotensive cats, the oral administration of cromakalim also produced a persistent decrease in blood pressure which for the oral dose of 60 µg/kg attained a nadir of over 60 % below the baseline value whereas heart rate increased only 25 % over baseline [71].

Overall, these studies indicate that K^+ channel openers produce tachycardia in conscious animals and this effect is strictly dependent upon the rate of onset of the blood pressure decrease.

Antivasoconstrictor effects

Ideally studies aimed to assess the antivasoconstrictor activity of a compound should be carried out on selected vascular beds (e.g. : isolated hind leg) by using constant flow (measure of changes in perfusion pressure) or constant pressure (measure of changes in blood flow) techniques. However, antivasoconstrictor studies are sometimes performed with rats in which the central nervous control to the vasculature has been removed either by destroying the spinal cord or by pharmacological blockade of autonomic ganglia. In these preparations, carotid artery blood pressure and, sometimes, blood flow to various organs are measured by Doppler flow-probes. Vasoconstrictor agents [noradrenaline, α_1-adrenoceptor (cirazoline, methoxamine, phenylephrine) and α_2-adrenoceptor

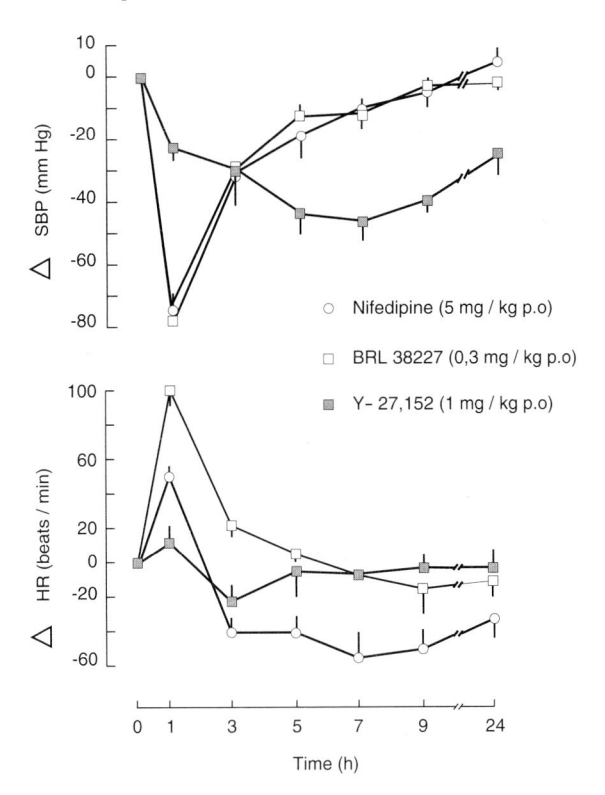

Fig. 13. Effects of nifedipine, BRL 38227 and Y-27,152 on systolic blood pressure (ΔSBP) and heart rate (ΔHR) in conscious spontaneously hypertensive rats. Redrawn from [21]

(B-HT 920, UK-14,304) agonists, angiotensin II, vasopressin, endothelin-1, 5-hydroxytryptamine, prostaglandin $F_{2\alpha}$] can be injected as i.v. doses singley or cumulatively. Electrical stimulation of the spinal cord (via the pithing rod) allows the pressor responses to endogenously released noradrenaline to be studied. The effects of bimakalim (SR 44866) and cromakalim were studied on the systemic and regional vascular responses evoked by cirazoline, UK-14,304 and electrical stimulation of the spinal cord in pithed spontaneously hypertensive rats [72]. At a dose which decreased carotid artery blood pressure by approximately 20 %, these compounds did not affect post-synaptic α_1-adrenoceptor mediated increases in systemic and regional (kidney, mesentery and hindquarter) vascular resistance but slightly attenuated those mediated by postsynaptic α_2-adrenoceptors. Bimakalim and cromakalim markedly reduced the pressor but not the tachycardic effects elicited by electrical stimulation of the spinal cord. This effect persists when the decrease in baseline blood pressure produced by K$^+$ channel openers is corrected with the pressor agent prostaglandin $F_{2\alpha}$. However, it disappears when vasopressin is used as vasoconstrictor agent. The authors [72] suggest that in spontaneously hypertensive rats, bimakalim and cromakalim inhibit the release of endogenous noradrenaline from sympathetic neurons innervating blood vessel without affecting noradrenaline release from neurons innervating the cardiac sinus node which control heart rate.

In another study using anaesthetized, ganglion-blocked spontaneously hypertensive rats [73], cromakalim dose-dependently antagonized the pressor responses to incremental i.v. infusions of noradrenaline and phenylephrine but did not change the vasoconstrictor effects of methoxamine, angiotensin II or vasopressin. By contrast, nifedipine antagonized all of these pressor agents, being most effective against responses to norepinephrine and angiotensin II. In pithed rats, both cromakalim and nifedipine, at a dose which reduced markedly (40-50 %) baseline diastolic blood pressure, depressed (over 40-60 %) the pressor response to electrical stimulation of the spinal cord. Restoration of the lowered diastolic blood pressure to within the control range with an i.v. infusion of vasopressin prevented the inhibitory effects of cromakalim but not those of nifedipine. In a study recently performed in our laboratory by treating pithed normotensive rats with a dose of aprikalim and levcromakalim that slightly decreases baseline blood pressure, we found that these K^+ channel openers and the calcium entry blocker nitrendipine inhibited similarly (approximately 40-60 %) the pressor effects to angiotensin II, noradrenaline and to electrical stimulation of the spinal cord. This indicates that K^+ channel openers like nitrendipine do not impair the release of endogenous noradrenaline.

In conclusion, further studies are needed to clarify the effects of K^+ channel openers on resistance vessel reactivity to various excitatory stimuli. Although, there is evidence for inhibition of pressor responses to spinal cord stimulation by K^+ channel openers in pithed rats, no convincing proof exists that this is due to a reduction in noradrenaline release from sympathetic nerve terminals. In our opinion, this is very unlikely since vascular neurotransmission studies under in vitro conditions and even in pithed rats have failed to reveal a direct inhibition of noradrenaline release [4, 6]. However, it is of interest to note that K^+ channel openers reduce the contractile responses evoked by electrical stimulation of the vagus nerve in guinea-pigs. This effect was attributed to inhibition of the release of acetylcholine and excitatory non-adrenergic non-cholinergic (NANC) neurotransmitters from neurons innervating airways [74].

Systemic and regional haemodynamic effects

The haemodynamic profile of K^+ channel openers depends on the animal species and particularly the experimental conditions (anaesthesia, conscious state, dose of K^+ channel opener, etc.). In normotensive anaesthetized rats cromakalim at a dose which markedly reduced mean carotid artery blood pressure (by approximately 40 mmHg) did not modify cardiac output, increased hindquarter blood flow and decreased the mesenteric blood flow slightly and the renal blood flow moderately [2, 32]. Similar effects were produced by bimakalim (EMD 52692, SR 44866) [2]. By contrast, in conscious normotensive rats cromakalim and bimakalim did not change renal, increased mesenteric and slightly lowered hindquarter blood flow [2]. Since the doses of cromakalim and bimakalim studied were notably hypotensive, the total and regional vascular resistances measured were markedly reduced. In another study of conscious rats, a brief (3 min) i.v. infusion of a moderately hypotensive dose of BRL 38227 also produced a substantial increase in mesenteric blood flow. However, it did not change hindquarter blood flow and decreased renal blood flow [75]. A study of blood flow distribution to various organs after the administration of cromakalim was carried out by injecting radioactive microspheres into anaesthetized rabbits [76]. Cromakalim and its active enantiomer caused dose-related increases in blood flow which were marked for the heart and stomach, less pronounced for the small intestine and slight for the brain. No notable change in blood flow occurred in the kidneys and the skeletal muscle. This blood flow profile of a K^+ channel opener in the rabbit differs from that of calcium antagonists, which are potent vasodilators of the coronary, cerebral and skeletal muscle regions [76]. In a cardiohaemodynamic study carried out in the pentobarbitone anaesthetized, open-chest dog, small to moderate doses of cromakalim and pinacidil decreased aortic blood pressure, but increased venous return and cardiac output. The last two effects also occurred in animals in which central reflex regulation had been abolished by surgical disruption of afferent reflex pathways (bilateral vagotomy and carotid sinus denervation). In intact dogs, the highest i.v. bolus dose of cromakalim (100 µg/kg) studied reduced transiently both cardiac output and venous return. This effect resulted from the similarly transient, concurrent fall in myocardial contractility, since in the cardiopulmonary bypass experiment (where the heart is excluded from the circulation), this dose of cromakalim still increased venous return [77]. In anaesthetized pigs, i.v. bimakalim (EMD 52692) decreased blood pressure dose-dependently. This effect was due to peripheral vasodilation since cardiac output was not changed. Heart rate increased, left ventricular end diastolic pressure decreased and myocardial contractility did not change. Blood flow to the brain increased substantially whilst blood flow to the kidneys decreased and that to various skeletal muscles did not change. In this preparation, intracoronary infusion of bimakalim at doses lacking hypotensive effects increased left anterior descending coronary artery flow by up to 128 %. The subepicardial area benefited almost twice as much as did the subendocardium from this vasodilator effect [78]. In anaesthetized dogs with an intact innervation to the kidney, an i.v. injection of cromakalim caused dose-related decreases in aortic blood pressure, did not change renal blood flow and increased slightly urine formation. However, an intrarenal artery injection of cromakalim at non hypotensive or slightly hypotensive doses, in preparations with denervated kidneys, increased renal blood flow, decreased renal vascular resistance and elevated the levels of urine production, sodium urinary excretion, fractional urinary excretion and urine osmolality. The diuretic effect appears to be due partly to the inhibition of sodium and water reabsorption at a side distal to the proximal tubules. All these effects were suppressed by pretreating the animals with glibenclamide [79].

In conclusion, K^+ channel openers are potent peripheral vasodilators virtually devoid of myocardial depressant effects even at doses producing large decreases in blood pressure. These compounds redistribute cardiac output and appear to preferentially cause dilation of the coronary circulation.

Effects on coronary blood flow

Aprikalim and nicorandil injected into the coronary artery of anaesthetized dogs immediately increased coronary blood flow without changing blood pressure [22, 32]. This effect was antagonized by glibenclamide. In the same preparation, an i.v. infusion of a non-hypotensive dose of aprikalim markedly elevated coronary blood flow [22]. Thus, it is possible to obtain coronary vasodilation with K^+ channel openers without concurrent hypotension. Similarly, in conscious dogs, cromakalim and pinacidil produced greater percentage increases in the coronary blood flow than decreases in blood pressure. In this preparation, cromakalim and pinacidil dose-dependently relaxed both conductance and resistance coronary vessels. The hypotensive and coronary vasodilator effects of cromakalim were not affected by prior combined ß-adrenoceptor and muscarinic receptor blockade which prevented K^+ channel openers from producing tachycardia. Thus, the coronary vasodilator activity of these compounds is independent of changes in myocardial metabolic demands due to sympathetic nerve activation. The dilation of epicardial large coronary arteries caused by cromakalim, unlike that caused by nitroglycerin, is partly mediated by a flow-dependent mechanism [80]. Finally, in conscious normotensive dogs Y-27,152 administered orally increased coronary blood flow. This effect was long lasting and occurred without changes in blood pressure [81]. Interestingly, pinacidil produced coronary steal only when the stenosis of the circumflex coronary artery was added to the occlusion of the left descending coronary artery and it was administered at a dose lowering markedly aortic blood pressure, and increasing strongly heart rate, cardiac output and left ventricle contractility [82]. K^+ channels appear to play a significant role in regulating the coronary circulation. In anaesthetized dogs, infusion of high doses of glibenclamide directly into the left circumflex coronary artery decreased dose-dependently the flow through this artery and therefore increased coronary vascular resistance. These effects were accompanied by signs of ischaemia and a reduction in contractility within the area of the left circumflex artery that disappeared upon removal of the drug [83]. In the dog, K^+ channels appear also to modulate the reactive hyperemia, that is, the increase of coronary blood flow above baseline value occurring after a brief occlusion of a coronary artery. Glibenclamide infused into the coronary artery reduced the flow debt repayment after a 30-s coronary artery occlusion [83]. At the dose studied, glibenclamide did not modify acetylcholine-induced vasodilation but reduced that caused by adenosine. Similar results were found after i.v. administration of glibenclamide to dogs, which attenuated the debt repayment and inhibited markedly vasodilator response to adenosine. In guinea-pig perfused heart, glibenclamide (1 µM) reduced reactive hyperemia, and debt repayment [84]. Finally, glibenclamide completely abolished the dilation of coronary arterioles less than 100 µm diameter which accompanies a graded decrease in coronary perfusion pressure in the beating dog heart. However, responses of larger arterioles (> 100 µm) were not modified. Thus, these vascular glibenclamide-sensitive K^+ channels are critically important in the dilation of coronary collateral vessels during ischaemia ([85], and KL Lamping, personal communication).

Antihypertensive activity of K^+ channel openers

A substantial amount of evidence exists that K^+ channel openers decrease blood pressure in several models of hypertension. Oral administration of levcromakalim, produces dose-related decreases in blood pressure not only in conscious spontaneously hypertensive rats but also in renal hypertensive cats and dogs. This effect was always accompanied by tachycardia, which in the rat was abolished by pretreatment with the $ß_1$-adrenoceptor antagonist atenolol, indicating that the increase in heart rate is mainly mediated by cardiac ß-adrenoceptors [71]. RP 49356, the racemic mixture containing RP 52891 (aprikalim), at a small oral dose (0.25 mg/kg) elicited notable and prolonged reductions in blood pressure which were of a similar magnitude one hour after its administration to renal hypertensive, DOCA-salt hypertensive, as well as spontaneously hypertensive, rats [22]. Ro 31-6930, a K^+ channel opener of the benzopyran series was a potent antihypertensive agent in spontaneously hypertensive rats. Its blood pressure effects were accompanied by pronounced tachycardia [86]. In the conscious renal hypertensive dog, Y-27,152 administered daily over 8-week, decreased blood pressure without causing tolerance to develop [21]. Similarly, no tolerance to the antihypertensive effects of RP 49356 during a 5-day dosing period [22] or to Ro 31-6930 during 22-day administration [86] to spontaneously hypertensive rats has been observed.

Interaction with the renin-angiotensin system

The humoral response to hypotension is an increased liberation of catecholamines, renin and aldosterone. Cromakalim, levcromakalim and diazoxide increased plasma renin activity in anaesthetized intact rats as well as in pithed rats in which the central reflex control has been removed by bilateral vagotomy and destruction of the spinal cord [69]. In conscious renal hypertensive cats, an oral dose of BRL 38227 producing a large decrease in blood pressure increased plasma renin activity 4-fold and plasma aldosterone concentration 2-fold [71]. There is also direct evidence that cromakalim (like calcium entry blockers) can increase renin secretion when added to primary cultures of juxtaglomerular cells [87]. Thus, K$^+$ channel openers can increase the plasma levels of renin either by a reflex mechanism triggered by their potent vasodilator action or directly by opening K$^+$ channels in the juxtaglomerular cells [2, 60]. In these cells, the hyperpolarizing effect of K$^+$ channel openers (like blockade of voltage-operated Ca^{2+} channels) produces a decrease in intracellular free Ca^{2+} which functions as an activation signal for renin secretion [88]. By contrast, in the vascular smooth muscle (from which the juxtaglomerular apparatus is derived), the increase in intracellular free calcium activates the cell (contraction). Increased plasma renin activity elevates the plasma concentration of angiotensin II, a potent vasoconstrictor octapeptide, which functionally opposes and, thus partially reduces, the vasorelaxant effects of K$^+$ channel openers. Indeed, in pentobarbitone-anaesthetized rats the hypotensive effects of cromakalim were significantly enhanced by enalapril, an inhibitor of angiotensin I converting enzyme (fig. 10) [2]. Furthermore, losartan (DUP 753), a blocker of the angiotensin II vascular receptor, also enhanced the blood pressure lowering effects of diazoxide, cromakalim and aprikalim (fig. 10) in both intact pentobarbitone-anaesthetized and pithed rats [69]. Evidence exists that pinacidil and cromakalim can also elevate plasma renin activity in man [14, 87]. However, this finding does not imply that K$^+$ channel openers will necessarily produce a pronounced activation of the renin-angiotensin system when administered repeatedly at antihypertensive doses which are absorbed slowly. Interestingly, the prodrug Y-27,152, which decreases blood pressure gradually in renal hypertensive dogs, only slightly increased plasma renin activity after the first dose but this effect did not persist during the 8-week repeated dose study despite the antihypertensive activity of the compound [21].

Effects on insulin secretion

ATP-sensitive K$^+$ channels play a crucial role in the regulation of membrane potential in pancreatic ß-cells. Intracellular ATP is an essential link between glucose metabolism and membrane depolarization. ATP generated by glucose metabolism mediates the closure of ATP-dependent K$^+$ channels which depolarizes the membrane causing voltage-dependent calcium channels to open and thus, trigger insulin secretion [89, 90]. Pancreatic ATP-sensitive K$^+$ channels are the site of action of pharmacological agents promoting and inhibiting insulin secretion. Antidiabetic sulphonylureas stimulate insulin secretion by causing electrical and ionic events similar to those induced by glucose. Specific high affinity binding sites for sulphonylurea receptors have been identified on the cell membrane of different insulin-secreting cell lines [60]. A pharmacological inhibitor of insulin secretion is diazoxide which hyperpolarizes the pancreatic ß-cell membrane by increasing its permeability to K^+. This effect is achieved by increasing the probability that ATP-dependent K^+ channels will assume the open state. Thus, diazoxide inhibits insulin secretion by a sequence of events opposite to that triggered by hypoglycemic sulphonylureas. Pinacidil, cromakalim and RP 49356 were also found to inhibit the release of insulin from pancreatic glucose-stimulated ß-cells; however, this effect occurred only with extremely high concentrations (100-500 µM) of these compounds. Therefore, it seems unlikely that these effects result from interactions with the same K^+ channels that mediate the vasorelaxation ($IC_{50} < 1$ µM, fig. 4). In a study designed to assess the effects of various K^+ channel openers in vivo pithed rats were infused continuously with glucose to raise baseline secretion of insulin. Diazoxide lowered insulin to levels similar to those measured before starting glucose infusion. However, cromakalim, nicorandil or aprikalim (RP 52891) even at doses 40-fold higher than those producing the same hypotensive effect as diazoxide in intact pentobarbitone anaesthetized, normotensive rats, did not have this effect. Glibenclamide pretreatment antagonized dose-dependently the hypoinsulinemic activity of diazoxide with a very low ED50 value (49 µg/kg i.v.). In contrast, a much higher dose of glibenclamide (approximately 10 µg/kg i.v.) is required to reduce the hypotensive effect of diazoxide by 50 % [69]. Therefore, among the numerous K^+ channel openers, diazoxide is the only one that potently inhibits insulin secretion in vitro and in vivo at doses which produce vasorelaxation.

K^+ channel openers have potential for treating cardiovascular diseases

The preclinical profile of K^+ channel openers presented clearly supports a clinical potential for the use of these drugs in cardiovascular pathologies which require a decrease in peripheral vascular resistance, an inhibition of vessel propensity to excessively vasocontract and a prolongation of the viability of the myocardial tissue undergoing transient oxygen deficiency.

Hypertension

K^+ channel openers can be functionally classified as direct vasodilators. Therapeutic experience with these agents indicates that they can reduce elevated blood pressure in patients (see chapter 16). A decrease in blood pressure achieved with a potent peripheral vasodilator is generally accompanied by adverse reflex counterregulation reactions which can manifest themselves as tachycardia, headache, flushing, increase in plasma renin activity, plasma aldosterone concentration and catecholamine secretion, and sodium and water retention leading to oedema and body weight gain. Therefore, the use of K^+ channel openers as monotherapy to reduce blood pressure may be accompanied by a series of undesirable effects which are no longer acceptable in clinical practice. In order to improve the poor profile of these drugs several suggestions can be made. Firstly, these agents could be rationally combined with established antihypertensive agents and, in particular, ACE inhibitors, ß-adrenoceptor antagonists and diuretics. Although diuretic agents are an apparently judicious choice for fighting the oedema produced by K^+ channel openers, in our opinion, they treat the symptoms and not the cause. Indeed, diuretics can increase plasma renin activity which leads to a rise in angiotensin II. The latter octapeptide is a very potent vasoconstrictor reducing partly the vasodilator activity of K^+ channel openers and causing sodium and water retention via the liberation of aldosterone. Furthermore, angiotensin II can produce vascular smooth muscle mitogenesis and cellular hypertrophy. For all these reasons, we believe that a better approach to improve the cardiovascular tolerance and efficacy of K^+ channel openers is their concurrent prescription with an ACE inhibitor and, when available, with an angiotensin II receptor antagonist. This suggestion is supported by experimental animal data presented in a previous section. It is important to stress that multidrug therapy is not the exception but rather the routine in the treatment of moderate and severe hypertension. A normalization of elevated blood pressure is often obtained with the use of doses of the combination components which are smaller than those generally prescribed when each single agent is taken as monotherapy. Furthermore, from a theoretical standpoint the simultaneous intake of small doses of antihypertensive drugs with distinct but complementary mechanisms of action will more effectively normalize elevated blood pressure, which is a physiological parameter tightly regulated by multiple independent but finely integrated neural and humoral mechanisms. Another possible avenue to reduce the adverse effects of K^+ channel openers is to smooth the absorption phase of these drugs. A fast entry of a vasodilator into the blood stream can rapidly lower peripheral vascular resistance, thereby causing headaches and flushes. Thus, appropriate controlled release formulations of rapidly bioavailable K^+ channel openers should be developed to attenuate reflex-mediated reactions. Experimental results previously detailed appear to support this point.

In conclusion, on the basis of these considerations, it would appear that K^+ channel openers are not ideal antihypertensive drugs for monotherapy. However, these agents could become useful if appropriately formulated and co-prescribed with selected established antihypertensive agents. Furthermore, small doses of K^+ channel openers could benefit the hypertensive patient since in animals these compounds appear to possess unique cardioprotective properties at doses which do not produce evident haemodynamic effects as it has been discussed in chapter 13 of this book. Therefore, small doses of K^+ channel openers could usefully provide needed myocardial protection to hypertensive patients since most of current antihypertensive drugs do not appear to substantially reduce cardiovascular mortality and morbidity, which is the ultimate goal of any treatment of hypertension.

Angina pectoris

K^+ channel openers have preclinical properties which are undoubtedly desirable for therapeutic agents aimed at treating patients with transient and chronic heart coronary disease: these agents can improve oxygen delivery and also reduce oxygen consumption within the ischaemic region. Indeed, K^+ channel openers relax coronary conductance arteries, increase selectively coronary blood flow and antagonize the vasoconstrictor activity of a large number of excitatory stimuli and do not appear to produce coronary steal. Furthermore, as presented in chapter 13, they can markedly reduce the left ventricular dysfunction which follows single or multiple ischaemic insults of brief duration (stunned myocardium) and substantially decrease also the loss of viable myocardium resulting from a prolonged ischaemic period followed by reperfusion. The latter beneficial myocardial effects can be observed in experimental animals with doses of K^+ channel openers which are devoid of notable general and coronary vasodilator activity. Thus, if preclinical findings can be reproduced in human subjects, K^+ channel openers may have antianginal activity at doses that do not provoke an undesirable, reflex-mediated activation of the sympathetic nervous system, an effect which would be a deleterious physiological reaction for a myocardium already ischaemic or lacking a safety margin of blood flow reserve.

Congestive heart failure

One of the presently used treatments for this disease, although it is aimed at alleviating symptoms, is to reduce peripheral vascular resistance. K^+ channels can produce this effect in a very effective manner. Since patients with congestive heart failure now almost routinely receive ACE inhibitors and diuretics, the adverse reflex mediated effects of K^+ channel openers due to peripheral vasodilation are probably not a limiting factor for using these drugs to treat this disease although it remains to be investigated whether an increase in cardiac sympathetic tone may be undesirable. Interestingly, the cardioprotective effects afforded by K^+ channel openers may be of particular benefit to patients with congestive heart failure which developed after a myocardial infarction since they are at elevated risk of suffering a secondary infarct.

Peripheral vascular diseases

This application is covered in chapter 18 of this book. Briefly, activation of K^+ channels can improve the energy metabolism and the mechanical performance of skeletal muscles suffering oxygen deficiency. This may be achieved partly by a selective dilator of collateral vessels supplying the ischaemic skeletal muscle and partly by a better utilization of high energy phosphates. These mechanisms are evidently of therapeutic potential for treating patients with peripheral vascular disease, a disabling disorder of elderly people characterized by poor blood supply to the lower limbs due mostly to atherosclerosis.

Differentiation of K^+ channel openers is a task for the futur

K^+ channel openers are a novel class of pharmacological agents. Well-designed clinical trials are now required to prove their theoretically promising clinical potential. The preclinical cardiovascular profile of these compounds appears to be even more interesting than that of calcium antagonists. The distinctive mechanism of K^+ channel openers is the activation of vascular K^+ channels which have the physiological task of driving to rest, clamping at rest and, also reinforcing the resting state of vascular myocytes. Hence, K^+ channel openers can reduce and prevent increases in vascular tone which are associated with pathological manifestations. A final remark appears to us essential. In this review we have presented the various K^+ channel openers as an uniform class of drugs sharing virtually the same pharmacological profile. Although we do not yet possess sufficient arguments to discriminate satisfactorily between the numerous available compounds, it is not an original prophecy to predict that K^+ channel openers are not all created equal and this is certainly true when chemistry is taken into consideration. Therefore, future pharmacological work, which is full of promise, is needed to clarify this point.

The authors wish to thank Evelyne Chazot for her admirable patience to type the numerous versions of this manuscript; Karen Pepper for restyling our English and Drs J. Evans, P. Hicks, T. Nakajima, P.E. Puddu, U. Quast, J. Randall, A. Roach and A. Weston for discussion and criticism of our manuscript.

References

1. Hirst GDS, Edwards FR (1989) Sympathetic neuroeffector transmission in arteries and arterioles. Physiol Rev 69: 546-604
2. Richer C, Pratz J, Mulder P, Mondot S, Giudicelli JF, Cavero I (1990) Cardiovascular and biological effects of K^+ channel openers, a class of drugs with vasorelaxant and cardioprotective properties. Life Sci 47: 1693-1705
3. Gopalakrishnan M, Janis RA, Triggle DJ (1993) ATP-Sensitive K^+ channels: Pharmacologic properties, regulation, and therapeutic potential. Drug Develop Res 28: 95-127
4. Weston AH, Edwards G (1992) Recent progress in potassium channel opener pharmacology. Biochem Pharmacol 43: 47-54
5. Cook NS (1990) Potassium channels. Structure, classification, function and therapeutic potential. Ellis Horwood Ltd, Chichester
6. Longman S, Hamilton TC (1992) Potassium channel activator drugs: mechanism of action, pharmacological properties, and therapeutic potential. Med Res Rev 12: 73-148
7. Quast U (1992) Potassium channel openers: pharmacological and clinical aspects. Fundam Clin Pharmacol 6: 279-293
8. Edwards G, Weston AH (1990) Structure-activity relationships of K^+ channel openers. Trends Pharmacol Sci 11: 417-422
9. Bower BD, Rayner PHW, Stimmler L (1967) Leucine-sensitive hypoglycemia treated with diazoxide. Arch Dis Child 42: 410-415
10. Quast U, Cook NS (1989) In vitro and in vivo comparison of two K^+ channel openers, diazoxide and cromakalim, and their inhibition by glibenclamide. J Pharmacol Exp Ther 250: 261-271
11. Trube G, Rorsman P, Ohno-Shosaku T (1986). Opposite effects of tolbutamide and diazoxide on the ATP-dependent K^+ channel in mouse pancreatic ß-cells. Pflügers Arch 407: 493-499
12. Meisheri KD, Cipkus LA, Taylor CJ (1988) Mechanism of action of minoxidil sulfate-induced vasodilatation: a role for increased K^+ permeability. J Pharmacol Exp Ther 245: 751-760
13. Buhl AE, Waldon DJ, Conrad SJ, Mulholland MJ, Shull KL, Kubicek MF, Johnson GA, Brunden MN, Stefanski KJ, Stehle RG, Gadwood RC, Kamdar BV, Thomasco LM, Schostarez HJ, Schwartz TM, and Diani AR (1992) Potassium channel conductance: A mechanism affecting hair growth both in vitro and in vivo. J Invest Dermatol 98: 315-319
14. Meisheri KD, Swirtz MA, Purohit SS, Cipkus-Dubray LA, Khan SA, Oleynek JJ (1991) Characterization of K^+ channel-dependent as well as-independent components of pinacidil-induced vasodilation. J Pharmacol Exp Ther 256: 492-499
15. Friedel HA, Brogden RN (1990) Pinacidil. A review of its pharmacodynamic and pharmacokinetic properties, and therapeutic potential in the treatment of hypertension. Drugs 39: 929-967

16. Yanagisawa T, Satoh K, Taira N (1979) Circumstantial evidence for increased potassium conductance of membrane of cardiac muscle by 2-nicotinamidoethyl nitrate (SG-75). Japan J Pharmacol 29: 687-694
17. Taira N (1989) Nicorandil as a hybrid between nitrates and potassium channel activators. Am J Cardiol 63: 18J-24J
18. Borg C, Mondot S, Mestre M, Cavero I (1991) Nicorandil: differential contribution of K^+ channel opening and guanylate cyclase stimulation to its vasorelaxant effects on various endothelin-1-contracted arterial preparations. Comparison to Aprikalim (RP 52891) and nitroglycerin. J Pharmacol Exp Ther 259: 526-534
19. Current drugs. Potassium channel modulators handbook (1991)
20. Mc Caully RJ (1991) Variations on the benzopyran nucleus. Current drugs: potassium channel modulators: KCM-B5-KCM-B19
21. Nakajima T, Shinohara T, Yaoka O, Fukunari A, Shinagawa K, Aoki, Katoh A, Yamanaka T, Setoguchi M, Tahara T (1992) Y-27,152, a long-acting K^+ channel opener with less tachycardia: antihypertensive effects in hypertensive rats and dogs in conscious state. J Pharmacol Exp Ther 261: 730-736
22. Cavero I, Aloup JC, Mondot S, Le Monnier de Gouville AC, Mestre M (1991) Cardiovascular pharmacology of the carbothioamide K^+ channel opener RP 49356 and its active enantiomer, aprikalim, RP 52891. Current drugs: potassium channel modulators: KCM-B70-KCM-B81
23. Aloup JC, Farge D, James C, Mondot S, Cavero I (1990) 2-(3-pyridyl)-tetrahydrothiopyran-2-carbothioamide derivatives and analogues: a novel family of potent potassium channel openers. Drugs Future 15: 1097-1108
24. Sturgess NC, Ashford MLJ, Cook DL, Hales CN (1985) The sulphonylurea receptor may be an ATP-sensitive potassium channel. Lancet 2: 474-475
25. Bray K, Quast U (1992) Differential inhibition by tedisamil (KC 8857) and glibenclamide of the responses to cromakalim and minoxidil sulphate in rat isolated aorta. Naunyn-Schmiedeberg's Arch Pharmacol 345: 244-250
26. Noack Th, Edwards G, Deitmer P, Greengrass P, Morita T, Andersson PO, Criddle D, Wyllie MG, Weston AH (1992) The involvement of potassium channels in the action of ciclazindol in rat portal vein. Br J Pharmacol 106: 17-24
27. Mc Pherson GA, Angus JA (1990) Characterization of responses to cromakalim and pinacidil in smooth and cardiac muscle by use of selective antagonists. Br J Pharmacol 100: 201-206
28. Cipkus-Dubray L, Swirtz M, Khan S, Humphrey S, Skaletzky L, Meisheri K. U-37,883A: a structurally novel antagonist of the vascular KATP openers. FASEB J 6: Abs 4869
29. Mc Cullough JR, Normandin DE, Lee Conder M, Sleph PG, Dzwonczyk S, Grover GJ (1991) Specific blook of the anti-ischemic actions of cromakalim by sodium 5-hydroxydecanoate. Circ Res 69: 949-958
30. Itoh K, Seki N, Suzuki S, Ito S, Kajikuri J, Kuriyama H (1992) Membrane hyperpolarisation inhibits agonist-induced synthesis of inositol 1, 4, 5-trisphosphate in rabbit mesenteric artery. J Physiol (Lond) 451: 307-328
31. Weir SW, Weston AH (1986) The effects of BRL 34915 and nicorandil on electrical and mechanical activity and on ^{86}Rb efflux in rat blood vessels. Br J Pharmacol 88: 121-128
32. Cavero I, Mondot S, Mestre M (1989) Vasorelaxant effects of cromakalim in rats are mediated by glibenclamide-sensitive potassium channels. J Pharmacol Exp Ther 248: 1261-1268
33. Cavero I, Pratz J, Mondot S (1991) K^+ channel opening mediates the vasorelaxant effects of nicorandil in the intact vascular system. Z Kardiol 80: 35-41
34. Cook NS, Weir SW, Danzeisen MC (1988) Anti-vasoconstrictor effects of the K^+ channel opener cromakalim on the rabbit aorta: comparison with the calcium antagonist isradipine. Br J Pharmacol 95: 741-752
35. Eltze M (1989) Glibenclamide is a competitive antagonist of cromakalim, pinacidil and RP 49356 in guinea-pig pulmonary artery. Eur J Pharmacol 165: 231-239

36. Hamilton TC, Weir SW, Weston AH (1986) Comparison of the effects of BRL 34915 and verapamil on electrical and mechanical activity in rat portal vein. Br J Pharmacol 88: 103-111
37. Cocks TM, King SJ, Angus JA (1990) Glibenclamide is a competitive antagonist of the thromboxane A2 receptor in dog coronary artery in vitro. Br J Pharmacol 100: 375-378
38. Haeusler G (1983) Contraction, membrane potential, and calcium fluxes in rabbit pulmonary arterial muscle. Fed Proc 42: 263-26839
39. Hermsmeyer K (1976) Electrogenesis of increased norepinephrine sensitivity arterial vascular muscle in hypertension. Circ Res 38: 362-367
40. Lawson K, Barras M, Zazzi-Sudriez E, Martin DJ, Armstrong JM, Hicks PE (1992). Differential effects of endothelin-1 on the vasorelaxant properties of benzopyran and non-benzopyran potassium channel openers. Br J Pharmacol 107: 58-65
41. Cain CR, Nicholson CD (1989) Comparison of the effects of cromakalim, a potassium conductance enhancer, and nimodipine, a calcium antagonist, on 5-hydroxytryptamine responses in a variety of vascular smooth muscle preparations. Naunyn-Schmiedeberg's Arch Pharmacol 340: 293-299
42. Mc Carron JG, Quayle JM, Halpern W, Nelson MT (1991) Cromakalim and pinacidil dilate small mesenteric arteries but not small cerebral arteries. Am J Physiol 261: H287-H291
43. Wilson C, Cooper SM (1989) Effect of cromakalim on contractions in rabbit isolated renal artery in the presence and absence of extracellular Ca^{2+}. Br J Pharmacol 98: 1303-1311
44. Taylor SG, Southerton JS, Weston AH, Baker JRJ (1988) Endothelium-dependent effects of acetylcholine in rat aorta: a comparison with sodium nitroprusside and cromakalim. Br J Pharmacol 94: 853-863
45. Doggrell SA, Smith JW, Downing OA, Wilson KA (1989) Hyperpolarizing action of cromakalim on the rat aorta. Eur J Pharmacol 174: 131-133
46. Bray KM, Weston AH, Duty S, Newgreen DT, Longmore J, Edwards G, Brown TJ (1991) Differences between the effects of cromakalim and nifedipine on agonist-induced responses in rabbit aorta. Br J Pharmacol 102: 337-344
47. Nakao K, Okabe K, Kitamura H, Kuriyama H, Weston AH (1988) Characteristics of cromakalim-induced relaxations in the smooth muscle cells of guinea-pig mesenteric artery and vein. Br J Pharmacol 95: 795-804
48. Videbæck LM, Aalkjær C, Mulvany MJ (1988) Pinacidil opens K^+-selective channels causing hyperpolarization and relaxation of noradrenaline contractions in rat mesenteric resistance vessels. Br J Pharmacol 95: 103-108
49. Leblanc N, Wilde DW, Keef KD, Hume JR (1989) Electrophysiological mechanisms of minoxidil sulfate-induced vasodilation of rabbit portal vein. Circ Res 65: 1102-1111
50. Furukawa K, Itoh T, Kajiwara M, Kitamura K, Suzuki H, Ito Y, Kuriyama H (1981) Vasodilating actions of 2-nicotinamidoethyl nitrate on porcine and guinea-pig coronary arteries. J Pharmacol Exp Ther 218: 248-259
51. Quast U, Baumlin Y (1988) Comparison of the effluxes of $^{42}K^+$ and $^{86}Rb^+$ elicited by cromakalim (BRL34915) in tonic and phasic vascular tissue. Naunyn-Schmiedeberg's Arch Pharmacol 338: 319-326
52. Edwards G, Henthorn M, Weston AH (1990) Some in vitro effects of potassium channel openers on rat bladder and portal vein. Eur J Pharmacol 183: 2408-2409
53. Lodge NJ, Cohen RB, Havens CN, Colatsky TJ (1991) The effects of the putative potassium channel activation WAY-120, 491 on ^{86}Rb efflux from the rabbit aorta. J Pharmacol Exp Ther 256: 639-644
54. Kreye VA, Lenz T, Theiss U (1991) The dualistic mode of action of the vasodilator drug, nicorandil, differentiated by glibenclamide in ^{86}Rb flux studies in rabbit isolated vacular smooth muscle. Naunyn-Schmiedeberg's Arch Pharmacol 343: 70-75
55. Bray K, Quast U (1991) Some degree of overlap exists between the K^+ channels opened by cromakalim and those opened by minoxidil sulphate in rat isolated aorta. Naunyn-Schmiedeberg's Arch Pharmacol 344: 351-359

56. Newgreen DT, Bray KM, Mc Harg AD, Weston AH, Duty S, Brown BS, Kay PB, Edwards G, Longmore J, Southerton JS (1990) The action of diazoxide and minoxidil sulphate on rat blood vessels: a comparison with cromakalim. Br J Pharmacol 100: 605-613
57. Bray K, Quast U (1992) A specific binding site for K^+ channel openers in rat aorta. J Biol Chem 267: 11689-11692
58. Gopalakrishnan M, Johnson DE, Janis RA, Triggle DJ (1991) Characterization of binding of the ATP-sensitive potassium channel ligand, [^{3}H] glyburide, to neuronal and muscle preparations. J Pharmacol Exp Ther 257: 1162-1171
59. Fosset M, De Weille JR, Green RD, Schmid-Antomarchi H, Lazdunski M (1988) Antidiabetic sulphonylureas control action potential properties in heart cells via high affinity receptors that are linked to ATP-dependent K^+ channels. J Biol chem 263: 7933-7936
60. Geisen K, Hitzel V, Ökomonopoulos R, Pünter J, Weyer R, Summ HD (1985) Inhibition of ^{3}H-Glibenclamide-binding to sulphonylurea receptors by oral antidiabetics. Arzneim Forsch / Drug Res 35: 707-712
61. Bernardi H, Fosset M, Lazdunski M (1988) Characterization, purification, and affinity labeling of the brain ^{3}H-Glibenclamide-binding protein, a putative neuronal ATP-regulated K^+ channel. Proc Natl Acad Sci USA 85: 9816-9820
62. Kajioka S, Nakashima M, Kitamura K, Kuriyama H (1991) Mechanisms of vasodilatation induced by potassium-channel activators. Clin Sci 81: 129-139
63. Noack T, Deitmer P, Edwards G, Weston AH (1992) Characterization of potassium currents modulated by BRL 38277 in rat portal vein. Br J Pharmacol 106: 717-726
64. Nelson MT, Patlak JB, Worley JF, Standen NB (1990) Calcium channels, potassium channels, and voltage dependence of arterial smooth muscle tone. Am J Physiol 259: C3-C18
65. Silberberg SD, van Breemen C (1992) A potassium current activated by lemakalim and metabolic inhibition in rabbit mesenteric artery. Pflügers Arch 420: 118-120
66. Clapp LH, Gurney AM (1992) ATP-sensitive K^+ channels regulate resting potential of pulmonary arterial smooth muscle cells. Am J Physiol 262: H916-H920
67. Post JM, Jones AW (1991) Stimulation of arterial ^{42}K efflux by ATP depletion and cromakalim is antagonized by glyburide. Am J Physiol 260: H 848-H854
68. Standen NB, Quayle JM, Davies NW, Brayden JE, Huang Y, Nelson MT (1989) Hyperpolarizing vasodilators activate ATP-sensitive K^+ channels in arterial smooth muscle. Science 245: 177-180
69. Pratz J, Mondot S, Montier F, Cavero I (1991) Effects of the K^+ channel activators, RP 52891, cromakalim and diazoxide, on the plasma insulin level, plasma renin activity and blood pressure in rats. J Pharmacol Exp Ther 258: 216-222
70. Von Beckerath N, Cyrys S, Dischner A, Daut J (1991) Hypoxic vasodilatation in isolated, perfused guinea-pig heart: an analysis of the underlying mechanisms. J Physiol 442: 297-319
71. Clapham JC, Hamilton TC, Longman SD, Buckingham RE, Campbell CA, Ilsley GL, Gout B (1991) Antihypertensive and haemodynamic properties of the potassium channel activating (-)- enantiomer of cromakalim in animal models. Arzneim Forsch/Drug Res 41: 385-391
72. Richer C, Mulder P, Doussau MP, Gautier P, Giudicelli JF (1990) Systemic and regional haemodynamic interactions between K^+ channel openers and the sympathetic nervous system in the pithed SHR. Br J Pharmacol 100: 557-563
73. Buckingham RE (1988) Studies on the anti-vasoconstrictor activity of BRL 34915 in spontaneously hypertensive rats ; a comparison with nifedipine. Br J Pharmacol 93: 541-552
74. Ichinose M, Barnes PJ (1990) A potassium channel activator modulates both exitatory non-cholinergic and cholinergic neurotransmission in guinea pig airways. J Pharmaco Exp Ther 252: 1207-1212
75. Gardiner SM, Kemp PA, Bennett T (1991) Effects of N^G-nitro-L-arginine methyl ester on vasodilator responses to adrenalin or BRL 38227 in conscious rats. Br J Pharmacol 104: 731-737
76. Hof RP, Quast U, Cook NS, Blarer S (1988) Mechanism of action and systemic and regioanl hemodynamics of the potassium channel activator BRL 34915 and its enantiomers. Circ Res 62: 679-686

77. Gotanda K, Yokoyama H, Satoh K, Taira N (1989) Cardiohemodynamic effects of cromakalim and pinacidil, potassium channel openers, in the dog, with special reference to venous return. Cardiovasc Drugs Ther 3: 507-515
78. Sassen LMA, Duncker DJGM, Gho BCG, Diekmann HW, Verdouw PD (1990) Haemodynamic profile of the potassium channel activator EMD 52692 in anaesthetized pigs. Br J Pharmacol 101: 605-614
79. Hayashi K, Matsumura Y, Yoshida Y, Ohyama T, Hisaki K, Suzuki Y, Morimoto S (1990) Effects of BRL 34915 (cromakalim) on renal hemodynamics and function in anesthetized dogs. J Pharmacol Exp Ther 252: 1240-1246
80. Giudicelli JF, Drieu la Rochelle C, Berdeaux A (1990) Effects of cromakalim and pinacidil on large epicardial and small coronary arteries in conscious dogs. J Pharmacol Exp Ther 255: 836-842
81. Sakamoto S, Liang CS, Stone CK, Hood WB (1989) Effects of pinacidil on myocardial blood flow and infarct size after acute left anterior descending coronary artery occlusion and reperfusion in awake dogs with and without a coexisting left circumflex coronary artery stenosis. J Cardiovas Pharmacol 263: H399-H404
82. Nakajima T (1991) Y-27,152: a long-acting K^+ channel opener with less incidence of tachycardia. Cardiovasc Drug Rev 9: 372-384
83. Imamura Y, Tomoike H, Narishige T, Takahashi T, Kasuya H, Takeshita A (1992) Glibenclamide decreases basal coronary blood flow in anesthetized dogs. Am J Physiol 263: H399-H404
84. Aversano T, Ouyang P, Silverman H (1991) Blockade of the ATP-sensitive potassium channel modulates reactive hyperemia in the canine coronary circulation. Circ Res 69: 618-622
85. Clayton FC, Hess TA, Smith MA, Grover GJ (1992) Coronary reactive hyperemia and adenosine-induced vasodilation are mediated partially by a glyburide-sensitive mechanism. Pharmacology 44: 92-100
86. Komaru T, Lamping KG, Eastham CL, Dellsperger KC (1991) Role of ATP-sensitive potassium channels in coronary microvascular autoregulatory responses. Circ Res 69: 1146-1151
87. Paciorek PM, Burden DT, Burke YM, Cowlrick IS, Perkins RS, Taylor JC, Waterfall JF (1990) Preclinical pharmacology of Ro 31-6930, a new potassium channel opener. J Cardiovasc Pharmacol 15: 188-197
88. Ferrier CP, Kurtz A, Lehner P, Shaw SG, Pusterla C, Saxenhofer H, Weidmann P (1989) Stimulation of renin secretion by potassium channel activation with cromakalim. Eur J Clin Pharmacol 36: 443-447
89. Kurtz A (1986) Intracellular control of renin release—An overview. Klin Vochenschr 64: 838-846
90. Petit P, Loubatières-Mariani MM (1992) Potassium channels of the insulin-secreting ß-cell. Fundam Clin Pharmacol 6: 123-134
91. Lebrun P, Antoine MH, Herchuelz A (1992) K^+ channel openers and insulin release. Life Sci 51: 795-806

CHAPTER 13
Potassium channel openers in the heart

D Escande and I Cavero

K^+ channel activation is a natural and effective cellular mechanism that mediates most of the biological effects of acetylcholine and adenosine on the heart muscle [1]. By binding onto atrial muscarinic (M1) and adenosine (A1) receptors, respectively, acetylcholine and adenosine activate K^+ channels of small unitary conductance through a G_i protein-regulated process. The consequences of K^+ channel activation in the atrium are decreased automaticity, decreased action potential duration and refractoriness, decreased contractility and slowed atrio-ventricular conduction. Apart from natural mediators, several synthetic K^+ channel openers have recently been identified (see chapter 12). These substances form a new class of chemically diverse drugs including benzopyran (e.g. cromakalim), guanidine (e.g. pinacidil), pyridine (e.g. nicorandil), pyrimidine (e.g. minoxidil), benzothiadiazepine (e.g. diazoxide) and thioformamide (e.g. aprikalim) derivatives. As a rule, all synthetic K^+ channel openers discovered so far are much more potent in smooth muscle than in striated muscle. Therefore, the primary effect of K^+ channel openers is smooth muscle relaxation in the vascular network, in the airways and in the urinary track. However, at much higher concentrations than those producing smooth muscle relaxation, K^+ channel openers are also able to open K^+ channels in the normally-oxygenated myocardium. Although of no therapeutic significance, this particularity has been advantageously exploited by cellular electrophysiologists to identify the category of K^+ channels opened by K^+ channel openers [2-4].

Under conditions of impaired cellular metabolism such as ischaemia, the situation is entirely different since striated muscle K^+ channels become much more sensitive to K^+ channel openers. This striking behaviour is the basis of the cytoprotective properties of K^+ channel openers observed experimentally both in the heart and in the skeletal muscle (see also chapter 18). Pharmacological protection against myocardial damage caused by ischaemia and reperfusion is currently a field of intensive research, partly as a consequence of the success of thrombolytic therapy in myocardial infarction. The prospect that potassium channel openers exert direct anti-ischaemic effects through a "naturally" operating protective mechanism [5] opens an exciting field of clinical research. This concept also provides a new approach for the treatment of ischaemic heart disease.

Only very high concentrations of K^+ channel openers activate K^+ channels in normally-oxygenated myocardium

As already stated above, concentrations of K^+ channel openers producing vasodilation exert virtually no effect on the normally-oxygenated myocardium. In dog papillary muscle perfused with blood, the threshold dose of cromakalim leading to decreased contractility is around 300 times that increasing coronary flow (fig. 1) [6]. In canine Purkinje and myocardial cells, at least 30 µM pinacidil is needed to abbreviate the action

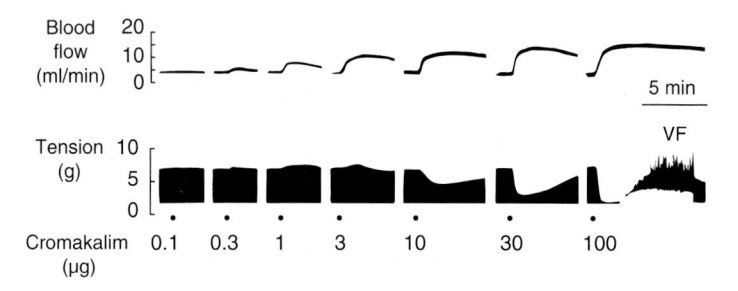

Fig. 1. Comparative effects of cromakalim on coronary blood flow through the anterior septal artery (top) and on developed tension (bottom) of a dog papillary muscle preparation perfused in vitro. Note that increased coronary blood flow occurred at much lower concentrations than decreased force. VF is ventricular fibrillation which ensued at 100 µg of cromakalim. Reproduced from [6] with permission

potential [7] whereas in clinical trials performed in hypertensive patients treated with maximal doses of pinacidil, only plasma levels between 0.3 - 1 µM could be achieved. In guinea-pig papillary muscle [8] and in canine Purkinje fibres [9], the minimal concentrations of nicorandil that shorten the action potential are 100 and 10 µM respectively (fig. 2), whereas in man antianginal effects of nicorandil are obtained with plasma concentrations ranging from 0.5 to 1.5 µM.

The predominant effect of high concentrations of K^+ channel openers in the heart muscle is shortening of the action potential and refractory period; i.e. the inverse of the effect of a K^+ channel blocker (see chapters 4 and 11). Since the plateau is abbreviated, the period of time available for Ca^{2+} to enter the cell diminishes and thus the developed peak force also diminishes. K^+ channel openers only slightly modify the other parameters of the myocardial action potential: (i) the resting membrane potential, V_m, hyperpolarizes by just a few millivolts because V_m is already close to the equilibrium potential for K^+ and; (ii) the upstroke velocity and the action potential amplitude are usually unchanged. When even higher concentrations of K^+ channel openers are used, the ultimate consequence of action potential shortening is the arrest of both the electrical and mechanical activities of the cardiac fibres. In spontaneously firing Purkinje fibres, there is a reduction in the frequency of the automatic activity and eventually its cessation because the activation of K^+ channels creates an additional outward current during diastole that opposes the depolarizing influence of the inward currents. By contrast, cromakalim, bimakalim and pinacidil produce only slight effects on pacemaker action potential and automaticity in sinoatrial node cells [10].

The target for K^+ channel openers in the heart is the ATP-sensitive K^+ channel

In cardiac cells, a series of reports demonstrate that K^+ channel openers selectively target ATP-sensitive K^+ channels (fig. 3) which are normally closed under normoxic conditions (i_{K-ATP} channels; see chapter 3). The increase in the outward current together with the modifications in the action potential produced by K^+ channel openers can be fully antagonized by sulphonylureas such as glibenclamide and tolbutamide, which are specific

Fig. 2. Effects of nicorandil on guinea-pig papillary muscle action potentials driven at various stimulation rates. The upper, middle and lower traces for each panel are the zero potential, the transmembrane potential and dV/dt. Same cell throughout. Note that action potential shortening was obtained at very high concentrations of nicorandil. Reproduced from [8] with permission

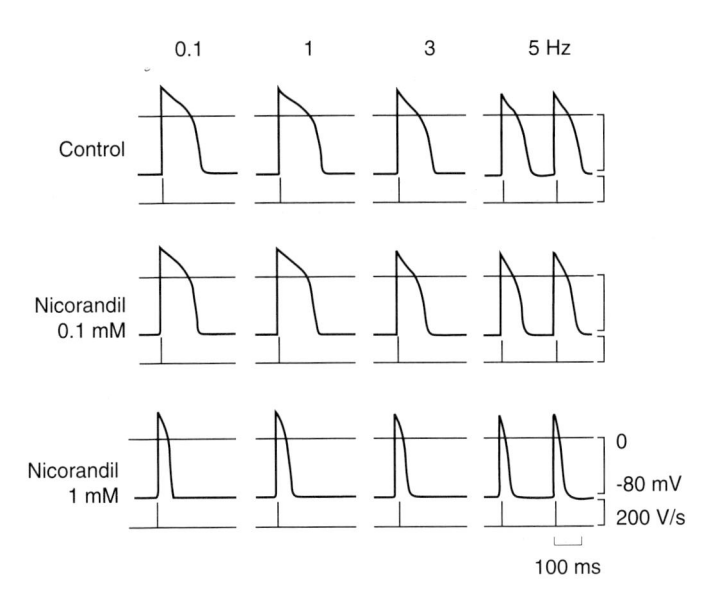

blockers of ATP-sensitive K$^+$ channels [11] (see chapters 10 and 11). At the single channel level, K$^+$ channel openers activate ATP-sensitive K$^+$ channels but fail to affect the inward rectifier K$^+$ channels, the Na$^+$-activated K$^+$ channels or the muscarinic K$^+$ channels [12, 13]. The fact that ATP-sensitive K$^+$ channels are activated in intact cells suggests that the drugs are able to relieve the natural block produced by intracellular ATP. Furthermore, in detached patches, the opening effects of K$^+$ channel openers depend strongly on the ATP concentration applied on the intracellular side of the membrane [12-17]; if no ATP is present, then the activity of ATP-sensitive K$^+$ channels is maximal and is not influenced by K$^+$ channel openers [12]. Although in patch-clamp experiments K$^+$ channel openers behave as if they competitively displace ATP from its binding site, this behaviour needs further biochemical demonstration. More recently, a specific binding site for K$^+$ channel openers has been found in rat aortic rings [18]; although the precise nature of the channel activated by K$^+$ channel openers in vascular smooth muscle cells may be somewhat different from that opened in cardiac cells (see chapter 12), it would be of interest to know whether the binding of K$^+$ channel openers to this receptor is influenced by tricyclic nucleotides. Nevertheless, in patch-clamp experiments, cardiac ATP-sensitive K$^+$ channels open at a higher ATP concentration in the presence of a K$^+$ channel opener than in its absence. The consequence of this mechanism is that the effectiveness of a K$^+$ channel opener on contractility and electrical activity critically depends on the intracellular ATP concentration and therefore on the metabolic state of the cells. Ripoll et al. [16] have shown that 100 µM RP 49356 (a racemate of which aprikalim is the active enantiomer) produces no effect on the mechanical activity of isolated rat ventricular cells in the presence of extracellular glucose. In contrast, the same concentration markedly diminishes contractility when extracellular glucose is replaced by 2-deoxyglucose, which inhibits the glycolytic metabolism.

Several investigators have observed that, in intact cells, the effect of K$^+$ channel openers on ATP-sensitive K$^+$ channels is strongly temperature-dependent, inasmuch as their

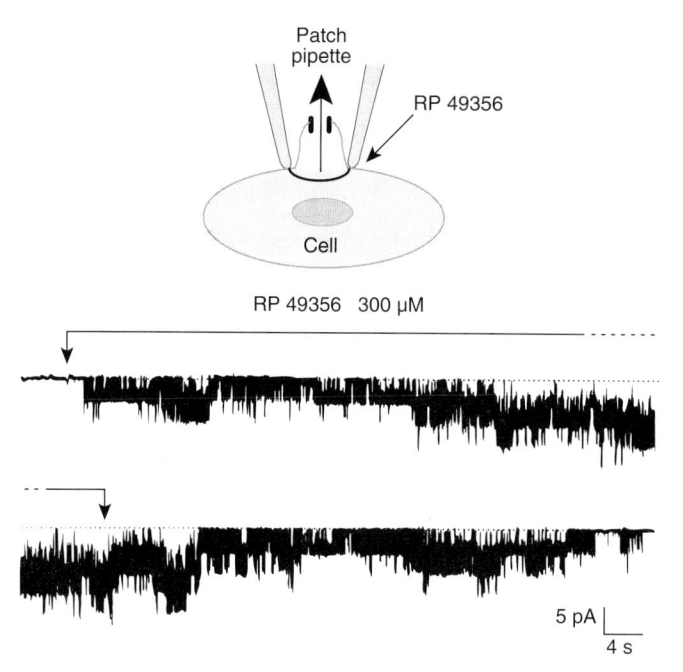

Fig. 3. Externally-applied RP 49356 (a racemate of which aprikalim is the active enantiomer) opened ATP-sensitive K⁺ channels in a cell-attached patch held at -60 mV. K⁺ concentration in the pipette and in the external medium was 140 mM. Downward deflections are inward single channel currents. Guinea-pig ventricular myocytes. Note the absence of channel activity before RP 49356 was applied. Reproduced from [2] with permission

opening is more marked and rapid at 37°C than at 20°C. The mechanism responsible for this particularity is not known at present. One possibility is that the binding site for K⁺ channel openers is located on the intracellular side of the membrane. Indeed, externally applied K⁺ channel openers open ATP-sensitive K⁺ channels recorded in the cell-attached configuration [12](fig. 3). The consequence is that the drug must penetrate the cell before activating the channels; this process may depend on the fluidity of the membrane, a temperature-dependent parameter. Another possibility is that the intracellular ATP concentration declines as the temperature increases thus increasing the effects of K⁺ channel openers.

K⁺ channel openers exert cardioprotection during acute ischaemia

Since K⁺ channel openers are more active in metabolically impaired myocardium, it is possible to determine concentrations that do not alter the baseline activity of the cardiac muscle but which will be operative during ischaemia.

Experimental evidence for cytoprotection has been obtained in vitro

In an in vitro perfused rat heart model of transient global ischaemia, nicorandil improves the mechanical function of the myocardium during reperfusion and prevents the loss of total adenine nucleotides that occurs as a result of ischaemia and reperfusion [19]. This cytoprotective property is also shared by pinacidil, cromakalim [20] and aprikalim, and

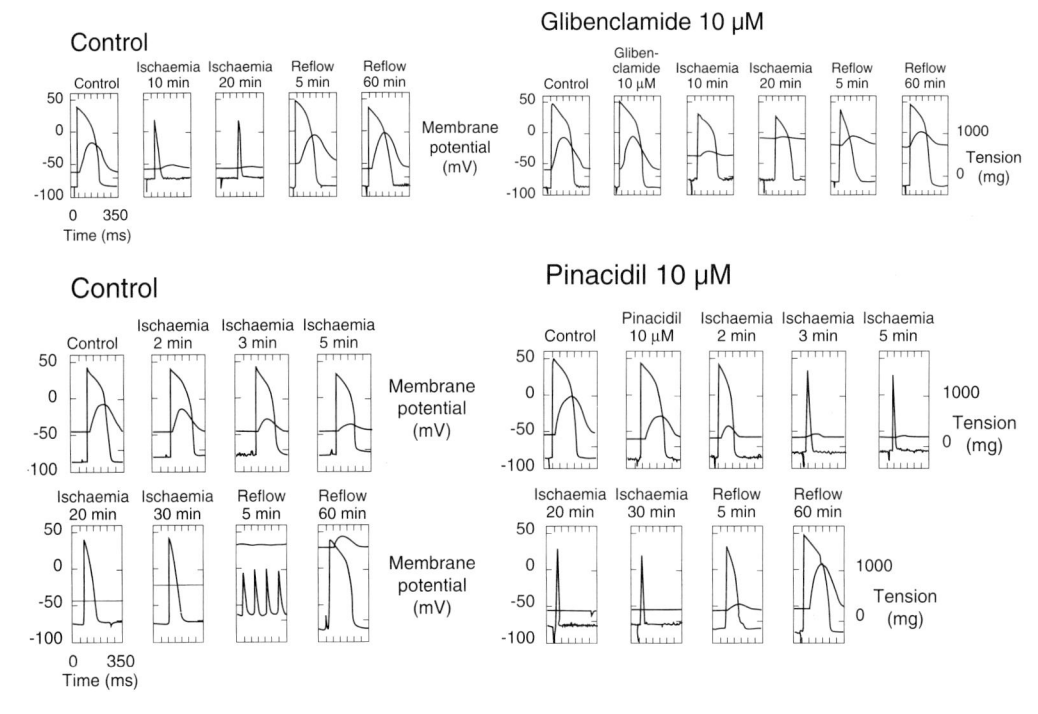

Fig. 4. Effects of glibenclamide 10 µM and pinacidil 10 µM on electrical and mechanical activity of guinea-pig right ventricular wall during 20 min (upper panel) and 30 min (lower panel) of ischaemia and reperfusion. Note: (i) the absence of a decline in action potential duration and the increased diastolic tension in the presence of glibenclamide; (ii) the accelerated action potential shortening during ischaemia and the improved recovery of mechanical function during reperfusion in the presence of pinacidil. Reproduced from [52] with permission

is observed both under constant pressure and constant flow conditions, making a major haemodynamic participation in cytoprotection unlikely. Additional experiments with cromakalim have demonstrated that, like its vasodilating properties, the cardioprotective effects are stereoselective [21]. Furthermore, cardioprotection produced by K^+ channel openers is completely reversed by glibenclamide or 5-hydroxydecanoate; stereoselectivity and reversal by K^+-ATP channel blockers strongly suggest the involvement of ATP-sensitive K^+ channels in the mechanism of cytoprotection. K^+ channel openers lead to cardioprotection in the absence of significant cardiodepression before ischaemia, making K^+ channel openers somewhat different from the calcium antagonists.

In the arterially perfused isolated ventricle of the guinea-pig, pinacidil reduces myocardial damage caused by ischaemia-reperfusion (fig. 4; see [52]). In contrast, glibenclamide alone worsens the consequences of ischaemia on the mechanical function of the myocardium.

Cytoprotection has also been demonstrated in vivo

In vivo preclinical models for assessing the therapeutic potential of compounds aimed to prevent or attenuate ischaemic damage attempt to replicate human pathological situations resulting from regional ischaemic insults of variable duration. Any ischaemic insult lasting from 5 to 20 min (the duration is a function of the animal species) is generally accompanied by reversible biochemical, functional, electrical, and ultra-structural alterations, although full recovery to normality can require from several hours to days. In contrast, ischaemic periods lasting over 30 min produce myocardial lesions of an irreversible nature (infarction) and sometimes fibrillation whether or not they are followed by reperfusion. It is also possible to cause cardiac damage by subjecting the heart to excessive energy expenditure (e.g. by overpacing or intensive exercise in the presence or absence of a critical coronary stenosis) which cannot be matched by energy production due to insufficient oxygen delivery.

K^+ channel openers have been studied in most of the available models of myocardial ischaemia in dogs, pigs, rats and rabbits. A number of results demonstrate that these compounds can afford cardioprotection which is characterized by reduced cardiac damage as assessed by functional, biochemical and anatomical parameters.

K^+ channel openers accelerate the recovery of contractility in the stunned myocardium

Exposure of the myocardium to a brief period of ischaemia produces reversible injury in that reperfusion can prevent the death of myocytes which would otherwise occur. The injury consists of reduced systolic shortening, early systolic bulging, ultrastructural changes and decreased high energy phosphate levels in the cell. The transient post-ischaemic myocardial contractile dysfunction is termed stunning and can last for over 24 hours [22]. The effects of pharmacological treatments on the stunned myocardium are generally studied in anaesthetized dogs and pigs, although conscious dogs have also been used. The stunning insult is produced by occluding the left circumflex or the left anterior descending coronary artery for a period of time spanning from 5 to 20 min, followed by reperfusion of the heart for several hours. The parameter mostly used to assess myocardial damage is the regional contractile activity which is determined with sets of two piezoelectric crystals inserted ~10 mm apart and 7-8 mm deep in the subendocardium of the normoxic and ischaemic regions. The difference between end diastolic and end systolic distance determined with each pair of crystals provides a measure of myocardial segment shortening which is a reliable index of myocardial contractile function. This parameter is expressed either as an absolute value or as a percentage of baseline (pre-ischaemia) shortening. In these experiments, aortic blood pressure, coronary blood flow, electrocardiogram and heart rate (kept constant by pacing) are routinely measured. Furthermore, collateral blood flow to the ischaemic region is measured particularly in dogs, since values of this parameter greater than 0.15 ml/min/g (fresh tissue) result in only a small amount of contractile dysfunction. In order to obtain this latter information myocardial blood flow in the subendocardial and subepicardial regions is assessed with radioactive microspheres. Dogs with a markedly developed coronary collateral circulation are discarded from data analysis. The area at risk is expressed as a percentage of total left ventricle wall weight [23].

The oldest available results on a possible cardioprotective effect of K^+ channel opening are those obtained with nicorandil. Pretreatment of dogs with nicorandil before the

Fig. 5. Segment shortening (expressed as percentages of baseline values) within a left ventricle area before, during and after a 15 min occlusion of the left anterior descending coronary artery in dogs treated with saline *(VE)*, nicorandil *(NIC)* (100 μg/kg i.v. bolus followed by 25 μg/kg/min for the duration of the experimental procedure) or glibenclamide *(GL:* 0.3 mg/kg i.v.) alone or followed by nicorandil. Note the improved mechanical recovery after pretreatment with nicorandil. Redrawn from [23]

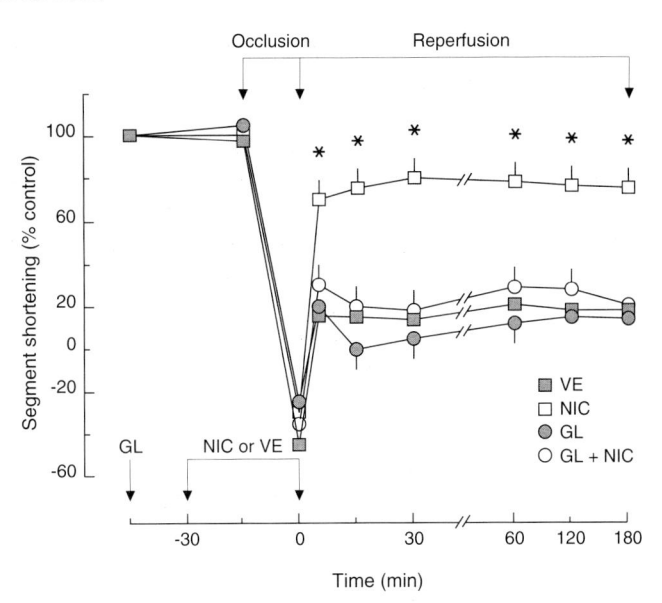

transient ischaemic insult produced greatly accelerated and impressive recovery of cardiac function toward normal on reperfusion (fig. 5) [23-25]. In matched experiments, nitroglycerin [24] and nifedipine [25] at doses producing similar hypotensive effects (20 - 30 mmHg) as nicorandil were also studied. Nifedipine did not reduce myocardial stunning whereas nitroglycerin was marginally beneficial (fig. 6). Nicorandil treatment attenuated the loss of high energy phosphates within the myocardial ischaemic region in anaesthetized dogs [26]. The efficacy of nicorandil in improving the tension-generating ability of the stunned myocardium was also demonstrated in conscious dogs [27], hence negating a possible contribution of barbiturate anaesthesia to this beneficial action. In these early publications, the possible involvement of K^+ channels in the cardioprotective action of nicorandil was not suspected. However, the recent finding that glibenclamide pretreatment abolished the beneficial effects of nicorandil, indicates that the improvement in myocardial function following a brief ischaemic insult after nicorandil treatment should be attributed to the opening of K^+ channels on cardiac myocytes (fig. 5) [23]. In addition, since in these experiments glibenclamide did not antagonize the moderate hypotensive effects produced by the dose of nicorandil studied, we cannot agree with the conclusion of Grover et al. [28] that the cardioprotective effects of nicorandil, but not of cromakalim, are mediated by a peripheral mechanism (decrease in peripheral resistance).

Studies with the novel potassium channel opener aprikalim in barbiturate anaesthetized dogs have also demonstrated that the stunned myocardial region can markedly recover its contractile function during the reperfusion period (fig. 6). In these experiments, the cardioprotective dose (10 μg/kg i.v. 5 min before ischaemia + 0.1 μg/kg/min i.v. throughout the experimental procedure) of aprikalim was entirely devoid of systemic and cardiac haemodynamic effects. However, this dose of aprikalim failed to reduce the contractile dysfunction due to the stunning insult when it was administered intravenously at the beginning of the reperfusion period. In addition, the beneficial action of aprikalim was entirely abolished by pretreating the dogs with a dose of glibenclamide

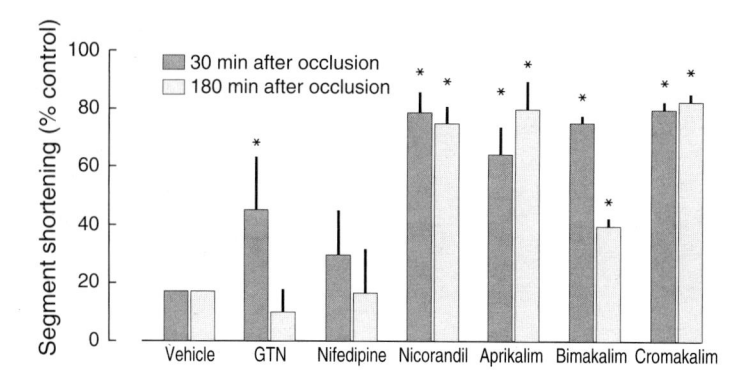

Fig. 6. Segment shortening (see fig. 5) values measured 30 and 180 min after starting reperfusion in dogs which were given (15 min before the stunning insult) vehicle, nitroglycerin *(GTN:* 10 µg/kg followed by 3.0 µg/kg/min) [24]; nifedipine (10 µg/kg followed by 3 µg/kg/min) [25]; cromakalim (1 µg/kg/min i.v. throughout the experimental procedure) [58], nicorandil (100 µg/kg i.v. bolus followed by 25 µg/kg/min [23] and bimakalim (3 µg/kg followed by 0.1 µg/kg/min). The vehicle value is the mean of control values reported in the references quoted in this legend. The value of nifedipine at 180 min has been adjusted to the mean of vehicle value reported in this figure in order to avoid the representation of individual control values of each study

(0.3 mg/kg i.v.) that by itself did not enhance ischaemic injury in this model of single stunning. However, a three-fold higher dose of glibenclamide as well as a high dose of tolbutamide worsened the contractile failure of the stunned region [29]. The low dose of glibenclamide also produced an aggravation of post-ischaemic ventricular dysfunction when it was studied in a multiple stunning model in dogs [30]. In the single stunning preparation, bimakalim (EMD 52692), a benzopyran K^+ channel opener, markedly accelerated the recovery of myocardial contraction within the first 30 min following reperfusion of the ischaemic heart [29]. However, after three hours of reperfusion this improvement was less impressive than that obtained with i.v. aprikalim, nicorandil or intracoronary cromakalim (fig. 6). The reason for this difference remains to be determined.

In conclusion, solid evidence exists in favour of a true cardioprotective activity of K^+ channel openers against a stunning insult when these compounds are administered before coronary artery occlusion. This protection is substantiated by a rapid recovery of contractile function during reperfusion of the myocardium that has been made ischaemic for a short period. Finally, the beneficial action of K^+ channel openers in the stunned myocardium is reminiscent of acute hibernation [22], which describes a quiescent, fully viable myocyte that has the potential for rapidly recovering its contractile function when it is reperfused.

K^+ channel openers have equivocal effects on infarct size

Ischaemic insults persisting for over 30 min produce death of cardiac myocytes. The size of the necrotic region is a function of several parameters such as the volume of tissue undergoing ischaemia (risk area), the duration of the ischaemic insult, the myocardial oxygen demand during ischaemia, and the degree of collateral blood flow to the ischaemic area. Furthermore, the extent of the infarct can be extended by the reperfusion process during which activated neutrophils liberate oxygen free radicals which speed up the necrotic process of severely injured myocytes.

Myocardial infarction can be produced by occlusion of the left circumflex or the left anterior descending coronary artery for periods longer than 30 min followed by prolonged reperfusion (3-6 h) in various species of experimental animals such as rats, rabbits, dogs and pigs. Generally, these studies are carried out under anesthesia, but have also been done in conscious animals. Several haemodynamic parameters (aortic blood pressure, coronary blood flow, regional blood flow) are measured. The collateral blood flow is determined from the regional blood flow results obtained with radioactive microspheres. As mentioned earlier, this is a very important parameter particularly in dogs since, if its value is elevated, it can result in a reduction in the size of the infarct. The area at risk is expressed as a percentage of the left ventricle and the infarct size is expressed as a percentage of the area at risk [31].

The effects of K^+ channel openers on infarct size are variable and range from exacerbation to reduction. Sakamoto et al. [32] reported that a dose of pinacidil which reduced blood pressure (25 mmHg) and increased heart rate (approximately 50 beats/min) and cardiac output (approximately 1.2 l/min) had no effect on the myocardial blood flow or infarct size which followed 4-h occlusion of the left anterior descending coronary artery and 20-h reperfusion in awake dogs. In contrast, the same dose of pinacidil in dogs, with a moderate stenosis (approximately 50%) of the proximal left circumflex coronary artery, lowered infarct zone myocardial blood flow and increased infarct sizes. The infarct size attained 69% of the zone risking necrosis (control group 54%) following the occlusion of the left anterior coronary artery. Thus, in this study, in the absence of moderate stenosis of the left circumflex artery, pinacidil, at a dose increasing myocardial oxygen consumption and decreasing coronary perfusion pressure, does not modify the infarct size produced by a prolonged occlusion of the left anterior descending coronary artery. In contrast, pinacidil worsens necrosis when stenosis is added to the coronary artery occlusion. This increase in infarct size is probably due to a reduction in collateral blood flow produced by pinacidil in the presence of coronary stenosis [32]. In another study carried out in anaesthetized dogs, cromakalim and celikalim infused into the coronary artery at a dose devoid of effects on coronary blood flow and blood pressure (starting 10 min before and then throughout the experimental procedure) markedly increased the size of myocardial infarct measured after 90 min of occlusion of the left circumflex coronary artery followed by 5-h reperfusion [33]. However, when the two compounds were infused i.v. at twice the intracoronary dose, infarct size was not modified. A very serious shortcoming of this study is the lack of collateral blood flow measurements. Indeed, the infarct size of the vehicle-treated group is rather low (28%) for a 90 min occlusion of the left circumflex coronary artery when it is compared with values (over 50%) obtained by other laboratories [31, 34] employing the same experimental procedure. In anaesthetized dogs, cromakalim and pinacidil, infused into the coronary artery throughout the experimental procedure starting 10 min before the coronary artery occlusion at a dose which did not change the haemodynamic status of the animals, markedly reduced (approximately 50%) the infarct size [34]. Similarly, i.v. administration of aprikalim at a dose lacking general and cardiac haemodynamic activity, 10 min before starting 60 or 90 min occlusion of the left circumflex coronary artery, pro-

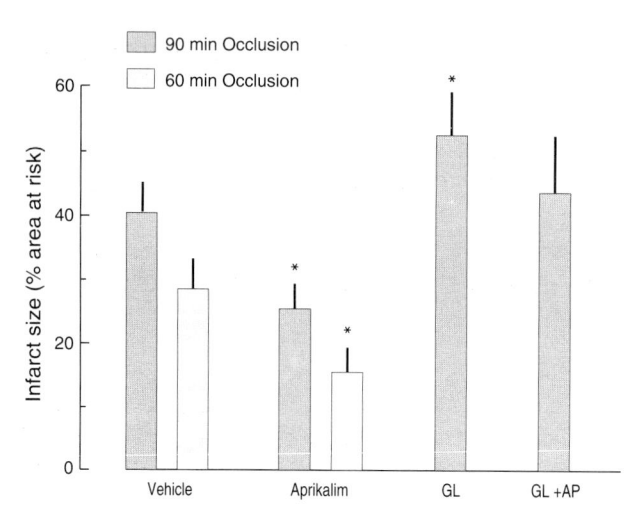

Fig. 7. Effects of aprikalim *(AP)* (10 µg/kg i.v. bolus followed by 0.1 µg/kg/min throughout the experimental procedure) on infarct size in dogs in which the left circumflex coronary artery was occluded for 60 or 90 min and then reperfused for 5 h. The effects of glibenclamide *(GL;* 1.0 mg/kg i.v.) alone or followed by aprikalim is also shown. Redrawn from [31] and [35]

duced a significant decrease (40-50%) in the infarct size (fig. 7) [31, 35]. This effect was prevented by glibenclamide pretreatment which, at the dose studied (1 mg/kg i.v.), significantly increased the size of the infarct. In these studies, animals with a sub-endocardial blood flow in the central ischaemic region greater than 0.15 ml/min/g were excluded from data analysis. Nicorandil has also been reported to decrease infarct size by approximately 50% in the anaesthetized dog subjected to 2-h occlusion of the left anterior descending coronary artery followed by 30 min reperfusion. This beneficial effect was obtained with an i.v. infusion of nicorandil started 10 min after the coronary artery occlusion and continued throughout the reperfusion period. It should be noted that nicorandil substantially decreased aortic blood pressure and myocardial oxygen consumption since the pressure-heart rate product declined by 30% at the end of 2-h occlusion, an effect which was not observed in the vehicle-treated group [36]. However, under the same experimental conditions, a dose of nifedipine which decreased the pressure-heart rate product by 23% did not significantly reduce infarct size [36]. Similarly, in the anaesthetized dog, a nicorandil infusion initiated 15 min after starting a 6-h occlusion of the left anterior descending coronary artery not followed by reperfusion, reduced the size of the myocardial infarct by 31%. Furthermore, cardiac output of the nicorandil-treated group remained at the preocclusion level whereas there was a 25% decrease in the control group. However, nicorandil, but not the vehicle, markedly reduced total peripheral vascular resistance (43%) and the pressure-heart rate product (49%) at the end of the occlusion period [37]. Anaesthetized pigs were subjected to 1-h occlusion of a distal branch of the left anterior descending coronary artery followed by 2-h reperfusion. Bimakalim (3 µg/kg bolus + 0.1 µg/kg/min) was infused i.v. starting (i) 15 min before reperfusion up to the end of the experiments, or (ii) 15 min before occlusion up to 1-h reperfusions, or (iii) 15 min before occlusion up to 15-min reperfusion. Pigs given the solvent (saline) served as control. Bimakalim reduced infarct size by over 50% only when it was administered before the occlusion period (ii & iii above). In contrast, bimakalim administered at reperfusion or before occlusion and then throughout the reperfusion period significantly reduced ventricular arrhythmia and tachycardia. These results indicate that the latter effect occurs independently of the decrease in infarct size and requires the presence of bimakalim throughout the reperfusion period [38].

Overall, K⁺ channel openers can reduce infarct size. However, this beneficial effect is not a constant finding and may be due to the experimental procedure as already discussed.

K⁺ channel openers afford cytoprotection by decreasing energy expenditure in the ischaemic zone

K⁺ channel openers do not increase extracellular K⁺ accumulation during acute ischaemia

It is well known that from the very early phase of ischaemia, the extracellular K⁺ concentration increases [39]. Several studies have documented increases in extracellular K⁺ concentration, $[K^+]_o$, in the ischaemic zone of up to 12 mM 10 min after coronary occlusion. Extracellular K⁺ accumulation during ischaemia partially depolarizes ischaemic cells and creates zones of slow conduction and altered refractoriness which eventually lead to re-entrant arrhythmias. Its time course is typically triphasic [39] with: (i) an initial rising phase lasting less than 10 min followed (ii) by a plateau or a falling phase and then (iii) by a late rising phase.

The cellular mechanism that governs extracellular K⁺ accumulation during ischaemia is not yet elucidated. A role for ATP-sensitive K⁺ channels has been hypothesized since these channels become opened under conditions of impaired intracellular metabolism and thus may represent an unusual pathway for K⁺ ions to leave the ischaemic cells and to accumulate extracellularly. Pharmacological tools have been used in various studies to investigate the participation of ATP-sensitive K⁺ channels during ischaemia [40-47]: the findings support the view that ATP-sensitive K⁺ channels play only a minor role in determining extracellular K⁺ accumulation during its initial phase because glibenclamide exerts weak effects on its occurrence. However, the use of glibenclamide as a tool to block ATP-sensitive K⁺ channels during ischaemia is controversial because: (i) on the one hand, the blocking activity of glibenclamide may increase as extracellular pH decreases [48] and; (ii) on the other hand, its activity may decrease as intracellular ADP increases [45].

Most recently, the effects of nicorandil, levcromakalim and cromakalim were also investigated on extracellular K⁺ accumulation and extracellular acidification induced by ischaemia [44, 46, 47] (Wilde et al., personal communication): the findings show that these agents do not modify extracellular pH, nor do they accelerate extracellular K⁺ accumulation (fig. 8) when employed at therapeutically relevant concentrations. In globally ischaemic rat hearts (Wilde et al., personal communication) or in arterially-perfused rabbit septa treated with cromakalim [47], there is even a slight decrease in the amount of accumulated extracellular K⁺ (fig. 8).

K⁺ channels are involved in the natural defence of the myocardium against an ischaemic insult

Ischaemia combines the deleterious effects of hypoxia, interruption of the delivery of energy substrates to the myocytes and arrest of the washout of by-products of the cell metabolism (e.g. protons) from the interstitial space.

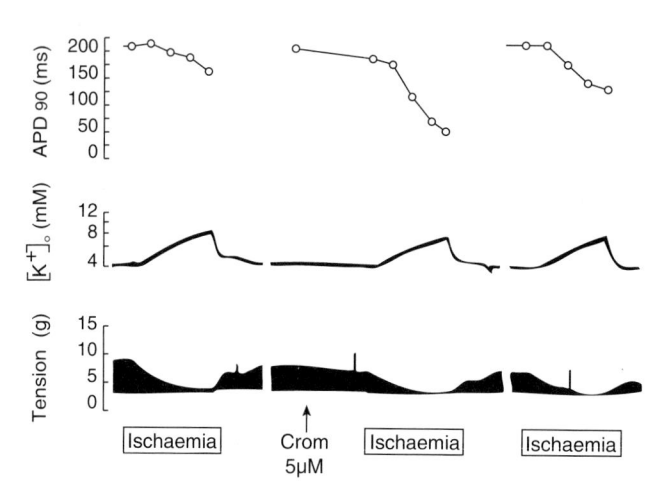

Fig. 8. Effects of cromakalim 5 μM *(Crom)* on action potential duration at 90% repolarization *(APD90)*, extracellular K⁺ concentration ($[K^+]_o$) and tension during total ischaemia in isolated rabbit interventricular septum. Note the accelerated action potential shortening and the unchanged extracellular K⁺ accumulation in the presence of cromakalim. Reproduced from [47] with permission

During hypoxia, it is well known that the action potential rapidly shortens. This decreases the time available for Ca^{2+} to enter the cell during the cardiac beat and consequently decreases the contractile force generated by the injured myocardium. Ultimately, action potential shortening leads to electrical and mechanical incompetence. When a "normal" action potential duration is restored by experimentally controlling the membrane potential, contraction is also restored [49]. Pretreatment of the myocardium with glibenclamide prevents the action potential shortening produced by hypoxia [3, 40], thus suggesting a role for ATP-sensitive K⁺ channels in its genesis. Furthermore, ATP-sensitive K⁺ currents can be directly recorded by means of the patch-clamp technique when either dinitrophenol [50] or hypoxia [51] are used to impair the cell metabolism. Thus the opening of ATP-sensitive K⁺ channels is likely to cause action potential shortening and mechanical failure during hypoxia.

During ischaemia, the action potential also shortens (figs. 4, 8 and 9) although less markedly than during hypoxia. The action potential plateau abbreviation produced by ischaemia can be prevented by glibenclamide [45, 46, 52] or tolbutamide [43], demonstrating the involvement of ATP-sensitive K⁺ channels in its genesis. The role of ATP-sensitive K⁺ channels during ischaemia has been a subject of controversy because during the early phase of ischaemia and also of hypoxia, the intracellular ATP concentration only decreases slightly and remains in any case much higher than the ATP concentration required to block the channels in excised membrane patches. This argument is not relevant for many reasons. First, as soon as a patch is excised, ATP-sensitive K⁺ channels undergo a process called "run-down": their activity spontaneously declines and their sensitivity to the blocking effects of ATP increases [12]. Since the amount of run-down in a membrane patch cannot be precisely assessed, a common bias of excised membrane patch experiments is to overestimate the ATP sensitivity of the channels. Second, it has been calculated that activation of less than 0.5% is sufficient to account for the observed action potential shortening [53, 54]. This is equivalent to the activation of the 125 channels less sensitive to ATP assuming a total of ≈ 25 000 channels per cell. Third, the activity of ATP-sensitive K⁺ channels during ischaemia is not only governed by intracellular ATP, which slightly declines as a consequence of reduced aerobic energy yield (fig. 10), but also by several other products such as protons and ADP, which accumulate.

Fig.9. Changes in monophasic action potential duration at 50% of the repolarization *(ADP$_{50}$)* measured before, during, and after a 15-min occlusion of the left anterior descending coronary artery (LAD) of dogs treated with either saline or cromakalim (1 µg/kg/min into the LAD for the period indicated in the figure). Note the accelerated action potential shortening in the presence of the K$^+$ channel opener also illustrated in fig. 4 and in fig. 8 under in vitro conditions. Redrawn from reference [58]

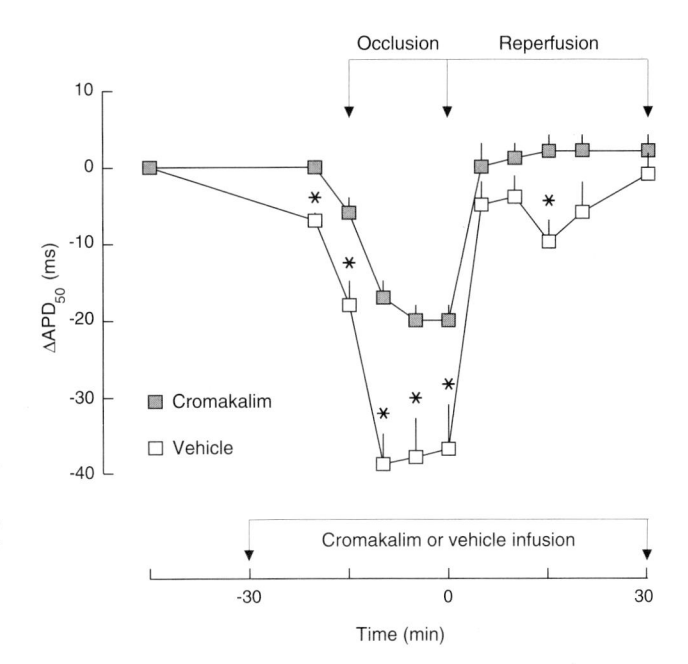

Increased ADP and to a lesser extent increased H$^+$ in the ischaemic cells increase the sensitivity of the channels to the small decrease in ATP [55]. Furthermore, adenosine, a by-product of ATP, accumulates extracellularly and promotes the activation of ATP-sensitive K$^+$ channels via a G-protein pathway in the myocardium [56].

As occurs during hypoxia, the consequence of ATP-sensitive K$^+$ channel activation during ischaemia is shortening of the action potential leading to a decreased amplitude of intracellular Ca^{2+} transients. Experiments measuring intracellular Ca^{2+} concentration and contraction during ischaemia showed an initial fall in the amplitude in intracellular Ca^{2+} transients (minimum achieved after ≈ 5 min) accompanied by a marked decline in muscle systolic shortening and then a secondary rise in intracellular Ca^{2+} which peaked at around 25 min [57]. Overloading of the ischaemic muscle with calcium plays an important role in the generation of ischaemic arrhythmias, irreversible muscle damage and finally cell death. The fall in intracellular Ca^{2+} transients that occurs during the initial phase of ischaemia is likely to be related to the ATP-sensitive K$^+$ channel activation and consequent action potential shortening. This event is beneficial for cell survival because: (i) it reduces and delays Ca^{2+} overload thus preserving cell integrity; (ii) it is responsible for the loss of contractile activity, a desirable mechanism which saves the energy necessary for the fundamental activities of the cells, e.g. active extrusion of calcium and sodium.

K$^+$ channel openers accelerate action potential shortening and mechanical failure in the ischaemic zone

In the presence of a K$^+$ channel opener, shortening of the action potential and contractile failure occur even more rapidly after the onset of ischaemia [44, 47, 52, 58] (see figs. 4, 8 and 9) and late contracture is delayed. Conversely, glibenclamide prolongs the time to

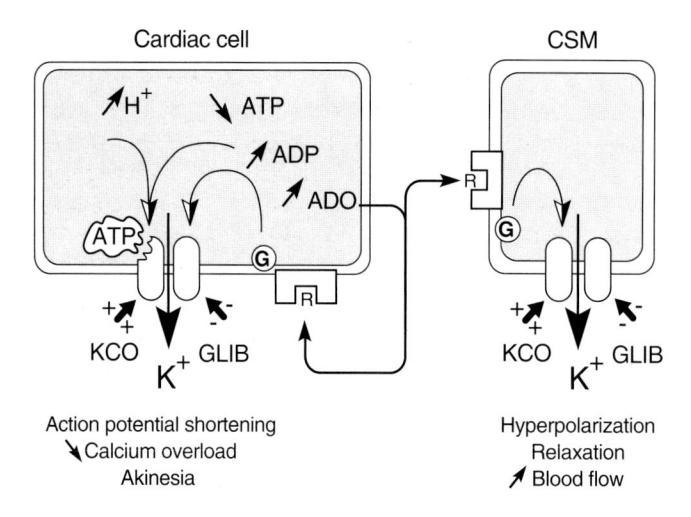

Fig. 10. Schematic drawing of the cellular events that lead to the opening of ATP-sensitive K$^+$ channels during ischaemia. During ischaemia, a limited decrease in the intracellular ATP concentration combines with decreased intracellular pH, increased intracellular ADP and increased extracellular adenosine to activate ATP-sensitive K$^+$ channels in cardiac cells and in coronary smooth mucle cells *(CSM)*. Adenosine, a by-product of ATP breakdown cascade is released extracellularly where it accumulates and activates an A1 receptor coupled to a G-protein. K$^+$ channel opener *(KCO)* and glibenclamide *(GLIB)* antagonistically modulate the opening of ATP-sensitive K$^+$ channels

mechanical arrest [44, 45]. In dogs, the epicardial ventricular monophasic action potential was measured before, during and after occlusion of the coronary artery. Ischaemia produced a clearcut shortening of the ventricular action potential [30, 58]. This effect was strongly accelerated and accentuated by a dose of cromakalim (fig. 9) or aprikalim, which produced excellent cardioprotective effects. Furthermore, the markedly shortened monophasic action potential recovered its pre-ischaemic duration within the first 3 min of the reperfusion period. The monophasic action potential effects of cromakalim and aprikalim were fully antagonized by glibenclamide, which was used in a relatively high dose (3 mg/kg i.v.) in the study of cromakalim and in a 10-fold lower dose in the study of aprikalim.

Data obtained with the patch-clamp technique revealed that ATP sensitive K$^+$ channels open at a higher ATP concentration in the presence of a K$^+$ channel opener than in its absence [12-17], an effect also produced by low pH or elevated intracellular ADP as described above. Therefore, in the presence of a low concentration of the drug, the number of activated channels at the onset of ischaemia is sufficient to accelerate the shortening of the action potential and to speed up the loss of contractile activity within the ischaemic zone even though, in the absence of ischaemia, these drugs at the same concentration fail to activate K$^+$ channels appreciably and thus fail to modify baseline electrical and mechanical function. A more rapid arrest of both electrical and mechanical activity during ischaemia would extend the survival time of the cells because it would save ATP. Since the main mechanism of myocardial ventricular dysfunction at reperfusion following the stunning insult has been attributed to myocyte calcium overload [22], the ATP saved can be used by the cell to maintain the critical processes which prevent calcium overload.

K$^+$ channel openers produce coronary vasodilation, and this effect can theoretically alleviate the imbalance between energy utilization and energy production which is ultimately responsible for the functional failure of the stunned myocardium. However, experimental results obtained with intracoronary infusion of cromakalim [58] or intravenous administration of aprikalim [30] in the stunned myocardium model clearly indicate that these compounds do not increase blood flow to the ischaemic endocardium or epicardium during the occlusion period, although blood flow to these tissues increases during reperfusion. Bimakalim and aprikalim have been reported to produce a marked reduction in neutrophil infiltration in the non-infarcted peri-ischaemic zone in a model of irreversible myocardial injury [30]. This could also contribute to the cardioprotective action of K^+ channel openers although, again, this cannot be the main mechanism since cardioprotection by K^+ channel openers is also found under in vitro conditions where polymorphonuclear leukocytes, the major source of oxygen-derived free radicals, are absent.

Are K^+ channel openers pro- or antiarrhythmic?

Whether K^+ channel openers exert proarrhythmic or antiarrhythmic effects remains controversial. Nicorandil has been shown to possess antiarrhythmic properties against arrhythmias caused by enhanced automaticity and re-entry, particularly in the presence of a decreased membrane K^+ conductance as occurs during hypokalaemia [59]. This drug has also been reported to antagonize arrhythmogenic early and late after-depolarizations arising in cardiac Purkinje fibres [60]. Similarly, pinacidil and cromakalim abolished early after-depolarizations and suppressed repetitive activities induced by quinidine and class III antiarrhythmic drugs [61], effects observed at low concentrations (3 µM pinacidil) which were compatible with clinically relevant plasma concentrations and which produced no effect per se on the control action potential. Likewise, cromakalim suppressed the after potentials secondary to increased extracellular Ca^{2+} [62]. In vivo, pinacidil antagonizes bradycardia-dependent polymorphous ventricular tachycardias resembling *torsades de pointe* induced either by cesium ions (fig. 11; [61]), or clofilium [63]. In a dog model of subacute myocardial infarction, pinacidil exhibited antiarrhythmic effects when examined 22-24 h after a two-stage coronary artery ligation [64]. In contrast with studies showing an antiarrhythmic activity of K^+ channel openers, other reports have pointed out the potential arrhythmogenic properties of these drugs. Pinacidil at doses that markedly decrease blood pressure and increase heart rate, has been demonstrated to have a profibrillatory effect in conscious dogs exposed to acute ischaemia superimposed on a previous infarction [65]. In isolated rat hearts exposed to global ischaemia, cromakalim 10 µM and pinacidil 30 µM enhanced the rate of tachycardia and shortened the time required for the hearts to develop fibrillation [66].

Studies showing either pro or antiarrhythmic effects are difficult to reconcile. From a theoretical standpoint, the effects of K^+ channel openers on the heart subjected to ischaemia are twofold: (i) on the one hand, they accelerate action potential shortening and thus may aggravate inhomogeneity in refractoriness between ischaemic and normal tissues, a classical proarrhythmic mechanism. Conversely, the same mechanism should also render the ischaemic area electrically silent more rapidly, thus lessening the probability of arrhythmic disturbance; (ii) on the other hand, K^+ channel openers limit calcium overload and exert a direct inhibitory effect on calcium overload-related arrhythmogenic

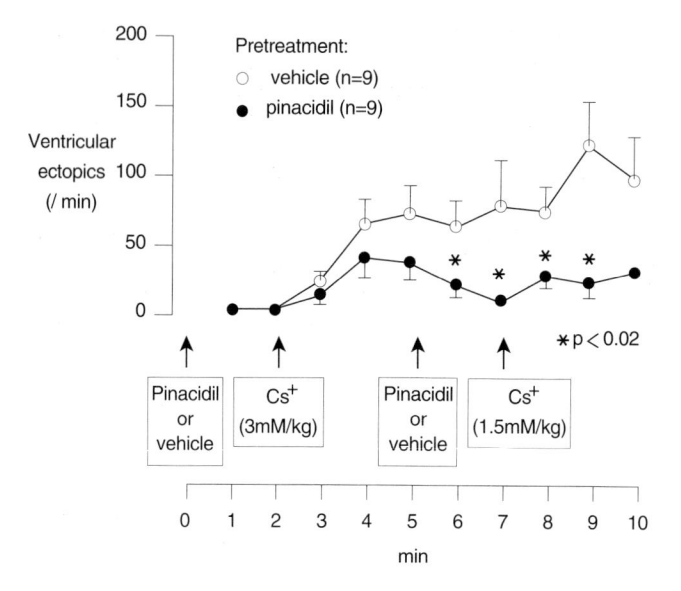

Fig. 11. Effect of pretreatment with vehicle or pinacidil in the ventricular ectopic beat frequency after intravenous cesium administration in anaesthetized rabbits. Significantly fewer arrhythmias were observed with pinacidil pretreatment. Reproduced from [61] with permission

after-potentials. This would be beneficial against reperfusion-induced arrhythmias. Clearly, further experimental investigations are needed to solve the problem. The experimental protocol that should be used for these studies is important: (i) the species chosen should have cardiac electrophysiological characteristics at the cellular level close to those of the human heart. For instance, the rat may not be the best choice because of its very particular ventricular action potential shape; (ii) the effects of K^+ channel openers on cardiac rhythm should be studied during regional rather than global ischaemia in order to mimic the clinical situation; (iii) finally, the doses of drug tested should be clinically realistic. When such guidelines are followed, as in the series of experiments performed at the Medical College of Wisconsin [31], the effects of K^+ channel openers like aprikalim appear to be essentially neutral vis-à-vis cardiac arrhythmias.

In summary, K^+ channel openers exert a striking cytoprotective effect whose mechanism remains incompletely understood. However, available data strongly support the view that a direct activation of cardiac ischaemic ATP-sensitive K^+ channels is the major determinant of cytoprotection. In the clinical setting, the challenge will be to define adequate methodology and protocols in order to prove the clinical relevance of these experimental findings.

References

1. Belardinelli L, Isenberg G (1983) Isolated atrial myocytes: adenosine and acetylcholine increase potassium conductance. Am J Physiol 244: H734-737
2. Escande D, Thuringer D, Le Guern S, Cavero I (1988) The potassium channel opener cromakalim (BRL 34915) activates ATP-dependent K^+ channels in isolated cardiac myocytes. Biochem Biophys Res Commun 154: 620-625

3. Sanguinetti MC, Scott A L, Zingaro GJ, Siegl PKS (1988) BRL 34915 (cromakalim) activates ATP-sensitive K$^+$ current in cardiac muscle. Proc Natl Acad Sci (USA) 85: 8360-8364
4. Escande D, Thuringer D, Le Guern S, Courteix J, Laville M, Cavero I (1989) Potassium channel openers act through an activation of ATP-sensitive K$^+$ channels in guinea-pig cardiac myocytes. Pflügers Arch 414: 669-675
5. Escande D, Cavero I (1992) K$^+$ channel openers and "natural" cardioprotection. Trends Pharmacol Sci 13: 269-272
6. Gotanda K, Satoh K, Taira N (1988) Is the cardiovascular profile of BRL 34915 charateristic of potassium channel activators. J Cardiovasc Pharmacol 12: 239-246
7. Smallwood JK, Steinberg MI (1988) Cardiac electrophysiological effects of pinacidil and related pyridylcyanoguanidines: relationship to antihypertensive activity. J Cardiovasc Pharmacol 12: 102-109
8. Kojima M, Ban T (1988) Nicorandil shortens action potential duration and antagonises the reduction of Vmax by lidocaine but not by disopyramide in guinea-pig papillary muscles. Naunyn-Schmiedeberg's Arch Pharmacol 337: 203-212
9. Yanagisawa T, Taira N (1981) Effect of 2-nicotinamidethyl nitrate (SG-75) on membrane potentials of canine Purkinje fibres. Japan J Pharmacol 31: 409-417
10. Gautier P, Bertrand JP, Guiraudou P (1991) Effects of SR 44866, a potassium channel opener on action potential of rabbit, guinea-pig, and human heart fibers. J Cardiovasc Pharmacol 17: 692-700
11. Fosset M, de Weille JR, Green RD, Schmid-Antomarchi H, Lazdunski M (1988). Antidiabetic sulphonylureas control action potential properties in heart cells via high affinity receptors that are linked to ATP-dependent K$^+$ channels. J Biol Chem 263: 7933-7938
12. Thuringer D, Escande D (1989) Apparent competition between ATP and the potassium channel opener, RP 49356, on ATP-sensitive K$^+$ channels of cardiac myocytes. Mol Pharmacol 36: 897-902
13. Takano M, Noma A (1990) Selective modulation of the ATP-sensitive K$^+$ channel by nicorandil in guinea-pig cardiac cell membrane. Naunyn-Schmiedeberg's Arch Pharmacol 342: 592-597
14. Nakayama K, Fan Z, Marumo F, Hiraoka M (1990) Interrelation between pinacidil and intracellular ATP concentrations on activation of the ATP-sensitive K$^+$ current in guinea-pig ventricular myocytes. Circ Res 67: 1124-1133
15. Nakayama K, Fan Z, Marumo F, Sawanobori T, Hiraoka M (1991) Action of nicorandil on ATP-sensitive K$^+$ channel in guinea-pig ventricular myocytes. Br J Pharmacol 103: 1641-1648
16. Ripoll C, Lederer WJ, Nichols CG (1990) Modulation of ATP-sensitive K$^+$ channel activity and contractile behavior in mammalian ventricle by the potassium channel openers cromakalim and RP 49356. J Pharmacol Exp Ther 255: 429-435
17. Tseng GN, Hoffman BF (1990) Actions of pinacidil on membrane currents in canine ventricular myocytes and their modulation by intracellular ATP and cAMP. Pflügers Arch 415: 414-424
18. Bray KM, Quast U (1992) A specific binding site for K$^+$ channel openers in rat aorta. J Biol Chem 267: 11689-11692
19. Pieper GM, Gross GJ (1989) Protective effect of nicorandil on postischaemic function and tissue adenine nucleotides following a brief period of low-flow global ischaemia on the isolated perfused rat heart. Pharmacology 38: 205-213
20. Grover GJ, McCullough JR, Henry DE, Lee Conder ML, Sleph PG (1989) Anti-ischaemic effects of the potassium channel activators pinacidil and cromakalim and the reversal of these effects with the potassium channel blocker glyburide. J Pharmacol Exp Ther 251: 98-104
21. Grover GJ, Newburger J, Sleph PG, Dzwonczyk S, Taylor SC, Ahmed SZ, Atwal KS (1991) Cardioprotective effects of the potassium channel opener cromakalim: stereoselectivity and effects on myocardial adenine nucleotides. J Pharmacol Exp Ther 257: 156-162

22. Stunning, hibernation, and calcium in myocardial ischaemia and reperfusion (1992) Opie LH (ed) Kluwer Academic Publishers, Boston
23. Auchampach JA, Cavero I, Gross GJ (1992) Nicorandil attenuates myocardial dysfunction associated with transient ischaemia by opening ATP-dependent potassium channels. J Cardiovasc Pharmacol 20: 765-771
24. Gross GJ, Pieper GM, Warltier DC (1987) Comparative effects of nicorandil, nitroglycerin, nicotinic acid, and SG-86 on the metabolic status and functional recovery of the ischaemic-reperfused myocardium. J Cardiovasc Pharmacol 10 (suppl 8): S76-S84
25. Gross GJ, Warltier DC, Hardman HF (1987) Comparative effects of nicorandil, a nicotinamide nitrate derivate, and nifedipine on myocardial reperfusion injury in dogs. J Cardiovasc Pharmacol 10: 535-542
26. Pieper GM, Gross JG (1987) Salutary action of nicorandil, a new antianginal drug, on myocardial metabolism during ischaemia and on postischaemic function in a canine preparation of brief, repetitive coronary artery occlusions: comparison with isosorbide dinitrate. Circulation 76: 916-928
27. Shimshak TM, Preuss KC, Gross GJ, Brooks HL, Warltier DC (1986) Recovery of contractile function in post-ischaemic reperfused myocardium of conscious dogs: influence of nicorandil, a new antianginal agent. Cardiovasc Res 20: 621-626
28. Grover GJ, Sleph PG, Parham CS (1990) Nicorandil improves postischaemic contractile function independently of direct myocardial effects. J Cardiovasc Pharmacol 15: 698-705
29. Auchampach JA, Maruyama M, Cavero I, Gross GJ (1992) Pharmacological evidence for a role of ATP-dependent potassium channels in myocardial stunning. Circulation 86: 311-319
30. Yao Z, Cavero I, Gross GJ (1993) Activation of cardiac K_{ATP} channels: an endogenous protective mechanism during repetitive ischaemia. Am J Physiol 264: H495-H504
31. Auchampach JA, Maruyama M, Cavero I, Gross GJ (1991) The new K^+ channel opener aprikalim (RP 52891) reduces experimental infarct size in dogs in the absence of haemodynamic changes. J Pharmacol Exp Ther 259: 961-967
32. Sakamoto S, Liang C, Stone CK, Hood WB (1989) Effects of pinacidil on myocardial blood flow and infarct size after acute left anterior descending coronary artery occlusion and reperfusion in awake dogs with and without a coexisting left circumflex coronary artery stenosis. J Cardiovasc Pharmacol 14: 747-755
33. Kitzen JM, McCallum JD, Harvey C, Morin ME, Oshiro GT, Colatsky TJ (1992) Potassium channel activators cromakalim and celikalim (WAY-120,491) fail to decrease myocardial infarct size in the anaesthetized canine. Pharmacol 45: 71-82
34. Grover GJ, Dzwonczyk S, Parham CS, Sleph PG (1990) The protective effects of cromakalim and pinacidil on reperfusion function and infarct size in isolated perfused rat hearts and anaesthetized dogs. Cardiovasc Drug Ther 4: 465-474
35. Gross GJ, Auchampach JA (1992) Blockade of ATP-sensitive potassium channels prevents myocardial preconditioning in dogs. Circ Res 70: 223-233
36. Lamping KA, Christensen CW, Pelc LR, Warltier DC, Gross GJ (1984) Effects of nicorandil and nifedipine on protection of ischaemic myocardium. J Cardiovasc Pharmacol 6: 536-542
37. Endo T, Nejima J, Kiuchi K, Fujita S, Kikuchi K, Hayakawa H, Okumura H (1988) Reduction of size of myocardial infarction with nicorandil, a new antianginal drug, after coronary artery occlusion in dogs. J Cardiovasc Pharmacol 12: 587-592
38. Schelling P, Becker KH, Lues I, Soei LK, Verdouw PD, Weygandt H (1992) Cardioprotection by the K^+ channel opener bimakalim in pigs under coronary artery occlusion and reperfusion. Eur Heart J 13 (Abstract suppl): 400
39. Kléber AG (1984) Extracellular potassium accumulation in acute myocardial ischaemia. J Mol Cell Cardiol 16: 389-394
40. Wilde AAM, Escande D, Schumacher C, Thuringer D, Mestre M, Fiolet J, Janse M (1990) Potassium channel accumulation in the globally ischaemic mammalian heart: a role for the ATP-sensitive potassium channel. Circ Res 67: 835-843

41. Bekheit SS, Restivo M, Boujdir M, Henkin R, Gooyandeh K, Assadi M, Khatib S, Gough WB, El-Sherif N (1990). Effects of glyburide on ischaemia-induced changes in extracellular potassium and local myocardial activation: a potential new approach to the management of ischaemia-induced malignant ventricular arrhythmias. Am Heart J 119: 1025-1033
42. Kantor PF, Coetze WA, Carmeliet EE, Dennis CC, Opie LH (1990) Reduction of ischaemic K^+ loss and arrhythmias in rat hearts. Effects of glibenclamide, a sulphonylurea. Circ Res 66: 478-485
43. Gasser RNA, Vaughan-Jones RD (1990) Mechanism of potassium efflux and action potential shortening during ischaemia in isolated mammalian cardiac muscle. J Physiol (Lond) 431: 713-741
44. Mitani A, Kinoshita K, Fukamachi K, Sakamoto M, Kurisu K, Tsuruhara Y, Fukumura F, Nakashima A, Tokunaga K (1991) Effects of glibenclamide and nicorandil on cardiac function during ischaemia and reperfusion in isolated perfused rat hearts. Am J Physiol 261: H1864-1871
45. Venkatesh N, Lamp ST, Weiss JN (1991) Sulphonylureas, ATP-sensitive K^+ channels, and cellular K^+ loss during hypoxia, ischaemia, and metabolic inhibition in mammalian ventricle. Circ Res 69: 623-637
46. Vanheel B, de Hemptinne A (1992) Influence of KATP channel modulation on net potassium efflux from ischaemic mammalian cardiac tissue. Cardiovasc Res 26: 1030-1039
47. Venkatesh N, Stuart JS, Lamp ST, Alexander LD, Weiss JN (1992) Activation of ATP-sensitive K^+ channels by cromakalim: effects on cellular K^+ loss and cardiac function in ischaemic and reperfused mammalian ventricle. Circ Res 71: 1324-1333
48. Findlay I (1992) Effects of pH upon the inhibition by sulphonylurea drugs of ATP-sensitive K^+ channels in cardiac muscle. J Pharmacol Exp Ther 262: 71-79
49. Stern MD, Silverman HS, Houser SR, Josephson RA, Capogrossi MC, Nichols CG, Lederer WJ, Lakatta EG (1988) Anoxic contractile failure in rat heart myocytes is caused by failure of intracellular calcium release due to alteration of the action potential. Proc Natl Acad Sci (USA) 85: 6954-6958
50. Escande D (1989) The pharmacology of ATP-sensitive K^+ channels in the heart. Pflügers Arch 414 (Suppl 1): S93-98
51. Friedrich M, Benndorf K, Schwalb M, Hirche HJ (1990) Effects of anoxia on K and Ca currents in isolated guinea-pig cardiocytes. Pflügers Arch 416: 207-209
52. Cole WC, McPherson CD, Sontag D (1991) ATP-regulated K^+ channels protect the myocardium against ischaemia-reperfusion damage. Circ Res 69: 571-581
53. Weiss JN, Venkatesh N, Lamp ST (1992) ATP-sensitive K^+ channels and cellular K^+ loss in hypoxic and ischaemic mammalian ventricle. J Physiol (Lond) 447: 649-673
54. Nichols CG, Lederer WJ (1991) Adenosine triphosphate-sensitive potassium channels in the cardiovascular system. Am J Physiol 261: H1675-H1686
55. Lederer WJ, Nichols CG (1989) Nucleotide modulation of the activity of rat heart ATP-sensitive K^+ channels in isolated membrane patches. J Physiol (Lond) 419: 193-211
56. Kirsch GE, Codina J, Birnbaumer L, Brown AM (1990) Coupling of ATP-sensitive K^+ channels to A1 receptors by G proteins in rat ventricular myocytes. Am J Physiol; 259: H820-H826
57. Lee JA, Allen DG (1992) Changes in intracellular free calcium concentration during long exposures to simulated ischaemia in isolated mammalian ventricular muscle. Circ Res 71: 58-69
58. D'Alonzo AJ, Darbenzio RB, Parham CS, Grover GJ (1992) Effects of intracoronary cromakalim on postischaemic contractile function and action potential duration. Cardiovasc Res 26: 1046-1053
59. Imanishi S, Arita M, Aomine M, Kiyosue T (1984) Antiarrhythmic effects of nicorandil on canine cardiac Purkinje fibres. J Cardiovasc Pharmacol 6: 772-779
60. Lathrop DA, Nanasi PP, Varro A (1990) In vitro cardiac models of dog Purkinje fibre triggered and spontaneous electrical activity: effects of nicorandil. Br J Pharmacol 99: 119-123

61. Fish FA, Prakash C, Roden DM (1990) Suppression of repolarization-related arrhythmias in vitro and in vivo by low dose potassium channel activator. Circulation 82: 1362-1369
62. Liu B, Golyan F, McCullough JR, Vassalle M (1988) Electrophysiological and antiarrhythmic effects of the K channel opener BRL 34915, in cardiac Purkinje fibres. Drug Dev Res 14: 123-139
63. Carlsson L, Abrahamsson C, Drews L, Duker G (1992) Antiarrythmic effects of potassium channel openers in rhythm abnormalities related to delayed repolarization. Circulation 85: 1491-1500
64. Kerr MJ, Wilson R, Shanks RG (1985) Suppression of ventricular arrhythmias after coronary artery ligation by pinacidil, a vasodilator drug. J Cardiovasc Pharmacol 7: 875-883
65. Chi L, Uprichard ACG, Lucchesi BR (1990) Profibrillatory actions of pinacidil in a conscious canine model of sudden coronary death. J Cardiovasc Pharmacol 15: 452-464
66. Wolleben CD, Sanguinetti MC, Siegl PKS (1989) Influence of ATP-sensitive potassium channel modulators on ischaemia-induced fibrillation in isolated rat hearts. J Mol Cell Cardiol 21: 783-788

From basic science to the clinic

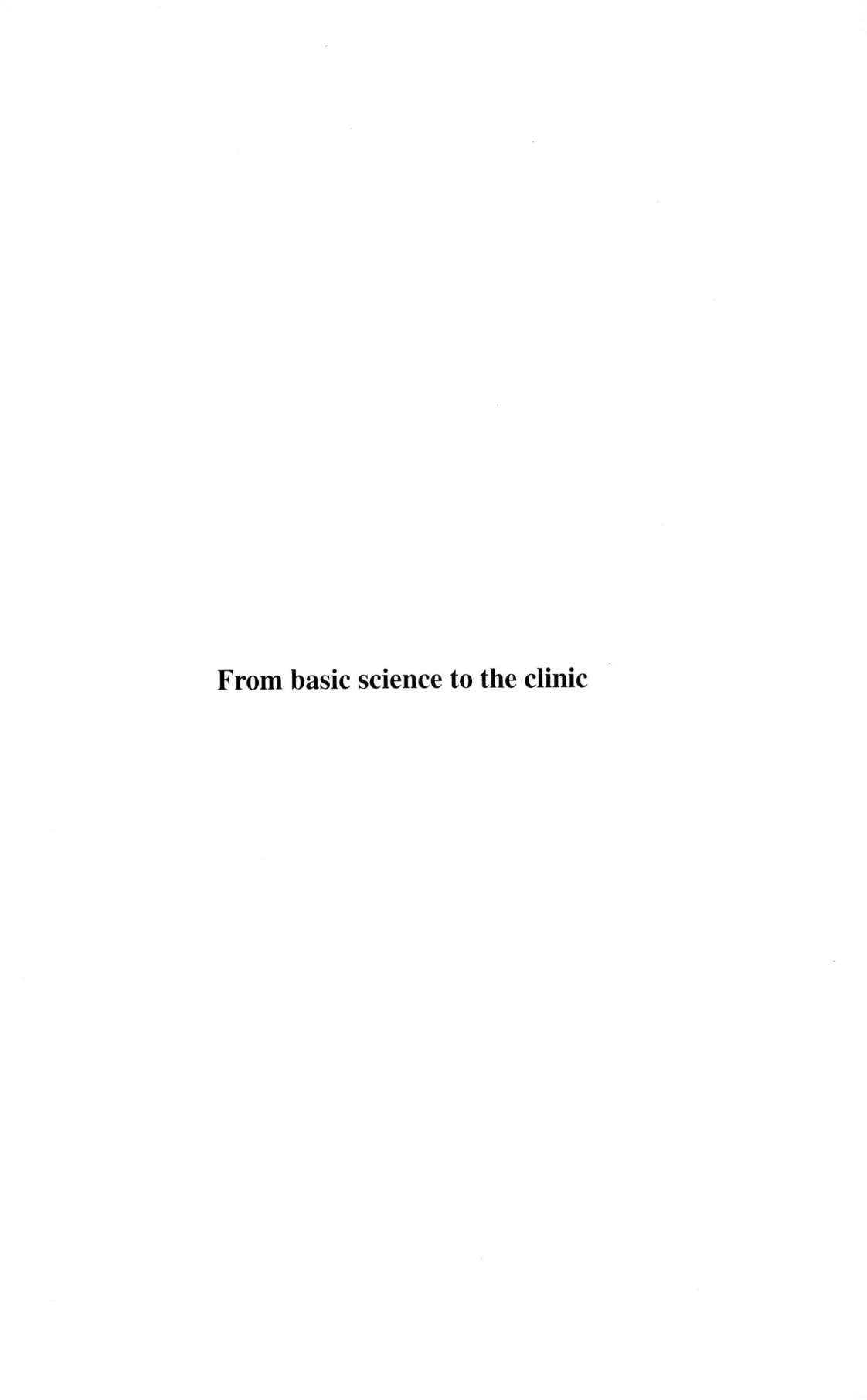

CHAPTER 14
Prolonging cardiac repolarization as an evolving antiarrhythmic principle

BN Singh, R Ahmed and L Sen

In the treatment of cardiac arrhythmias, there now is a growing consensus that antiarrhythmic compounds that act essentially by inhibiting the fast sodium-channel activity might not be as effective as those that produce their beneficial effects by prolonging repolarization [1, 2]. Such a belief recently came into sharp focus with the report of the results of the Cardiac Arrythmia Suppression Trials [3, 4]. In this placebo-controlled study in survivors of acute myocardial infarction with frequent premature ventricular contractions (PVC), it was found that flecainide and encainide, *two agents which markedly delay conduction, increased mortality and sudden death rate despite markedly suppressing PVCs.*

There now is extensive clinical and experimental evidence which suggests that this might be applicable to most, if not all class I agents. Meta-analytic studies [5] with a variety of class I agents (Table 1) have indicated that the deleterious effect of class I compounds might be a "class action" effect. Conversely, it was found that specific adrenergic blockers such as ß-receptor antagonists which reduce heart rate, while being relatively weak suppressants of PVCs, consistently reduced sudden death in the survivors of acute myocardial infarction [1]. In the last decade, increasing data have indicated that two compounds, sotalol and amiodarone, which not only reduce sympathetic excitation but also prolong cardiac repolarization with little or no effect on myocardial conduction, exhibit a potential to reduce sudden death in patients with manifest ventricular tachycardia and fibrillation [2, 6]. The purpose of this chapter is twofold. The first is to summarize briefly the data which suggest that agents which act essentially by delaying fast-channel mediated conduction may be more proarrhythmic than antiarrhythmic. They may prevent ventricular fibrillation less often than they induce potentially lethal arrhythmias in patients at risk from sudden death in the context of significant structural heart disease. The second objective is to critically discuss the evolving data which suggest that agents which act predominantly by lengthening repolarization and refractoriness during concomitant adrenergic modulation have a major antifibrillatory potential. They may prolong survival in patients at high risk for sudden death.

Agents that reduce conduction may increase rather than decrease mortality during the course of antiarrhythmic therapy

There is now much data, summarized at length elsewhere [7], indicating that a varying spectrum of proarrhythmic effects might result from all class I agents in a variety of experimental and clinical settings. A few salient features will be emphasized here. In the canine model of sudden death described by Lynch and Lucchesi [8], none of the class I agents exerted an antifibrillatory action. In this ischaemic canine post-infarct model, programmed electrical stimulation produces ventricular tachycardia (VT) in nearly all dogs;

Table 1. Overview of mortality data from randomized trials of antiarrhythmic drugs in myocardial infarction. Adapted from Yusuf and Teo [5]

Class of agent	N°. dead/ active	N°. randomized/ control	Odds ratio (95% confidence interval)
Class I			
Class IA	142/1,936	131/1,847	1.01 (0.79-1.30)
Class IB	306/7,068	275/6,945	1.06 (0.89-1.26)
Class IC	97/1,278	65/1,117	1.41 (1.02-1.95)
Quinidine/disopyramide/ mexiletine	4/49	5/47	0.75 (0.19-2.94)
Total class I	549/10,331	476/9,956	1.11 (0.97-1.26)
Class II (ß blockers)			
Early i.v. + short-term	530/14,511	603/14,415	0.87 (0.77-0.98)
Late + long-term	934/12,438	1,124/11,860	0.77 (0.70-0.84)
Class III			
Amiodarone	54/577	72/575	0.71 (0.48-1.03)
Class IV			
Calcium channel blockers	968/9,746	944/9,786	1.03 (0.94-1.13)

the VT almost invariably deteriorates into ventricular fibrillation (VF). The pretreatment of groups of dogs with different antiarrhythmic agents permits the evaluation of their potential antifibrillatory and their antiarrhythmic effects.

In this model, flecainide [9] provided no protection against the development of VF compared with the controls (fig. 1). A similar lack of protective effects against VF in this model has been reported for quinidine [10] as well as lidocaine [11]. Antiarrhythmics with class I properties also exert significant proarrhythmic effects in other experimental models [12]. These findings are consistent with a number of clinical observations. Furukawa et al. [13] found that a lesser slowing of conduction and a greater prolongation of refractoriness induced by procainamide tended to abolish re-entry in patients with inducible ventricular tachycardia. The converse tended to stabilize the re-entrant circuit and promoted the continued inducibility of ventricular tachycardia. Similar findings have been reported recently by Kus et al. [14] for the class Ic agent propafenone. Numerous lines of clinical evidence are also consistent [7].

In a series of meta-analyses [5, 15] of randomized clinical trials with various class I agents in the survivors of acute myocardial infarction, it was found that without exception these agents either had no effect on mortality or increased it (Table 1). The results of the Cardiac Arrythmia Suppression Trial 1 [3] are well known and widely appreciated. The recent results of the CAST II trial involving the effects of moricizine and a placebo [4], provide data of much concern. Electrophysiologically, moricizine is a class I agent but differs somewhat from flecainide and encainide. In the first 14 days of treatment with moricizine after myocardial infarction there was an excess mortality (17 of 665 patients died of cardiac arrests) as compared with the placebo (fig. 2) or no treatment (3 of 660

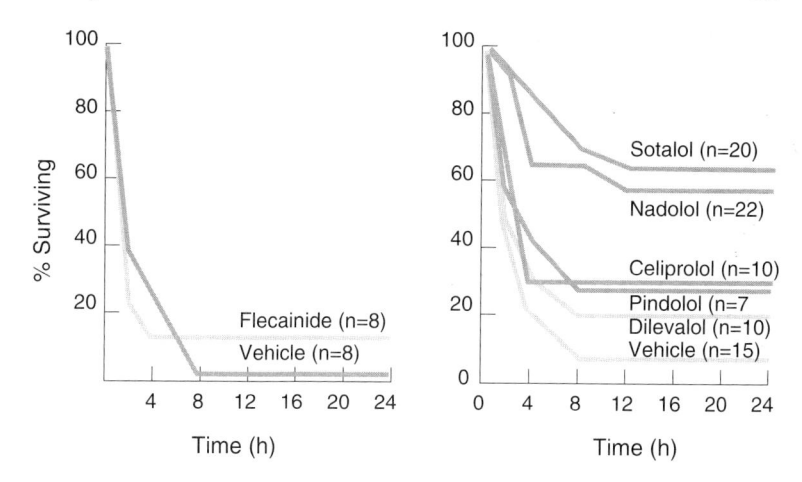

Fig. 1. Effects of flecainide and various types of ß-blockers on survival in the canine sudden death model (see text for details). Note that flecainide, a class I agent, has no significant effect; sotalol with a class III action provides 70% protection, nadolol a little less, and substantially less protection is provided by ß-blockers that have little bradycardic effect. Data reproduced from [8] with the permission of the authors

patients); a deleterious trend necessitating premature termination of the trial was also evident during the longterm segment of the study.

CAST II data are of much theoretical and practical importance. They confirm the results of CAST I and further invalidates the PVC hypothesis. Moreover, they support the concept of a "class action" being deleterious since moricizine is a class I agent but differs from flecainide and encainide. This is also emphasized by the recent report from Hallstrom et al. [16] in 941 patients with aborted sudden cardiac death treated between 1970 and 1985. Compared with no treatment, procainamide and quinidine appeared to worsen survival, although the difference did not reach statistical significance. In contrast, ß-blockers prolonged survival significantly ($p < 0.001$) compared with the cohort of the population never exposed to ß-blockers (fig. 2). Again, the data emphasize the reality of a class action effect with respect to the salutary influence of ß-blockers, the converse in the case of class I drugs. The possibility that the negative impact of class I agents on mortality pervades to a varying extent the entire spectrum of cardiac arrhythmias is real. It should engender considerable concern in view of their continuing and widespread use in the treatment of disorders of rhythm.

Quinidine, a class Ia agent, appears also to increase mortality in other subsets of patients. For example, compared with the placebo after one year of therapy to maintain stability of the sinus rhythm in patients with atrial fibrillation converted to sinus rhythm, therapy with the drug was associated with a mortality of 2.9% (0.8% on placebo; $p < 0.05$). This derived from data from meta-analytic studies in over 800 patients [17]. This apparent adverse effect on mortality has been supported by the data derived from an analysis of mortality in patients given antiarrhythmic therapy (essentially class I agents) in the Stroke Prevention Trial on Atrial Fibrillation [18]. Compared with the group not given antiarrhythmic therapy, there was a 2.5 fold increase in mortality. In a meta-analysis of 4 studies in ventricular arrhythmias (mainly PVCs), Morganroth and Goin [19],

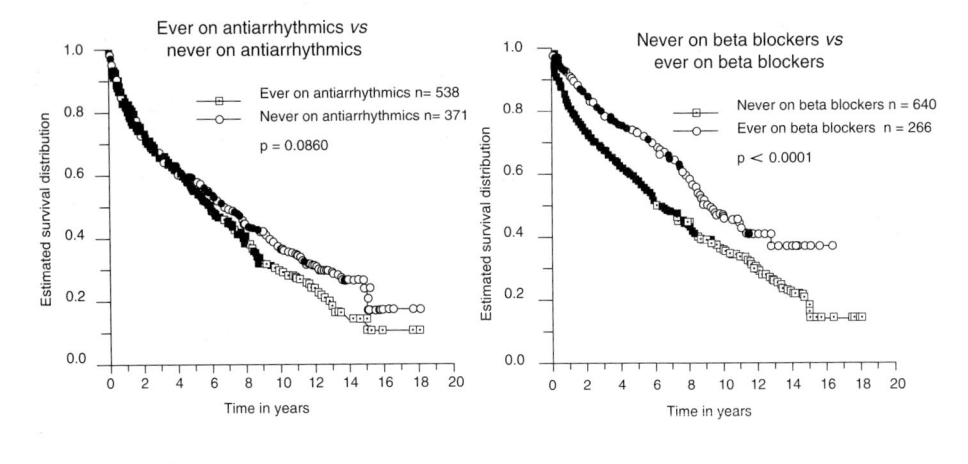

Fig. 2. Effects of ß-blockade on patients surviving cardiac arrest compared with patients either given no treatment or given conventional antiarrhythmic therapy (procainimide or quinidine), empirically ß-blockers were superior. From [16] with the permission of the authors and the American Journal of Cardiology (see text for details)

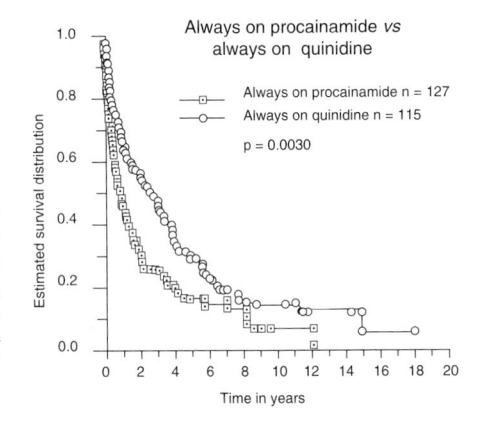

found an excess of mortality with quinidine compared with the placebo in patients with relatively well-preserved ventricular function. Salerno [20] attributed such an impact on mortality to proarrhythmic reactions and logically questioned whether the most commonly used antiarrhythmic agent (quinidine) in the USA was "worse than adverse". The case against the prophylactic use of class I agents in different subsets of patients with cardiac disease continues to grow as the data from clinical trials using mortality rather than surrogate end-points increase. Interestingly, an excess in mortality is also suggested by the data on lidocaine given to patients with acute myocardial infarction [21-23]. Similarly, in the IMPACT study, mexiletine also showed a trend to increase mortality when compared with the placebo in the post-infarction survivors [24]. It would seem that the data are consistent with the class I effect of lidocaine.

For these reasons, it appears unlikely that Na^+ channel blockers as a class will survive the close scrutiny of stringently controlled clinical trials with end-points such as sudden death and total mortality. Since the adverse effects of these agents are in those with significant structural heart disease (the very subset of patients in whom the antiarrhythmic therapy for the prevention of sudden death is needed the most) the overall role of class I agents is likely to shrink. It is possible their use in the future will be confined to the relief of symptoms from arrhythmias in patients with essentially normal hearts structurally. The issue is a matter of much practical importance to the clinician and to those involved in antiarrhythmic drug development. Available clinical and experimental data suggest a compelling need to shift from Na^+ channel blockers to agents that block sympathetic influences on the heart and to those which have the added property of homogeneously prolonging cardiac repolarization and refractoriness without a major effect on conduction [25]. This is the focus of the rest of this chapter.

Agents that delay repolarization offer the best scope for mortality reduction by preventing VF

To date there are three types of antiarrhythmic agents (ß-blockers, sotalol, and amiodarone) which appear to offer the best scope for mortality reduction in patients with cardiac disease by preventing the occurrence of ventricular fibrillation. Their effects are in sharp contrast to those of class I agents and Ca^{2+} channel blockers in different subsets of patients with respect to mortality reduction. All three attenuate sympathetic excitation; sotalol and amiodarone share the common property of lengthening the action potential duration and refractoriness, amiodarone having additional electrophysiologic effects and exceedingly complex pharmacokinetic and membrane effects [25]. The clinical profiles of these agents, which have been studied for a few decades, may permit tentative conclusions regarding which components of their electrophysiological properties are linked to their clinical antifibrillatory and profibrillatory actions. The synthesis and characterization of the so-called pure class III compounds are likely to provide insights into the precise significance of lengthening the action potential duration in preventing atrial and ventricular fibrillation.

ß-blockers are antifibrillatory agents

The evidence now is secure that drugs of this class reduce sudden death in survivors of acute infarction [1]. They also reduce mortality in the congenital long QT interval syndrome [26]. There is recent evidence that they may also be effective in controlling symptomatic sustained monomorphic ventricular tachycardia [27]. As already mentioned, they may be superior to class I agents in prolonging survival in those with aborted sudden death [16]. In the animal ischaemia model of sudden death, ß-blockers prevent the deterioration of ventricular tachycardia to ventricular fibrillation [8]. However, the precise mechanism of the overall salutary effect of ß-blockers in these various settings is not clear. It appears multifactorial. It is probably due to the reversal of the arrhythmogenic consequences of sympathetic stimulation. There is no evidence that the overall effect is due solely to the effect on myocardial ischaemia or the suppression of premature ventricular contractions [1]. These agents do however increase the ventricular fibrillation threshold [28]. By reducing heart rate, ß-blockers may reduce dispersion of refractory period in the ischaemic myocardium as has been reported during vagal stimulation [29]. The latter effect may be critical in their action as their beneficial effect in the survivors of acute infarction correlates well with decreases in heart rate [30]. Although the data are controversial, it has been suggested that their observed effect in the post-infarct patients may be related in part to the lengthening of the action potential duration during drug administration [31]. Such an observation in experimental animals has not been verified in humans [1]. Whatever the mechanisms that are finally established as mediating the salutary effects of ß-blockers on mortality in patients with ischaemic heart disease, it is clear that antifibrillatory agents of the future cannot ignore the importance of incorporating an anti-adrenergic action as an integral component of their overall molecular property [32-37]. Both sotalol and amiodarone exert significant anti-adrenergic actions, the former by an associated competitive ß-receptor blocking effect [36], the latter by non-competitive anti-adrenergic actions [37], which in part may be mediated via a cardiospecific thyroid hormone antagonism [37]. The fundamental role of adrenergic excitation in the genesis of cardiac arrhythmias must be remembered when dealing with compounds that act by interfering with the sources of sympathetic transmitters at the level of the adrenergic receptor or by other means.

Sotalol led to the development of the class III antiarrhythmic concept

Numerous lines of evidence from experimental and clinical studies have confirmed that lengthening of the action potential duration with an accompanying prolongation of the ERP constitutes an important antiarrhythmic principle. It is an effective approach to end and prevent the development of atrial and ventricular fibrillation even in the absence of any change in conduction velocity [25, 35-40].

Sotalol, like amiodarone, verapamil and propranolol, was synthesized in 1962. In 1968, Kaumann and Olsson [41] reported that when kitten papillary muscle was exposed to large concentrations of dl-sotalol, there was a marked increase in the action potential duration accompanied by an increase in tension development. In some preparations, accompanying the marked delay in repolarization was an after-depolarization (EAD) associated with an after-contraction. The authors did not consider the possibility that such a phenomenon might be associated with an antifibrillatory action on the one hand and a profibrillatory action on the other. Confirming their observations on the effects of sotalol on myocardial repolarization in 1970, Singh and Vaughan Williams [33] drew attention to the fact that such an electrophysiological effect was associated with a marked antifibrillatory action in the setting of digitalis intoxication. The lengthening of the action potential duration was also found to occur in the atria [33, 42]. The fact that the drug lengthened the action potential duration in ventricular muscle without a concomitant change in conduction velocity has been the focus of the so-called class III action. The principle is relatively simple in electrophysiological terms. There are a number of clinical correlates of such an antifibrillatory phenomenon. For example, atrial fibrillation is common in the case of hyperthyroidism, a state in which the atrial action potential duration is markedly shortened. The very converse occurs in the case of hypothyroidism [35, 37, 39]. Many years ago, it was found that in patients with atrial fibrillation converted to sinus rhythm by electrical conversion, relapses occurred more frequently in those whose atrial action potential (hence effective refractory period) was much shorter than in those whose APD was longer [39]. It was reasonable to presume that the longer action potential contributed to the electrical stability of the atrial tissue, which reduced the probability of fibrillation [35, 37]. The issue has not been as clear in the case of ventricular arrhythmias. However, the electrophysiological principle is compelling [25]. The "isolated" increase in the ERP produces widening of the cycle length of the sustained tachycardia and prevents it from deteriorating into fibrillation; the increase in the excitation wavelength of a re-entrant circuit makes the arrhythmia difficult to sustain by precluding the establishment of a stable circuit. The lengthening of the action potential duration elevates the ventricular fibrillation threshold and decreases the defibrillation threshold. An aggregate of such actions has been found to be associated with chemical defibrillation [43].

The most important hallmark of a class III antiarrhythmic agent is its antifibrillatory action. It is equally characteristic of this class of antiarrhythmic agents that, given the appropriate clinical circumstances, they will produce a variable incidence of *torsades de pointe* as a reasonably specific proarrhythmic effect [25].

Sotalol has both class III and ß-blocking properties

As discussed at length elsewhere [6, 36], sotalol exhibits a broad spectrum of antiarrhythmic actions in supraventricular and ventricular arrhythmias. The drug is known to exert a potent ß-blocking action and prolongs the action potential duration and refractoriness in virtually all cardiac tissues. These two properties are unrelated and the action potential duration is not prolonged by ß-blockade per se. Kato et al. [44] demonstrated that the levo- and the dextro-isomers of sotalol were equipotent in lengthening repolarization while the dextro-isomer was only 1/50th as potent as the levo-compound in blocking ß-receptors.

The effects of a number of antiarrhythmic drugs have been compared with sotalol in a standardized clinically relevant model of sudden arrhythmic death due to ventricular fibrillation in the dog described by Lucchesi and his colleagues [8]. Fig.1 shows the effects of intravenous flecainide, a "pure" class I agent that markedly depresses myocardial conduction, and prevents ventricular tachycardia (VT) deteriorating into ventricular fibrillation (VF). Compared with the vehicle serving as the control, there was no significant effect, but the drug appeared to shorten the time to the development of VF. In the Lucchesi model of sudden death, the effects of other classes of antiarrhythmic agents have yielded somewhat different results which are more in line with the expanding clinical data. For example, it was found that the ß-blocker, nadolol [45], given prior to programmed electrical stimulation prevented the development of ventricular fibrillation in a significant number of dogs; the effects of sotalol, the prototype class III antiarrhythmic agent with a potent ß-blocking action, was somewhat greater than that of nadolol (fig. 1). In contrast, ß-blockers which had vasodilator properties due to their α-blocking actions and lacked bradycardic actions provided only a modest protective effect, indicating the importance of inhibiting sympathetic excitation in preventing VF. The effects of sotalol appeared to be greater than those of nadolol, emphasing the importance of the so-called class III action as an antifibrillatory mechanism in the case of VF. However, the difference was not large and the possibility remains that the most significant antifibrillatory effect of sotalol may be mediated via its ß-adrenergic blocking actions.

It is nevertheless reasonable to assume that the overall spectrum of antiarrhythmic activity and the profile of adverse reactions of sotalol represents the summated effects of its ß-blocking action and its class III properties. A series of blind studies have established that sotalol suppresses premature ventricular contractions (PVCs) significantly more than a placebo, and at least one study revealed that the drug is more potent in this regard than the reference ß-blocker [46]. Confirmation of such a difference must, however,

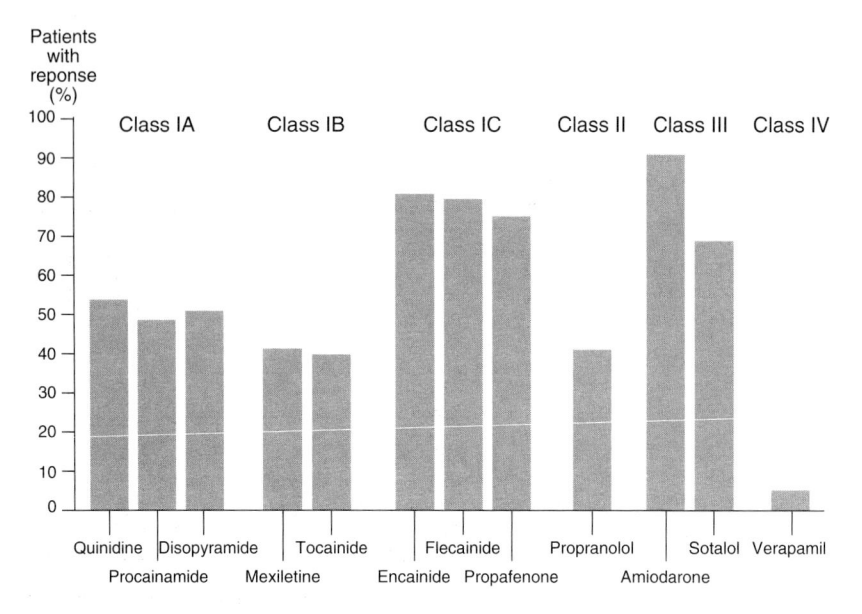

Fig. 3. PVC suppressant effects of various classes of antiarrhythmic drugs. The suppressant effects of sotalol is comparable to that of class IA and IB agents (see text for details). Based on the meta analysis data of various clinical trials reported by Salerno et al. [47]

await further controlled studies. The PVC suppressant effect of sotalol compared with that of various electrophysiological classes of antiarrhythmic compounds is shown in fig.3 [47]. The drug has a suppressant effect comparable to class Ia and Ib agents; it is less potent than class Ic agents and amiodarone. At present, it is unclear whether the suppressant effect of sotalol on PVCs stems essentially from its ß-blocking action or whether it is influenced additionally by its class III properties.

Of the greatest importance is the drug's effect on life-threatening ventricular arrhythmias, especially in patients with symptomatic sustained ventricular tachycardias, ventricular fibrillation or those surviving cardiac arrest. In these patients, both intravenous and chronic administration of the drug have been shown to prevent the re-induction of VT/VF by programmed electrical stimulation (PES) of the heart. The efficacy has ranged from 30 to 40% in recent studies using triple stimuli [36] and from 45 to 67% in the earlier studies using double extra-stimuli [48, 49]. The data have been critically reviewed by Deedwania and Singh [50] and by Singh [36]. The overall effectiveness of the drug on VT/VF-induced by PES is compared with that produced by electrophysiological classes of compounds in fig. 4 [51]. Sotalol appears to be superior to other agents.

A major recent advance has been the availability of the preliminary data from the Electrophysiologic versus Electrocardiographic Monitoring (ESVEM) study in determining the relative merits of PES versus Holter-guided therapy (HM) of VT/VF [52]. From the over 2000 patients screened, 486 patients satisfied the entry criteria of the study; 242 patients were to allocated at random to the PES limb and 244 to the HM limb. Six class I agents and sotalol were tested in the study (drug trials were positive in 14% by PES and in 38% by Holter) in each limb, resulting in 297 responders (45% in the PES limb and 77% in the Holter limb) who were followed for six years. There were two important

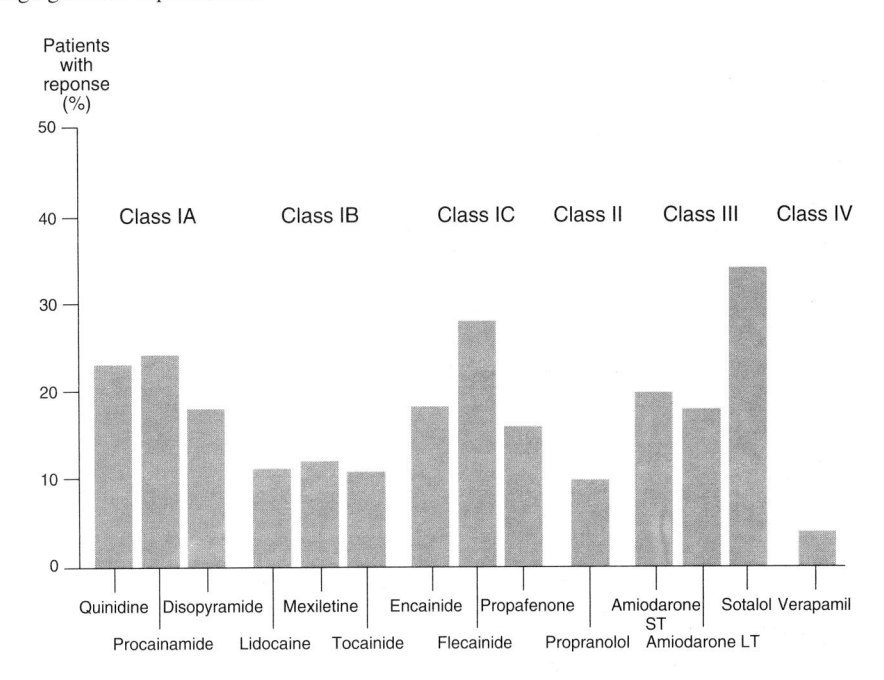

Fig. 4. The comparative efficacy of various electrophysiological classes of antiarrhythmic agents in preventing inducibility of VT/VF by PES. Data based on those reported in a review by Nattel [51]. See text for details

findings. First, there was no significant difference between PES and HM in predicting arrhythmia recurrence, sudden death, cardiac death or all-cause mortality. Second, sotalol was superior to class I agents individually or collectively (5 agents) with respect to total mortality, sudden death, cardiac death and especially for VT recurrence. For example, at one year, arrhythmia recurrence was found in 44% of the patients taking class I agents and in 21% of the patients taking sotalol ($p < 0.0007$). Two conclusions can be drawn from these results. First, the responses might be judged to be drug-specific rather than technique-specific, sotalol being more effective than class I agents. From these studies it cannot be concluded that class I agents are superior to a placebo as no control limb was used in ESVEM. Since sotalol was significantly more effective than class I agents, it is increasingly difficult to justify the continued use of Na^+-channel blockers in VT/VF as in the case of their use in other subsets of patients with cardiac disease. Second, although HM appeared to have a greater clinical applicability for selecting drug therapy of VT/VF, the possibility cannot be ignored that both techniques may lack intrinsic scientific validity [7]. The resolution of this issue of technique-specificity versus drug-specificity of the responses in antiarrhythmic therapy in VT/VF will require carefully controlled studies with the use of ICD back-up for safety in the non-treatment (equivalent to placebo) limb of randomized studies.

Amiodarone has complex antifibrillatory actions

The enigma of amiodarone action continues to puzzle as well as baffle experimental and clinical pharmacologists and cardiologists. It is overly simplistic to continue classifying the drug as a class III antiarrhythmic compound even though the most striking feature of its pharmacology is the consistent and marked lengthening of the action potential duration that the drug produces as a function of time at a constant dose. An extensive discussion of its varied electropharmacology has been presented elsewhere [37, 39]. A few key features will be mentioned here to emphasize that while the drug exhibits all four electrophysiological classes of action, its overall effects during chronic therapy cannot be mimicked simply by combining Na^+-channel blockers, ß-blockers, "pure" K^+-channel blockers and Ca^{2+}-channel blockers.

The main features may be summarized as follows:

(i) it is now widely appreciated that amiodarone exerts markedly different effects when it is given intravenously during acute administration and when it is given orally for varying periods of time. During acute drug administration, there is evidence for Ca^{2+}-channel blockade and non-specific anti-adrenergic effect but with little or no change in heart rate, the major effect being on the atrio-ventricular node. There is also evidence of Na^+-channel blockade at fast stimulus frequencies. There is a striking absence of effect on repolarization, which lengthens markedly on oral dose as a function of time, probably reaching a peak after many months, at least in humans. Only in part, the latter effect is accounted for by the accumulation of its metabolite, desethylamiodarone [37];

(ii) the drug has a long plasma elimination half - life, which appears to be between 30 and 110 days, although the pharmacodynamic half-life may be considerably longer [53]. No adequate explanation has been found for such a property;

(iii) amiodarone exhibits an unusual propensity for interacting pharmacokinetically with a variety of cardio-active agents;

(iv) the drug was originally designed and synthesized as a coronary dilator and an anti-ischaemic agent. It reduces heart rate almost to the same extent as a ß-blocker, an extent that is not accounted for by the drug's non-specific anti-adrenergic properties. Thus additional as yet unidentified factors are involved. The bradycardiac effects of the drug may play a crucial role not only by reducing myocardial oxygen consumption, but also by reducing temporal dispersion of the refractory period in the myocardium [53];

(v) in line with the general property of class III agents, amiodarone does not appear to significantly depress ventricular function. The drug does not appear to aggravate existing heart failure or to induce cardiac decompensation in those with a markedly reduced left ventricular ejection fraction in the majority of patients. Indeed, the drug is being tested in blind studies against a placebo to determine an impact on mortality in patients with manifest congestive heart failure [54];

(vi) the combination of the drug's complex electropharmacology, pharmacokinetics and pharmacodynamics is unique; it may hold the key to its unusual efficacy and the range of its well-known side effects. A number of its features as an antiarrhythmic agent should be emphasized. First, it is an unusually potent anti-ectopic agent which produces a near complete suppression of nonsustained ventricular tachycardia (fig. 5); it is exceedingly potent in reducing simple and complex PVCs (fig. 3). Yet its potency for suppressing inducible VT/VF after varying periods of time is inconsistent, unpredictable and generally of a low order of effectiveness. Despite the lack of suppression of VT/VF inducible by PES, the clinical outcome in the majority of so-called PES non-responders is favorable [54]. Second, when the drug is given empirically to patients with aborted sud-

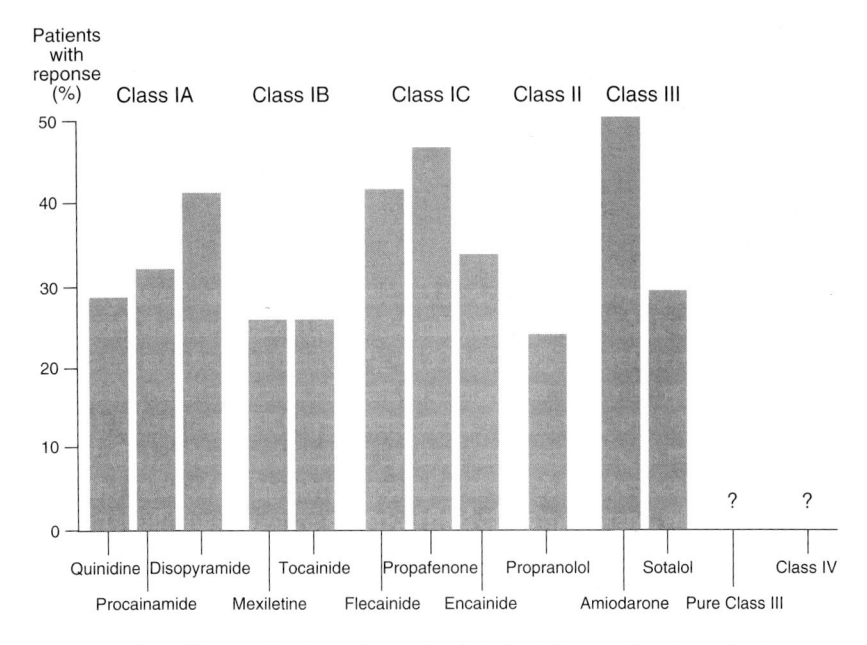

Fig. 5. The comparative efficacy of various electrophysiological classes of antiarrhythmic agents in suppressing nonsustained ventricular tachycardia. Reproduced with permission from [47] and the American Heart Association

den death and to those with manifest VT/VF, sudden death rate appears to be markedly reduced [53, 54], although no controlled trials have been carried out. Perhaps of much interest has been the supportive data from the experimental laboratory. In the case of chronic amiodarone administration, there is not only a significant bradycardia but also a marked prolongation of the ventricular ERP. Patterson et al. [57] showed that this compound was the most potent in preventing VT deteriorating into VF after being administered chronically for three weeks. The drug was found to be more effective than sotalol (fig. 6). It is therefore not surprising that a number of recent editorials and reviews have emphasized that amiodarone therapy is comparable to the implantable devices in their respective effects on sudden death [58, 59]. Based on these data, several clinical trials are being conducted to determine relative merits of the two forms of therapy. Clearly there is a growing consensus that amiodarone treatment might represent the best medical therapy for patients with manifest VT/VF and those surviving cardiac arrest. There is also a growing consensus that the antifibrillatory effects of the drug evident in the case of ventricular tachyarrhythmias, may be equally applicable to atrial fibrillation. In this setting, a low maintenance dose of the drug has been shown to maintain sinus rhythm after cardioversion in a higher proportion of patients than quinidine [60]. Here too, the drug may prove to be the most effective antifibrillatory compound during chronic therapy.

Less extensive data are available in the case of patients at high risk from sudden death. However, a relatively small study in heart failure has shown an improvement in mortality in patients with associated high density PVCs [61]. Similarly, an uncontrolled study in hypertrophic cardiomyopathy, has indicated benefit in relation to the suppression of ventricular tachycardia documented on Holter recordings [62]. Two recent clini-

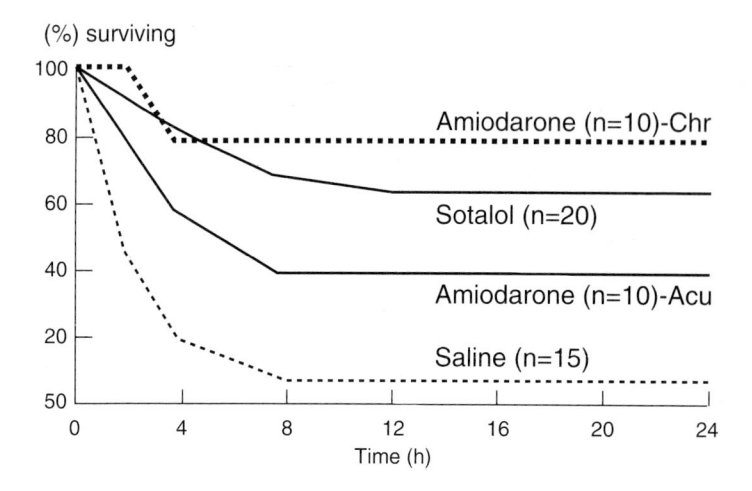

Fig. 6. Antifibrillatory effects of amiodarone in the conscious model of sudden death. Note that survival is the highest for the group of animals treated chronically with amiodarone, higher than that for sotalol and that after the acute intravenous administration of amiodarone. Data based on [57] and [8] with the permission of the authors and of the American Heart Association

cal studies are worthy of mention as they emphasize that amiodarone in relatively low doses exerts a beneficial effect in the survivors of acute myocardial infarction. The first was reported by Burkhardt et al. [63]. They randomized 312 survivors of acute infarction in a group given low dose amiodarone (n = 98), no antiarrhythmic treatment (n = 100) and individualized drug therapy (n = 114). At the end of one year, death rate in the no treatment group was 13%, in the individualized therapy group 10%, and 5% in the case of the amiodarone-treated group (p < 0.05 compared with the no treatment group). In the other study, Ceremuzynski et al. [64] conducted a double-blind study in 305 patients given a amiodarone and 308 patients given placebo who were followed for one year. The patients enrolled were recent survivors of acute myocardial infarction who could not be given ß-blockers because of specific contra-indications. In this study, amiodarone reduced cardiac mortality (p < 0.048); there were 21 deaths in the amiodarone group (6.9%) and 33 (10.7%) in the placebo group (fig. 7). Thus remarkably consistent data are emerging attesting to the high degree of efficacy of amiodarone in numerous subsets of patients with cardiac arrhythmias. Again, as in the case of sotalol in the ESVEM trial, a drug-specific response in the treatment of VT/VF is also supported by the data on amiodarone. The therapeutic responses of amiodarone bear little relationship to PES responses, indicating the importance of the nature of the match between the vulnerable substrate and the complexity of the drug's electropharmacological characteristics [7].

What is the rationale for developing a "pure" class III antifibrillatory agent?

Given the premise that Na^+-channel blockers are unlikely to retain their previous pre-eminent role in the treatment of the majority of cardiac arrhythmias, there is an increasing

Fig. 7. Effect of chronic amiodarone administration compared with a placebo in post-infarct survivors in whom ß-blockade could not be instituted. Amiodarone produced a significant reduction in mortality at 12 months. See text for details. Reproduced from [64] with the permission of the authors and of the Journal of American College of Cardiology

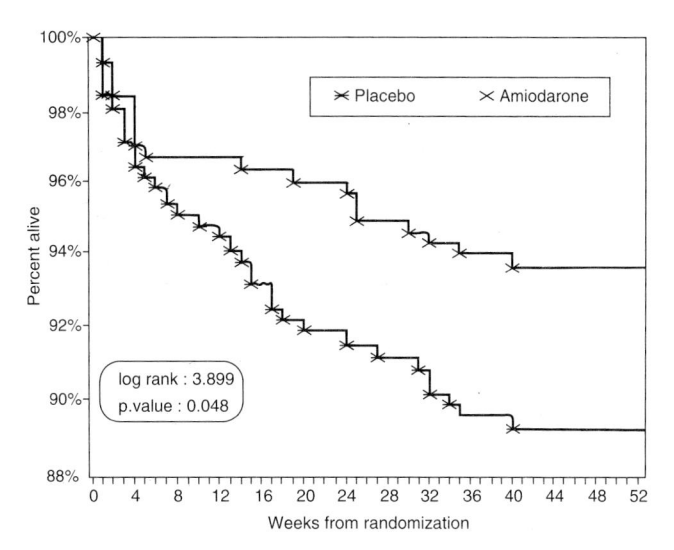

log rank : 3.899
p.value : 0.048

impetus to develop newer class III compounds as substitutes or alternatives to the prototype agents - sotalol and amiodarone. The limitations of these two agents are well defined. The major shortcomings of sotalol are its high degree of ß-blocking activity and the incidence of *torsades de pointe* which has been reported to be between 3 and 5% in the usual or the higher end of therapeutic doses [36]. Amiodarone has neither the ß-blocker side effects nor does it produce a significant incidence of *torsades de pointe*. On the other hand, the drug produces a wide range of side effects, especially at high doses, some of which (e.g. pulmonary fibrosis and hepatotoxicity) may be potentially lethal. The question therefore arose whether some or all of the limitations of sotalol and amiodarone might be overcome by creating compounds that exhibited profiles of a pure class III agent i.e. one in which the electrophysiological effect was confined to the propensity to produce an isolated lengthening of the action potential duration and the effective refractory period.

Numerous such compounds now have been synthesized and are undergoing clinical and experimental evaluation. The structures of some of these compounds are shown in fig. 8. Most are para-substituted benzamides (dofetilide, sematilide, ibutilide, E-4031), others [42] are derivatives of Inpea (e.g. almokalant) and some have more complex structures (e.g. MS-551, ambasilide). These differences have a bearing on their precise electrophysiological profiles. In the case of the para-substituted benzamides, the electrophysiological profile is reasonably predictable. In isolated tissues, these compounds produce concentration-related increases in the action potential duration and in the effective refractory period. There is no effect on conduction velocity in atrial or ventricular muscle or in Purkinje fibres. Similarly, a lack of slow-channel blocking action, or interactions with autonomic transmitters, is associated with no change in the atrio-ventricular node under the action of these compounds. Thus on the surface electrocardiogram these agents produce no effect on the AH, HV, or QRS intervals, but they may slightly widen the RR intervals by virtue of their prolonging the action potential duration in the sinus node fibres, thereby delaying the onset of the next action potential. However, this is usually a modest effect; the net effect on heart rate is small.

A number of "pure" class III agents have been studied in the Lucchesi model of sud-

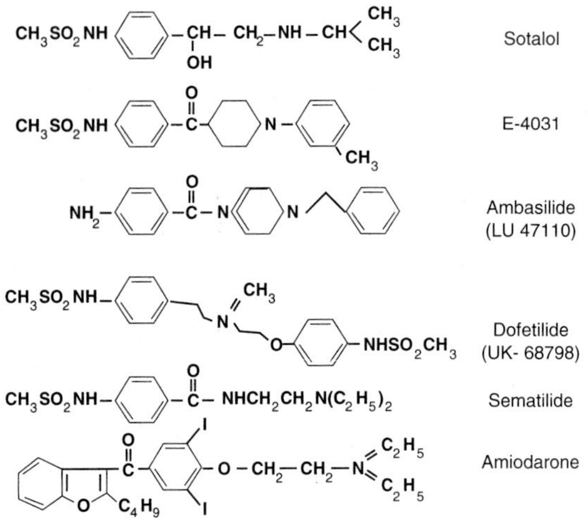

Fig. 8. Chemical structure of some newer class III agents compared with those of amiodarone and sotalol. Note that most newer agents with close structural similarities are likely to have reasonably similar electrophysiological profiles ("pure" class III actions)

den death in the experimental animals. For example, recently Lynch et al. [65] reported that 300 µg/kg of E-4031 administered intravenously prior to programmed electrical stimulation prevented the development of VF in 70% of dogs compared with 30% in the control series. Similar data have been reported for the dextro-isomer of sotalol [66], a compound with a weak ß-blocking activity [44]. D-sotalol functions essentially as a pure class III agent, although it has somewhat greater bradycardic action.

As the list of these so-called class III agents and the data on their antifibrillatory and proarrhythmic actions increase, it is becoming clear that there is a variable spectrum for not only their antifibrillatory actions but also for their proclivity to induce *torsades de pointe* for a given degree of prolongation of repolarization. Thus it is clearly important to define the circumstances under which these agents as a class produce antifibrillatory or antiarrhythmic actions, and under which they exert proarrhythmic actions relative to the fundamental electrophysiological correlates of these effects [40]. Therefore, the challenge is to define the basis for the observed differences in the antifibrillatory and proarrhythmic effects of the older and newer class III compounds in terms of measurable electrophysiological parameters in vitro and in vivo experimental models.

The available data indicate that ventricular fibrillation can be prevented by either ß-blockers or by pure class III agents individually, but their effects may summate as indicated by a potentially greater action in the case of sotalol, a compound exhibiting both actions. Finally, the fact that the effects of amiodarone given chronically are the most striking of those of all the compounds studied indicates that the prolongation of the action potential in an uniform manner merely provides the substrate for the class III action. The precise antifibrillatory and proarrhythmic *(torsades)* potentials of such compounds may therefore depend critically on associated properties. Some of these merit further discussion.

Class III agents exert rate- and use-dependent effects on repolarization and refractoriness

It has long been known that the refractory period in the heart decreases as the interval between beats decreases. An increase in the effective refractory period (ERP) in cardiac muscle is considered the critical determinant of an antifibrillatory action both in atrial [38] as well as ventricular tissue [2, 25, 35]. The influence of stimulation frequency on repolarization and the voltage-dependent refractory period against the background of class III actions is therefore clearly of major practical significance. An important characteristic of an ideal antifibrillatory agent might be the property to increase the effective refractory period, in a relative or an absolute sense, as the rate of the tachycardia accelerates. As indicated by Hondeghem and Snyders [67], Na^+-channel block is most pronounced at fast rates but the lengthening of the action potential duration and hence the voltage-dependent ERP is most striking at slow rates. Thus quinidine at slow rates has little Na^+-channel blocking effect but may markedly increase the action potential duration, a "pure" class III effect. Indeed, nearly all class III agents including n-acetylprocainamide (NAPA) and sotalol exhibit such a phenomenon of reversed use dependence [67]. They increase the ERP most markedly when the heart rate is slow. The newer class III agents (fig. 8), whether they are methanesulphonyl benzamides (such as sematilide, dofetilide, E-4031) or not (e.g. ambasilide), all appear to exhibit such a reversed use dependency with respect to decreases in the APD and ERP as the stimulus frequency is increased. However, quantitatively the frequency-dependent effect of all class III agents is not identical. Ambasilide [68, 69] is of particular interest in that it lengthens the APD in ventricular muscle as well as in Purkinje fibres, but it produces a greater shortening of the APD in Purkinje fibres than in ventricular muscle at frequencies between 30 and 120 beats/minute. However, the differential effect may be less striking at higher frequencies.

There are however two instances in which lengthening of the repolarization time and ERP, compared to baseline, increase or fail to decrease as the stimulus frequency is increased. The first is the case with the chronic administration with amiodarone. Anderson et al. [70] administrered 30 gr of amiodarone over three weeks to dogs and measured its rate-dependent effects at the end of this period and compared them to the results in a control untreated series of animals. They used the method of Sarma et al. [71] to determine the repolarization interval, defined as the interval between the activation time and the repolarization time in a unipolar electrogram. There was a strong correlation between the test-site ERP and the repolarization interval. The increases in the repolarization interval and the refractory periods resulting from amiodarone treatment did not vary with the cycle length over a wide range of stimulation frequencies (200 - 1000 ms). This effect has now been confirmed in humans [72]. The second set of relevant observations are those of Wang et al. [73]; they deal with the class Ic drug, flecainide, which in ventricular tissue exerts a markedly depressant effect on conduction but with a minimal effect on the ERP [74] despite producing a powerful block of the delayed rectifier K^+ current on isolated ventricular myocytes [75]. Wang et al. [73] showed using the atrial transmembrane potentials recorded by the standard microelectrode technique that flecainide and quinidine both demonstrated use dependency in blocking the fast sodium channel function as indicated by the Vmax of the rising phase of the action potential. In contrast, *flecainide markedly increased the action potential duration (and the ERP) as the rate was increased, whereas quinidine exhibited the characteristic reversed use dependency* previously reported in the ventricle. This appears to be an example of the class III action confined to the atrial myocardium at high stimulation frequencies; it provides an unusual electropharmacological probe to determine the ionic current profile of an ideal antiarrhythmic agent with a wider application. Such an effect is consistent with in vivo data in experimental animals in which increases in the refractory period were found to be more predictable determinants of conversion of atrial flutter than decreases in conduction velocity [38].

Proarrhythmia correlates with prolonged cardiac repolarization

It is now well recognized that delaying conduction markedly by class I agents produces a spectrum of what have been considered proarrhythmic effects. These include increase in the numbers of PVCs, runs of non-sustained ventricular tachycardia, ventricular tachycardia, ventricular fibrillation and sudden death [32]. This aggregate of effects has been considered reasonably specific for class I agents. The precise incidence of such proarrhythmic effects varies with individual agents, ranging from possibly 8 to 20% depending on the clinical status of the patient and the degree of slowing of myocardial conduction induced by a particular compound. Undoubtedly, such an incidence of potentially serious proarrhythmic effects must confound the accurate evaluation of the beneficial effect, especially that on mortality.

In the case of class III agents, the pattern of proarrhythmic effects resulting from lengthened cardiac repolarization differs from that of class I agents. The characteristic proarrhythmia complicating prolonged repolarization (long QT or QTc on the surface electrocardiogram) is *torsades de pointe*. Experimental data have indicated that the basis for *torsades de pointe* in most cases is likely to be early after-depolarization (EAD). The overall incidence ranges from less than 1% (e.g. amiodarone) to 5-8% or greater (quinidine, sotalol, n-acetylprocainamide); the figures for the newer so-called pure class III agents are not available, but preliminary data indicates that the overall incidence of *torsades* induced by these agents might be similar to that found with sotalol. The incidence of *torsades* varies not only with individual agents but also with the bradycardia and electrolyte disturbances such as hypokalaemia and hypomagnesaemia. In contrast to the life-threatening VT/VF produced by class I agents, *torsades* due to class III agents usually present with frank or incipient syncope and in only a small number of cases does it deteriorate into ventricular fibrillation. However, it is likely that cases of sudden death that are reported during therapy with class III agents are due to *torsades* deteriorating into ventricular fibrillation.

It is now clear that class III agents produce a spectrum of antifibrillatory and proarrhythmic activity which cannot be related in a systematic or quantitative manner to the simple lengthening of cardiac repolarization (QT or QTc intervals). However, the prolongation of repolarization must provide the substrate for the expression of either of these effects. Which effect prevails in a particular instance must depend on the modulation of the substrate by the clinical setting and the associated fundamental properties of the individual agent [7]. An issue that has become of much interest is whether the known and the emerging differences among various class III agents might be accounted for by their differing selectivity for the block of individual repolarizing currents which determine the time course and the shape of the action potentials in various myocardial tissues. Such currents and their channel kinetics can now be measured reliably in isolated myocytes from a variety of animal species using the whole-cell and patch-clamp techniques. The effects of a large number of antiarrhythmic agents on these channels have been determined by these techniques [76-86].

In the case of class I and class IV antiarrhythmic agents, the electrophysiological changes that depend essentially on the influence of these compounds on a single ion channel are those affecting the fast Na$^+$-channel (i_{Na}) and the slow-Ca^{2+} channel (i_{Ca}) respectively. In the case of class III agents, the issue is exceedingly complex for a number of reasons. First, and as emphasized above, the voltage-dependent refractoriness (the most important correlate of the class III action) is determined essentially by the length of the action potential duration which, in turn, is modulated importantly and on a beat-to-beat basis by the autonomic nervous system. Second, the overall duration of the action potential is determined by at least seven, perhaps discrete, K$^+$ channels (the most important being i_K, i_{K1}, i_{to}, i_{K-ATP}, i_{K-ACh}, see chapter 1). In addition there are contributions from inward currents during phase 2 of the action potential carried by Na through the Na$^+$ "window" current or a slowly inactivating Na$^+$ current, by Ca^{2+}, through Ca^{2+} channel and a possible Ca^{2+} "window" current, as well as by the electrogenic Na/Ca exchange [87]. Most of the newer class III agents appear to exert their principal effects on repolarization by inhibiting potassium channels [86]. However, the voltage and time-dependencies, and the activation and deactivation kinetics of these currents as well as the interactions among them and with the autonomic transmitters during repolarization are extremely complex. Thus it is inherently unlikely that arrhythmias occurring as an abnormality of repolarization are due to the perturbations of a single (what may be called) "arrhythmogenic current". This is suggested by the preliminary data on the effects of class III agents on repolarizing currents [85].

The delayed rectifier K$^+$ current (i_k) appears to be the principal target for most of the newer class III agents. The major issue is to determine the balance between the antifibrillatory and the proarrhythmic effects of class III agents relative to their electrophysiological influences on major repolarizing ionic currents. An attempt has been made to summarize such a correlation in Table 2. In Table 3, a correlation between the occurrence of *torsades de pointe* (clinical instances) as the specific proarrhythmic effect and the degree of protection from the development of VF in the Lucchesi canine ischaemia model of sudden death has been used.

The data need to be interpreted with some caution as they are not all derived from the same animal species under the same experimental conditions and may not have been correlated with changes in action potential duration. For example, in the study dealing with flecainide [75] a powerful inhibitory effect on i_K was found in cat ventricular myocytes, whereas it is known that over a very wide range of drug concentrations in the dog the drug exerted only a trivial effect on the APD [74]. Similarly, the effects of amiodarone in acute superfusion studies showed that the drug had a powerful inhibitory effect on i_K [85]. On the other hand, in such a situation, there is little or no increase in the APD, which only occurs consistently during chronic drug administration. It is, however, noteworthy that clofilium and amiodarone acutely affect both the i_K and i_{K1}, but their antiarrhythmic and proarrhythmic effects are vastly different. Similar comparisons for other agents also indicate that it might be exceedingly difficult to correlate ion-channel data with the clinically relevant in vivo observations dealing with the antiarrhythmic actions of various class III agents [25, 89]. The issue may be equally applicable to the proarrhythmic actions of these agents. No single ion-channel activity has been established as the basis for the development of *torsades,* although it is widely believed that the arrhythmia may

Table 2. Specificity of antiarrhythmic drugs in blocking cardiac ion channels*

Class	Agents	i_K	i_{K1}	i_{to}	i_{Ca}	i_{Na}
IA	Quinidine	Yes	Yes	Yes	Yes	Yes
	Disopyramide	Yes	Yes	Yes	No	Yes
IB	Lidocaine	No	No	No	No	Yes
IC	Flecainide	Yes	No	No	Yes	Yes
	Encainide	Yes	No	No	No	Yes
III	Sotalol	Yes	Yes	Yes	Yes	No
	Amiodarone**	Yes	Yes	No	Yes	Yes
	Clofilium	Yes	No	Yes	No	No
	Risotilide	Yes	No	No	No	No
	E-4031	Yes	No	No	No	No
	Dofetilide	Yes	No	No	No	No
	Tedisamil	Yes	No	Yes	No	No
	Ambasilide	Yes	No	No	No	No
	Terikalant	No	Yes	Yes	No	No
	Almokalant	Yes	No	No	No	No
	MS-551	Yes	No	No	No	No

* Adapted from Colatsky et al. [88]. ** The data on amiodarone apply to acute studies; its relevance to the chronic effects of the drug remains unclear since the antifibrillatory actions of the drug are largely confined to chronic administration (see text)

be due to early after-depolarizations on the basis of the slow-channel activity or that of the Na window current [87]. No clear pattern of correlation emerges.

The antiarrhythmic puzzle of amiodarone is key towards the development of the ideal antifibrillatory compound

While the experimental and clinical data for the expanding list of class III agents continue to reveal moderate efficacy as antifibrillatory agents with an incidence of *torsades* of 5-8%, the experience with chronic amiodarone administration consistently demonstrates marked efficacy with an unusually low arrhythmogenic potential [25, 89]. Amiodarone is effective in controlling VT/VF in over 60-70% of cases when conventional agents (especially class I) fail [55, 56]. It suppresses ectopy, and markedly suppresses non-sustained VT, prevents inducible VT/VF in some cases and slows VT rate; it prolongs survival in post-infarct and in cardiac arrest survivors [63, 64]. Thus, barring its side effect profile, amiodarone is a desirable prototype of a broad-spectrum antifibrillatory and an antiarrhythmic compound. The precise understanding of its cellular action may therefore provide the basis for the development of the ideal agent. On nearly all counts it differs from conventional as well as newer agents. Amiodarone has a complex EP profile that is unique, being characterized by relatively minor effects on Na^+ and Ca^{2+} channels with a broad array of extracardiac actions; its chronic action is dominated by marked prolonga-

Table 3. Specificity of antiarrhythmic drugs in blocking potassium channels relative to antiarrhythmic effects and development of *torsades de pointe****

Class	Agents	i_K	i_{K1}	i_{to}	TDP (Clinical)	SD Protection (Lucchesi experimental model)
IA	Quinidine	Yes	Yes	Yes	5-8%	< 10%
	Disopyramide	Yes	Yes	Yes	5%	< 10%
IB	Lidocaine	No	No	No	0	< 10%
IC	Flecainide	Yes	No	No	±	< 10%
	Encainide	Yes	No	No	±	< 10%
II	ß-Blockers	No	No	No	0	60%
III	Sotalol	Yes	Yes	Yes	5%	70%
	Amiodarone	Yes*	Yes*	No*	< 1%*	80%**
	Clofilium	Yes	No	Yes	10%	< 50%
	Risotilide	Yes	No	No	?	?
	E-4031	Yes	No	No	?	65%
	UK-68,798	Yes	No	No	?	?
	Tedisamil	Yes	No	Yes	?	?
	Ambasilide	Yes	No	No	?	?

* Data refers to acute effects only; ** refers to chronic effects. *** Adapted from Singh et al. [25]. Abbreviations: *TDP* = *torsades de pointe*; *SD* = sudden death

tion of APD and refractoriness with a modest effect on conduction evident at fast frequencies [8, 89]. Unlike other class III agents, the drug lengthens the action potential duration more in the ventricular muscle than in the Purkinje fibres, which may be the locus of origin of early after-depolarizations. Other class III agents lengthen the APD a great deal more in the Purkinje fibres than in ventricular muscle. Like ß-blockers, amiodarone reduces heart rate but with a slower onset and as a function of time, reaching a peak in three months [53]. Recent data from our laboratory indicate that the drug reduces temporal [53] as well as spatial [90] dispersion of repolarization in man. There are many features of the drug's action that suggest a metabolic basis of cellular and subcellular action. An intriguing possibility is that it might in part act by cardiospecifically antagonizing thyroid hormone action at a receptor level [34, 42, 89].

Amiodarone is structurally unique, resembling thyroxine [89]. It is iodinated; many of its actions overlap with those of myocardial hypothyroidism [89]. Thus the question arose whether a major component of amiodarone action is mediated via selective myocardial T_3-receptor inhibition as originally suggested [34]. Now there is substantial evidence in support of this assumption [89]. This effect is not due to the iodine contained in the amiodarone molecule; iodine alone, in doses equivalent to those contained in the effective dose of amiodarone, had no effect on atrial action potentials [34]. However, the concomitant administration of amiodarone and thyroid hormone prevented the development of repolarization changes evident after amiodarone alone [34]. This indicated that the electrophysiological effect of amiodarone may in part be mediated by a selective T_3 block in cardiac muscle as the inhibition of the peripheral conversion of T_4 to T_3 [91], resulting in a decrease in T_3, an increase in reverse T_3 and a minimal increase in T_4 in the plasma due to the blockade of 5'-monodeiodinase [92], could account for the observed EP changes. Thus, a direct inhibition of T_3 nuclear binding by amiodarone and/or its metabolite desethylamiodarone [93] has been postulated to result in a hypothyroid state at a cellular level [89]. Since the electrophysiological effects of hypothyroidism [94] on repolarization are nearly identical to those found after long-term amiodarone treatment, this phenomenon appears to exhibit cardiospecificity [34]. The drug generally does not produce generalized hypothyroidism [89]. Recently, Talajic et al. [95] reported a marked attenuation of amiodarone class III and sinus node effects in hypothyroid guinea-pigs. The pattern of ß-receptor density in the case of chronic amiodarone administration and in hypothyroidism also appears nearly identical [89]. In the rat, the changes in heart myosin isoenzymes following chronic amiodarone treatment has been shown to be identical to those found in hypothyroidism [96]. This was confirmed by Nag et al. [97] in studies involving rat cardiac muscle cell culture. Cardiac myocytes exposed to amiodarone in the absence of T_3 showed predominant isomyosin V_1. When they were exposed to amiodarone in the presence of T_3, they expressed prevalent isomyosin V_3, or both V_3 and V_1 equally. Supraphysiological concentrations of T_3 counteracted amiodarone effects, showing the expression of predominant isomyosin V_1. These findings are also consistent with a direct receptor interaction between amiodarone and T_3. A competitive interaction between T_3 and amiodarone in acute studies has recently been found by Cui et al. [98]. Hensley et al. [99] recently studied the effects of chronic amiodarone administration on Na/K-ATPase α-2 and ß-2 expression in the rat ventricle. As in the case of hypothyroidism, the α-2 isoform was significantly depressed by amiodarone. The lack of effect on α-1 and ß-1 expression in heart at six weeks, and the lack of effect on skeletal Na/K-ATPase expression, all seen in hypothyroidism, indicated that amiodarone's actions are isoform and tissue specific.

Amiodarone prevents electrically-induced early and late after-depolarizations [100]; it prevents Ba^{2+}-induced EADs [101], this may in part be related to the drug's associated Ca^{2+}-channel blocking effects inhibiting the propensity for EADs to develop [101]. It is of interest that in hyperthyroidism, i_{Ca} is increased, the converse in hypothyroidism [102]; the electrophysiological effects of hypothyroidism closely resembles those of chronic amiodarone administration [34]. In hyperthyroid ventricular muscle EADs develop (blocked by verapamil) when the muscle is driven at low frequencies [102]. Finally, the rate-dependent effects on the action potential duration in the case of hypothyroidism, at least in the atria [94], are identical to those following chronic amiodarone administration [70-72].

Thus there is compelling data suggesting that amiodarone might act fundamentally by competing with T_3 effect. If this were the case, a number of conclusions about the properties of amiodarone can be drawn: (i) the effects of the drug on repolarization, as with hypothyroidism [94], will be homogeneous from cell to cell with little or no dispersion of refractoriness. This will minimize focal re-excitation and re-entry; (ii) the long refractory period with little change in conduction will be antifibrillatory in all heart tissues; (iii) since the overall myocardial effects will parallel the time course of hormone action, the net steady-state effect of the drug will have a significant latency and the pharmacological half-life of the drug will outlast the plasma half-life of the drug, there being little likelihood of a relationship between serum, tissue or sarcolemmal drug levels and effect as shown in our laboratories [87]; (iv) the attenuation of i_{Ca} in amiodaronized or hypothyroid muscle will lead to a reduced proclivity for the development of EADs and hence *torsades de pointe;* (v) "cardioselective" hypothyroidism created by amiodarone and its own intrinsic non-competitive anti-adrenergic effect will summate to produce bradycardia and reduce oxygen consumption, properties of obvious utility in patients with coronary artery disease experiencing VT/VF. In the absence of the hypothyroid effect, the degree of observed bradycardia induced by amiodarone is unexplained. Finally, there are preliminary data from radio-ligand binding studies which have indicated that amiodarone and its principal metabolite, desethylamiodarone, may indeed function as thyroid hormone antagonists in different tissues. Particularly convincing are the data on rat pituitary cells reported recently by Norman and Lavin [104]. If such an effect occurred in the heart with a significant measure of tissue specificity, an entirely novel approach to effective class III antifibrillatory and antiarrhythmic actions might be the synthetic design of cardiospecific T_3 antagonists.

Increasing concern in recent years regarding the seriousness of the proarrhythmic effects of class I agents has drawn attention to the antifibrillatory compounds that act essentially by prolonging myocardial repolarization. The available data allow a number of tentative conclusions. First, the ß-blocker data in a variety of subset of patients emphasize the importance of ß-blockade as a major antiarrhythmic mechanism in its own right while being an integral component in more complex molecules such sotalol and amiodarone. Second, there is sufficient evidence that such drugs exert a varying spectrum of antifibrillatory and proarrhythmic (characterized by *torsades de pointe*) actions for a given degree of prolongation of repolarization. These differences are not readily accountable in terms of specificity of their actions for single or multiple repolarizing myocardial ion currents. Third, there are differences between the so-called pure class III agents such as sematilide, dofetilide and E-4031 on the one hand and more complex compounds such as sotalol and amiodarone, which also exert anti-adrenergic actions, on the other. It is unclear whether one should aim at the development of antifibrillatory compounds that are relatively simple molecules with clearly defined electrophysiological profiles in terms of actions on ion channels, currents, receptors and pumps, or those with complex electropharmacological profiles. The available data suggest the latter approach holds a greater promise for the synthetic design of agents with potent antiarrhythmic and antifibrillatory actions and minimal proarrhythmic reactions. Amiodarone appears to be a prototype.

Supported in part by Medical Research funds of the Veterans Administration (Washington, DC) and the American Heart Association, the Greater Los Angeles Affiliate, Los Angeles, California. In preparing this chapter, we have relied substantively on two papers [25, 36] on a similar subject by one of us (BNS).

References

1. Singh BN (1990) Advantages of ß-blockers versus antiarrhythmic drugs and calcium-channel antagonists in secondary prevention in survivors of myocardial infarction. Am J Cardiol 66: 9-20

2. Singh BN (1988) Control of cardiac arrhythmias by lengthening repolarization. Futura Publishing Co Inc., Mount Kisco, NY
3. The Cardiac Arrhythmia Suppression Trial (CAST) Investigators (1989) Preliminary report: effect of encainide and flecainide on mortality in a randomized trial of arrhythmia suppression after myocardial infarction. N Eng J Med 321 406-412
4. The Cardiac Arrhythmia Suppression Trial II Investigators (1992) Effect of the antiarrhythmic agent moricizine on survival after myocardial infarction. N Eng J Med 327: 227-233
5. Yusuf S, Teo KK (1991) Approaches to prevention of sudden death: need for fundamental reevaluation. J Cardiovasc Electrophysiol 2 (3): S233-S239
6. Singh BN (1990) Control cardiac arrhythmias with sotalol, a broad-spectrum antiarrhythmic with ß- blocking effects and class III activity. Am J Cardiol 765: 1A-84A
7. Singh BN (1993) Controlling cardiac arrhythmias by delaying conduction and predicting responses by Holter monitoring and electrophysiologic testing. A point counterpoint. Submitted to Circulation
8. Lynch JJ, Lucchesi BR (1987) How are animal models best used for the study of antiarrhythmic drugs? In: Hearse DJ, Mouring AS, Janse MJ (eds) Life-threatening arrhythmias during ischaemia and infarction. Raven Press, New York, NY, pp 169-196
9. Kou WH, Nelson SD, Lynch JJ, Montgomery DG, DiCarlo L, Lucchesi BR (1987) Effect of flecainide acetate of prevention of electrical induction of ventricular tachycardia and occurrence of ischaemic ventricular fibrillation during the early postmyocardial infarction period. Evaluation in a conscious canine model of sudden death. J Am Coll Cardiol 9: 359-365
10. Patterson E, Lucchesi BR (1983) Quinidine gluconate in chronic myocardial ischaemic injury. Differential effects in response to programmed stimulation and acute myocardial ischaemia in the dog. Circulation 68: 111-118
11. Wallace A, Gehret J, Heaney L, Stupienski R, Stein R, Lynch JJ (1991) Failure of lidocaine to prevent lethal ischaemic ventricular arrhythmias in a canine model of previous myocardial infarction. FASEB J (in press)
12. Nattel S, Pedersen DH, Zipes DP (1981) Alterations in regional myocardial distribution and arrhythmogenic effects of aprindine produced by coronary artery occlusion in the dog. Cardiovasc Res 15: 80-86
13. Furukawa T, Rozanski JJ, Monroe K, Gosselin AJ, Lister JR (1989) Efficacy of procainamide on ventricular tachycardia: relation to prolongation of refractoriness and slowing of conduction. Am Heart J 118: 702-708
14. Kus T, Dubue M, Lambert C, Shenasa M(1990) Efficacy of propafenone in preventing ventricular tachycardia, with insights from analysis of resetting response patterns. Am J Cardiol 15: 1229-1237
15. Furberg CD (1983) Effects of antiarrhythmic drugs on mortality after myocardial infarction. Am J Cardiol 53: 32C-36C
16. Hallstrom AP, Cobb LA, Yu BH, Weaver WD, Fahrenbruch CE (1991) An antiarrhythmic drug experience in 941 patients resuscitated from an initial cardiac arrest between 1970 and 1985. Am J Cardiol 68: 1025-1031
17. Coplen SE, Antman EM, Berlin JA, Hewitt P, Chalmers TC (1990) Efficacy and safety of quinidine therapy for the maintenance of sinus rhythm after cardioversion. A meta-analysis of randomized control trials. Circulation 82: 1106-1116
18. Flaker GC, Blackshear JL, McBride R, Kronmal RA, Halperin JL, Hart RG (1992) Antiarrhythmic drug therapy and cardiac mortality in atrial fibrillation. J Am Coll Cardiol 20: 427-532
19. Morganroth J, Goin JE (1991) Quinidine-related mortality in the short-to-medium-term treatment of ventricular arrhythmias: A meta-analysis. Circulation 84: 1977-1983
20. Salerno DM (1991) Quinidine: worse than adverse? Circulation 84: 2196-2198
21. Antman EM, Berlin JA (1992) Declining incidence of ventricular fibrillation in myocardial infarction: implications or the prophylactic use of lidocaine. Circulation 86: 764-773

22. Singh BN (1992) Routine prophylactic lidocaine administration in acute myocardial infarction: an idea whose time is all but gone? Circulation 86: 1033-1035
23. MacMahan S, Collins R, Peto R, Koster RW, Yusuf S (1988) Effects of prophylactic lidocaine in suspected acute myocardial infarction. JAMA 20: 1910-1916
24. IMPACT Research Group (1984) International mexiletine and placebo antiarrhythmic coronary trial. 1. Report on arrhythmias and other findings. J Am Coll Cardiol 4: 1148-1156
25. Singh BN, Sarma, JSM, Zhang ZH, Takanaka C (1992) Controlling cardiac arrhythmias by lengthening repolarization: rationale from experimental findings and clinical considerations. Ann NY Acad Sci 644: 187-209
26. Schwartz PJ (1985) The idiopathic long QT syndrome: progress and questions. Am Heart J 109: 399-411
27. Steinbeck G, Andresen D, Bach P, Haberl R, Oeff M, Hoffmann E, Von Leitner E-R (1992) A comparison of electrophysiologically-guided antiarrhythmic drug therapy with ß- blocker in patients with symptomatic, sustained ventricular tachyarrhythmias. N Eng J Med 327: 987-992
28. Anderson JL, Rodier HE, Green LS (1983) Comparative effects of ß-adrenergic blocking drugs on experimental ventricular fibrillation threshold. Am J Cardiol 51: 1196-202
29. Kent KM, Smith ER, Redwood DR, Epstein SE (1973) Electrical stability of acutely ischaemic myocardium. Influences of heart rate and vagal stimulation. Circulation 67: 291-298
30. Kjekshus JK (1986) Importance of heart rate in determining ß-blocking efficacy in acute and long-term myocardial infarction trials. Am J Cardiol 57: 43F-49F
31. Raine AEG, Vaughan Williams EM (1981) Adaptation to prolonged ß-blockade on rabbit atrial, Purkinje fibres and ventricular potentials and papillary muscle contractions. Circ Res 48: 804-815
32. Singh BN, Courtney K (1990) On the classification of antiarrhythmic mechanisms: experimental and clinical correlations. In: Zipes DP, Jalife J (eds) Cardiac electrophysiology: from the cell to the bedside. WB Saunders, Philadelphia, pp. 882-897
33. Singh BN, Vaughan Williams EM (1970) A third class of antiarrhythmic action. Effects on atrial and ventricular intracellular potentials, and other pharmacological actions on cardiac muscle, of MJ 1999 and AH 3474. Br J Pharmacol 39: 675-687
34. Singh BN, Vaughan Williams EM (1970) The effect of amiodarone, a new anti-anginal drug, on cardiac muscle. Br J Pharmacol 39: 657-667
35. Singh BN, Nademanee K (1985) Control of arrhythmias by selective lengthening of cardiac repolarization: theoretical considerations and clinical observations. Am Heart J 109: 421-430
36. Singh BN (1992) Antiarrhythmic action of DL-sotalol in ventricular and supraventricular arrhythmias. J Cardiovasc Pharmacol 20: 575-590
37. Singh BN, Venkatesh N, Nademanee K, Josephson MA, Kannan R (1989) The historical development, cellular electrophysiology and clinical pharmacology of amiodarone. Prog Cardiovasc Dis 31: 249-280
38. Feld GK, Venkatesh N, Singh BN (1986) Pharmacologic conversion and suppression of experimental canine atrial flutter. Circulation 74: 147-204
39. Singh BN (1983) Amiodarone; historical development and pharmacologic profile. Am Heart J 106: 788-796
40. Singh BN (1988) When is QT prolongation antiarrhythmic and when is it pro-arrhythmic? Am J Cardiol 63: 867-869
41. Kaumann AJ, Olsson C (1968) Temporal relationship between long-lasting aftercontractions and action potentials in cat papillary muscles. Science 163: 293-29
42. Singh BN (1992) Pharmacologic actions of certain drugs and hormones: focus on studies of antiarrhythmic mechanisms. D Phil Thesis, Hertford College, University of Oxford, 1971 Futura Publishing Co., Mt Kisco, NY
43. Bacaner MB (1968) Treatment of ventricular fibrillation and other acute arrhythmias with bretylium tosylate. Am J Cardiol 21: 530-538

44. Kato R, Yabek S, Ikeda N, Kannan R, Singh BN (1986) Electrophysiologic effects of dextro and levo-isomers of sotalol in isolated cardiac muscle and their in vivo pharmacokinetics. J Am Coll Cardiol 7: 116-125
45. Patterson E, Lucchesi BR (1982) Antifibrillatory actions of nadolol. J Pharmacol Exp Ther 223: 144-152
46. Deedwania PC (1990) Suppressant effects of conventional ß- blockers and sotalol on complex and repetitive ventricular premature complexes. Am J Cardiol 65: 43A-50A
47. Salerno D, Gillingham KJ, Berry DA, Hodges M (1990) A comparison of antiarrhythmic drugs for the suppression of ventricular ectopic depolarization. A meta-analysis. Am Heart J 120: 340-353
48. Senges J, Lengfelder W, Jauernig R, et al. (1984) Electrophysiologic testing of therapy with sotalol for sustained ventricular tachycardia. Circulation 69: 577-583
49. Nademanee K, Feld G, Hendrickson JA, Singh PN, Singh BN (1985) Electrophysiologic and antiarrhythmic effects of sotalol in patients with life-threatening ventricular tachycardia. Circulation 72: 555-563
50. Deedwania PC, Singh BN (1992) Sotalol: a unique ß-blocker with class III antiarrhythmic properties. In: Deedwania PC (eds) ß-Blockers and cardiac arrhythmias. Marcel Dekker Inc., NY, pp 253-292
51. Nattel S (1991) Antiarrhythmic drug classifications. A critical appraisal of their history, present status and clinical relevance. Drugs 41: 672-701
52. Harrington C (1992) ESVEM Findings: Holter equal to EP testing for predicting VT drug efficacy; sotalol triumphs. Cardio 6: 10-14
53. Antimisiaris M, Sarma JMS, Schoenbaum M, Pollard LM, Venkataraman K, Singh BN (1993) Amiodarone abolished circadian variability of QTc but not that of heart rate: significance for prevention of sudden death. J Am Coll Cardiol (Submitted)
54. Singh S, Fletcher RD, Fisher S, Deedwania PC, Lewis D, Massie B, Singh BN, Colling RP, CHF Stat Investigators (1992) Congestive heart failure: survival trial of antiarrhythmic therapy (CHFSTAT). Cont Clin Trials 13: 339-350
55. Herre JM, Sauve MJ, Griffin JC, Helmy L, Langberg JJ, Goldberg H, Scheinman MM (1989) Longterm results of amiodarone therapy in patients with recurrent sustained ventricular tachycardia or ventricular fibrillation. J Am Coll Cardiol 13: 442-449
56. Weinberg BA, Miles WM, Klein LS, Bolander JE, Dusman RE, Stanton MS, Heger JJ, Langefeld C, Zipes DP (1993) 5 year follow-up of 589 patients treated with amiodarone. Am Heart J (in press)
57. Patterson E, Eller BT, Abrams GD, Vasilades J, Lucchesi BN (1983) Ventricular fibrillation in conscious canine preparation of sudden coronary death. Prevention by short and longterm amiodarone administration. Circulation 68: 857-864
58. Connolly S, Yusuf L (1992) Evaluation of the implantable cardioverter defibrillator in survivors of cardiac arrest: The need for randomized trials. Am J Cardiol 69: 959-962
59. Kim SG (1993) Implantable defibrillator therapy: does it really prolong life? How can we prove it? Am J Cardiol (in press)
60. Middlekauff HR, Wiener I, Saxon LA, Stevenson WG (1992) Low-dose amiodarone for atrial fibrillation: time for a prospective study? Ann Int Med 116: 1017-1020
61. Cleland JG, Dargie HJ, Ford L (1987) Morbidity in heart failure: clinical variables of prognostic value. Br Heart J 58: 572-582
62. McKenna WJ, Oakley CM, Krikler DM, Goodwin JR (1985) Improved survival with amiodarone in patients with hypertrophic cardiomyopathy and ventricular tachycardia. Br Heart J 53: 412-420
63. Burkart F, Pfisterer M, Kiowski W, Follath F, Burckhardt D (1990) Effect of antiarrhythmic therapy on mortality in survivors of myocardial infarction with asymptomatic complex ventricular arrhythmias: basel antiarrhythmic Study of Infarct Survival (BASIS). J Am Coll Cardiol 16: 1711-1718

64. Ceremuzynski Y, Kleczar E, Kreminska-Pakula M, et al. (1992) Effect of amiodarone on mortality after myocardial infarction. J Am Coll Cardiol 20: 1056-1062
65. Lynch JJ Jr., Heaney LA, Wallace AA, Gehret JR, Selnick HG, Stein RB (1990) Suppression of lethal ischaemic ventricular arrhythmias by the class III agent E4031 in a canine model of previous myocardial infarction. J Cardiovasc Pharmacol 15: 764-775
66. Lynch JJ, Coskey LA, Montgomery DG, Lucchesi DR (1985) Prevention of ventricular fibrillation by dextro-rotatory sotalol in a conscious canine model of sudden coronary death. Am Heart J 109: 949-958
67. Hondeghem LM, Snyders DJ (1990) Class III antiarrhythmic agents have a lot of potential but a long way to go: reduced effectiveness and dangers of reverse use dependence. Circulation 81: 686-690
68. Takanaka C, Sarma JSM, Singh BN (1992) Electrophysiologic actions of ambasilide (LU47110), a novel class III agent, on the properties of isolated rabbit and canine cardiac muscle. J Cardiovasc Pharmacol 19: 290-298
69. Zhang Z-H, Follmer CH, Sarma JSM, Chen F, Singh BN (1992) The effects of ambasilide (LU47110), a new class III agent, on outward plateau currents in isolated guinea-pig ventricular myocytes. J Pharmacol Exp Therap 253: 40-48
70. Anderson KP, Walker R, Dustman T, Lux RL, Ershler PR, Kates RE, Urie PM (1989) Rate-related electrophysiologic effects of longterm administration of amiodarone on canine ventricular myocardium in vivo. Circulation 79: 948-958
71. Sarma JSM, Bilitch M, Melinte SG (1983) Ventricular refractory periods in relation to rate and test-site VT intervals in anaesthetized and conscious dogs: a canine model for conscious state measurements. Pace 6: 735-745
72. Sager PT, Uppal P, Follmer C, Antimisiaris M, Pruitt C, Singh BN (1993) The frequency-dependent electrophysiologic effects of amiodarone in humans. Circulation (in press)
73. Wang Z, Pelletier LC, Talajic M, Nattel S (1990) Effects of flecainide and quinidine on human atrial action potentials. Circulation 82: 274-283
74. Ikeda N, Singh BN, Davis LD, Hauswirth C (1985) Effects of flecainide on the electrophysiologic properties of isolated canine and rabbit myocardial fibers. J Am Coll Cardiol 5: 303-312
75. Follmer CH, Colatsky TJ (1990) Block of the delayed rectifier potassium current Ik by flecainide and E-4301 in cat ventricular myocytes. Circulation 82: 289-293
76. Rasmussen HS, Allen MJ, Blackburn KJ, Butrous GS, Dalrymple HW (1992) Dofetilide, a novel class III antiarrhythmic agent. J Cardiovasc Pharmacol 20: S96-S105
77. Roden DM, Bennett PB, Snyders DJ, Balser JR, Hondeghem LM (1988) Quinidine delays IK activation in guinea pig ventricular myocytes. Circ Res 62: 1055-1058
78. Arena JP, Kass RS (1988) Block of heart potassium channels by clofilium and its tertiary analogs: relationship between drug structure and type of channel blocked. Mol Pharmacol 34: 60-66
79. Hiraoka M, Sawada K, Kawano S (1986) Effects of quinidine on plateau currents of guinea pig ventricular myocytes. J Mol Cell Cardiol 18: 1097-1106
80. Salata JJ, Wasserstrom JA (1988) Effects of quinidine on action potentials and ionic currents in isolated canine ventricular myocytes. Circ Res 62: 324-337
81. Imaizumi Y, Giles WR (1987) Quinidine-induced inhibition of transient outward current in cardiac muscle. Am J Phyiol 253: H704-H708
82. Carmeliet E (1985) Electrophysiologic and voltage-clamp analysis of the effects of sotalol on isolated cardiac muscle and Purkinje fibres. J Pharmacol Exp Ther 232: 817-825
83. Berger F, Borchard U, Hafner D (1989) Effects of (+)- and (±)- sotalol on repolarizing currents and pacemaker current in sheep cardiac Purkinje fibres. Naunyn Schmiedeberg's Arch Pharmacol 340: 696-704
84. Balser JR, Hondeghem LM, Roden DM (1987) Amiodarone reduces time dependent Ik activation Circulation 76 (Suppl IV): IV-151
85. Sato R, Hisatome I, Singer D (1987) Amiodarone blocks an inward rectifier K^+ channel in gui-

nea pig ventricular myocytes. Circulation (Suppl IV): IV-150
86. Sawada K (1989) Depression of the delayed rectifier K^+ current by a novel class III antiarrhythmic agent E-4031 in guinea pig single ventricular myocytes (abstract). J Mol Cell Cardiol 21: S20
87. El-Sherif N, Craelius W, Boutjdir M, Gough WB (1990) Early afterdepolarizations and arrhythmogenesis. J Cardiovasc Electrophysiol 1: 145-160
88. Colatsky TJ, Follmer CH, Starmer CF (1990) Channel specificity in antiarrhythmic drug action. Mechanism of potassium-channel block and its role in suppressing and aggravating cardiac arrhythmias. Circulation 82: 2235-2242
89. Singh BN (1988) Electropharmacology of amiodarone. In: BN Singh (ed) Control of arrhythmias by lengthening of repolarization. Futura Publishing Co., Mt. Kisco, NY, pp 367-400
90. Cui G, Sen L, Uppal P, Sager PT, Singh BN (1992) Comparison of QT dispersion and RR interval induced by sematilide, amiodarone and sotalol. Circulation 86: I-393
91. Hershman JM, Nademanee K, Sugawara M, Pekary A, DiSteffano DG, Singh BN (1986) Thyroxine and triiodothyronine kinetics in cardiac patients taking amiodarone. Acta Endocrin 111: 193-199
92. Sogol PB, Hershman JM, Reed AW, Dillmann WH (1983) The effects of amiodarone on serum thyroid hormones and hepatic thyroxine 5'-monodeiodnination in rats. Endocrinol 113: 1464-1469
93. Latham KR, Selletti DF, Goldstein RE (1970) Interaction of amiodarone and desethylamiodarone with solubilized nuclear thyroid hormone receptors. J Am Coll Cardiol 1987: 872-881.
94. Freedberg AS, Papp GJ, Vaughan Williams EM (1970) The effects of altered thyroid state on atrial intracellular potentials. J Physiol (Lond) 207: 357-368
95. Talajic M, Nattel S, Davies M, McCann J (1990) Attenuation of class 3 and sinus node effects of amiodarone by experimental hypothyroidism. J Cardiovasc Pharmacol 13: 447-450.
96. Bagchi N, Brown TR, Schneider DS, Banerjee SK (1990) Effect of amiodarone on rat heart isoenzymes. Circ Res 67: 51-60
97. Nag AC, Lee Mei Li, Shepherd D (1990) Effect of amiodarone and the expression of myosin isoforms and cellular growth of cardiac muscle cells in culture. Circ Res 67: 51-60
98. Cui, G, Sen L, Singh BN (1992) Interaction of amiodarone with T_3 on Na^+ current in guinea-pig cardiac myocytes. J Am Coll Cardiol 19: 347A
99. Hensley CB, Bersohn MM, Sarma JSM, Singh BN, McDonough AA (1992) Amiodarone decreases Na, K- ATPase α-2 and ß- 2 expression specifically in cardiac ventricle. Submitted to Circ Res
100. Ohta M, Karaguezian HS, Mandel WJ, Peter TC (1987) Acute and chronic effects of amiodarone on delayed afterdepolarization and triggered automaticity in rabbit ventricular myocardium. Am Heart J 1214: 289-298
101. Takanaka C, Singh BN (1990) Barium-induced non-driven action potentials as a model of triggered automaticity and early afterdepolarizations: differing effects of amiodarone, and quinidine and significance of slow channel activity. J Am Coll Cardiol 15: 213-221
102. Binah O, Bernstein I, Gilat E (1970) Effects of thyroid hormone on the action potential and membrane currents of guinea pig ventricular myocytes. Pflügers Arch 409: 214-216
103. Sharp NA, Neel DS, Parsons RL (1985) Influence of thyroid hormone levels on the electrical and mechanical properties of rabbit papillary muscle. J Mol Cell Cardiol 17: 119-132
104. Norman MF, Lavin TA (1989) Competition of thyroid hormone action by amiodarone in rat pituitary tumor cells. J Clin Invest 83: 306-313

CHAPTER 15
Adenosine, ATP-sensitive potassium channels and myocardial preconditioning

K Mullane

> *"Twas brillig, and the slithy toves did gyre and gimble in the wabe; All mimsy were the borogroves, and the mome raths outgrabe"*
>
> *"It seems very pretty"*, she said when she had finished it, but it's rather hard to understand. (You see she didn't like to confess even to herself, that she couldn't make it out at all). *"Somehow it seems to fill my head with ideas - only I don't exactly know what they are."* [1]

Blockade of a major coronary artery renders a portion of the myocardium ischaemic. If unresolved that portion of the heart will become irreversibly damaged (infarcted). It is well documented that the size of a myocardial infarct depends on the duration and severity of the ischaemic insult; where severity is dictated by the area of myocardium at risk (i.e. the area of the heart served by the occluded blood vessel), the degree of residual or collateral blood flow and the rate of oxygen consumption at the time of coronary occlusion [2]. The introduction of thrombolytic therapy to limit the extension of myocardial infarct has resulted in a decrease in cardiac morbidity and mortality [3]. However, the restoration of blood flow to the ischaemic myocardium can also result in "reperfusion injury", where the ultimate degree of tissue damage is greater than would be expected if the reintroduction of blood and nutrients arrested all damage [4]. Consequently, scientists continue to seek new therapeutic strategies to decrease further the extent of myocardial damage associated with acute myocardial infarction.

Although some of the factors that contribute to myocardial infarction have been established, the exact biochemical mechanisms of cell death remain unknown. In an attempt to separate the deleterious consequences of the build-up of toxic metabolites formed during ischaemia (e.g. lactate, hydrogen ions, potassium) from the loss of high energy phosphates, Reimer and colleagues [5] used a model of four brief periods of ischaemia with intervening periods of reperfusion to permit the washout of the toxic products. These investigators [5] observed that ATP levels did not drop beyond the level achieved after the first bout of ischaemia, and no infarct was found in six of the seven dogs studied. Subsequently, this group found that these brief periods of ischaemia and reperfusion protected the heart from a more prolonged period of ischaemia [6]. This paradoxical induction of tolerance to a prolonged ischaemic insult by a prior brief exposure to ischaemia has been called "ischaemic preconditioning"[6].

Preconditioning is an all-or-nothing response

Various protocols to induce ischaemic preconditioning have been used, from multiple occlusions to a single period of ischaemia. Preconditioning is not provoked by 2.5 min of ischaemia, but it can be triggered by a single 5 min ischaemic episode, and is no greater after 4 x 5 min of ischaemia [7]. A reperfusion phase between the preconditioning ischaemia and the prolonged ischaemic insult of greater than 1 min [8] but less than 120 min is required [7]. After 120 min the response has decayed and any protective benefits derived from the conditioning stimulus are lost [7]. The duration of the final ischaemic insult is also important. In dogs preconditioning protects against 60 min ischaemia, but not 90 min ischaemia [9]. Thus it appears that preconditioning improves the heart's ability to tolerate ischaemia, but this protection is not absolute and myocellular injury is only delayed.

While ischaemic preconditioning has been described in a number of species including dogs, pigs, rabbits and rats, the precise temporal pattern can vary [7, 10]. Preconditioning can also be triggered by hypoxia [11] and intermittent thrombotic occlusions [12], but not by rapid atrial pacing [13] suggesting that a quite severe ischaemic insult is required to trigger the response.

There are two other important features of ischaemic preconditioning. One is that it has been demonstrated universally, with no known studies which fail to show a reduction in infarct size in preconditioned hearts. Secondly, the magnitude of the protection is impressive. For example, in rabbits subjected to 30 min ischaemia and reperfusion, infarct size is reduced from ~40% to ~10% of the area at risk [7].

Ischaemic preconditioning is reported to protect the heart from other consequences of ischaemia-reperfusion, including reperfusion arrhythmias [14] and post-ischaemic contractile failure [15, 16]. These phenomena have been termed arrhythmia preconditioning and functional preconditioning, respectively, to differentiate them from the infarct size-limiting preconditioning. Because these events are more controversial and may occur by mechanisms independent of those involved in infarct size-limiting preconditioning, they will not be considered in this review.

Finally, a number of mechanisms for preconditioning the heart with ischaemia have been proposed (see the excellent reviews by Walker and Yellon, [10]; Downey et al. [7]), but the strongest evidence to date favours a central role for adenosine, possibly acting to increase the open-state probability of ATP-sensitive K^+ channels. The critical evaluation of this hypothesis represents the basis of this review.

Preconditioning is linked to adenosine

The proposal that adenosine might be the mediator of preconditioning was originated by Downey and co-workers [17]. Based upon the rapid accumulation of adenosine in the ischaemic myocardium [18] coupled with reports on the cardioprotective effects of adenosine by an action on A_1-receptors [19], Downey et al. [17] examined the ability of two adenosine receptor antagonists, 8-sulphophenyltheophylline or PD 115,199 to block preconditioning. Both agents completely blocked the protection provided by preconditioning, while having no effect on infarct size in naive (unconditioned) hearts.

Moreover, in isolated blood-perfused hearts, a 5 min infusion of adenosine could substitute for the preconditioning ischaemia to protect the heart [17]. Because adenosine is extremely labile in blood with a half-life of seconds [18], there was no adenosine present during the second period of ischaemia. Consequently, the adenosine must have triggered some secondary mechanism(s) of protection, not dependent upon the continued presence of the purine.

Classically, two types of adenosine receptors have been described: namely the A_1 and the A_2 receptors [18]. In the heart the activation of A_1 receptors depresses automaticity and conduction within the sinus and atrioventricular nodes to decrease heart rate and propagation through the conducting system, while the A_2 receptors are thought to mediate coronary vasodilation and inhibit neutrophil activation [18]. Pretreatment with an A_1-selective agonist, R-phenylisopropyladenosine (R-PIA), could also protect the heart from the prolonged ischaemic insult, whereas the A_2-selective agonist, CGS 21680, could not [17]. Other studies have confirmed the protective effects of A_1-specific agonists [20, 21].

While the evidence implicating A_1 receptor activation in preconditioning is extremely suggestive, it is not definitive. Most studies have used R-PIA as the ligand [17, 20, 21]. Another A_1-specific agonist, cyclopentyladenosine, was ineffective in providing protection in a porcine model, despite prevention of preconditioning with an adenosine receptor antagonist [22]. R-PIA also has a much longer half-life than adenosine, consequently it would still be present in the circulation during the prolonged period of ischaemia. Thus it has the opportunity to continue to act and produce cardioprotection by a mechanism that is perhaps separate from the secondary effects triggered by preconditioning. A short-acting A_1 agonist, or the temporal characteristics of A_1-mediated protection, have not been studied to date.

Despite questions concerning the mechanism by which adenosine promotes tolerance to ischaemia, the evidence that adenosine is the endogenous mediator is quite convincing. Further support for adenosine-mediated preconditioning is provided by studies with acadesine, an agent which augments extracellular adenosine levels locally within the heart during an ischaemic event [23, 24]. Acadesine may enhance the tolerance to ischaemia and decrease the time required for preconditioning to occur, consistent with its ability to augment adenosine levels [25].

The mechanisms by which adenosine mediates enhancement of ischaemic tolerance may involve a G-protein pathway

Adenosine receptor purification by agonist affinity chromatography yields complexes of the A_1 receptor with pertussis toxin-sensitive GTP binding proteins, G_o and G_i [26]. Ischaemic preconditioning is suppressed when animals are pretreated with pertussis toxin [27] which ADP-ribosylates and inhibits specific G proteins. This observation [27] is consistent with the hypothesis that preconditioning results from activation of an A_1 receptor-G protein coupled response, although stimulation of A_1 receptors may also result in an increased or decreased coupling of other receptors to their G proteins [26]. A_1 receptor-G protein linked events which have been described to date include activation of Na/Ca exchange [28], activation of muscarinic K^+ channels in the atria [29], and inactivation of voltage-dependent Ca^{2+} channels [30]. Intracellular calcium levels may also be decreased by at least two other mechanisms. Activation of A_1 receptors can decrease phospholipase C activity, resulting in diminished inositol trisphosphate formation and intracellular calcium release [26]. The second mechanism by which adenosine can reduce intracellular calcium levels is via activation of ATP-sensitive K^+ channels in the heart, also via an A_1-G protein-dependent mechanism [31]. Activation of ATP-sensitive K^+ channels results in hyperpolarization of the cell which, in turn, decreases the open probability of the voltage-dependent Ca^{2+} channels, thereby attenuating the entry of calcium into the cell. This latter sequence involving the ATP-sensitive K^+ channel is an attractive mechanism for preconditioning and has received the most attention to date.

Does a sub-type of the A_1 receptor mediate preconditioning?

R-PIA is the predominant A_1 agonist used successfully to mimic preconditioning in various species[17, 20, 21]. R-PIA shows about 100 fold selectivity for the A_1 receptor over the A_2 receptor [32]. Protection is also observed in rabbits with 2-chloro-N6-cyclopentyl adenosine, which is 1500-fold A_1 selective, whereas another A_1 selective agonist cyclohexyladenosine, which exhibits a 400-fold selectivity for the A_1 receptor [32], does not evoke the preconditioning phenomenon in pigs [22]. Likewise, while the nonselective adenosine receptor antagonists 8-SPT and PD 115,199 block preconditioning, the A_1 selective antagonist DPCPX (with 740-fold selectivity for the A_1 receptor [32]) fails to prevent preconditioning in rabbits [7]. This apparent conundrum might infer either additional properties of the active compounds apart from recognition of the A_1 receptor, or the presence of A_1 receptor sub-types. While at first pass the idea that structurally diverse compounds which either activate or block adenosine receptors, such as R-PIA, 8-SPT or PD 115,199, could equally activate or block respectively another system appears remote, the seemingly close structural requirements for a glucose transporter(s) with adenosine receptors has led to the suggestion that activity at a glucose transportation site could contribute to preconditioning [33]. However, a more likely explanation for the differential activities of various agonists and antagonists is that there is a particular sub-type of an A_1 receptor with affinities for the agonists and antagonists different from those of the classical A_1 receptor upon which the selectivity and binding affinities were determined [32]. Data compiled from structure-activity relations with various agonists and antagonists, together with studies in which adenosine receptors have been cloned and sequenced, indeed corroborate the idea that sub-types of A_1 receptors exist [26]. To date these have been tentatively labelled as A_{1a}, A_{1b}, and A_3, although the complete repertoire of possibilities is not yet known. Moreover, some of these A_1 receptor sub-types are also linked to a pertussis toxin-sensitive G binding protein [26] and thus their involvement would be consistent with the finding that inhibition of G_i or G_o with pertussis toxin prevents preconditioning [27]. The A_1 receptor or sub-type responsible for preconditioning has yet to be identified. Receptor binding studies appear to be too insensitive to define A_1 receptors in the myocardium, so that an alternative approach to the identification of a receptor subtype, perhaps involving application of the polymerase chain reaction technology to amplify the RNA transcripts encoding the receptor, is probably required.

> "Alice sighed and gave it up. "It's exactly like a riddle with no answer" she thought. [1]

Is adenosine both the inducer and the mediator of preconditioning?

The finding that adenosine receptor blockade either before or after the conditioning ischaemia is equally effective at attenuating the preconditioning response [34] indicates that occupancy of adenosine receptors is required for both inducing preconditioning and mediating the protective response during the prolonged ischaemia. This raises an apparent paradox. Why is occupancy of adenosine receptors in the conditioning ischaemia a requisite for cardioprotection when occupancy of the same receptors in naive hearts subjected to the prolonged ischaemic period alone, does not result in the same protection? An alternative way of asking the same question is: what is different about adenosine activity in the brief conditioning ischaemia from that in the prolonged ischaemic insult? One suggestion proposed by Kitakaze and colleagues [35] is that preconditioned hearts release more adenosine during ischaemia than virgin hearts, due to conditioning-enhanced activity of 5' nucleotidase which converts AMP to adenosine. Consequently the higher adenosine levels in the preconditioned hearts result in greater cardioprotection. Unfortunately, direct measurements of interstitial adenosine levels using microdialysis probes show reduced adenosine levels during ischaemia in preconditioned hearts, probably because of reduced ischaemic stress [36].

Another potential explanation for the finding that adenosine is both the inducer of preconditioning and the mediator, is that the transducing mechanism of adenosine effects is somehow "activated" or "primed", such that a second release of adenosine during the prolonged ischaemic insult now elicits an enhanced response. Whether this could occur at the level of the G protein or some regulatory unit, the ATP-sensitive K$^+$ channel, or an intracellular phosphorylation step for example, is not known. However, elucidation of the mechanism merits some attention, wherein exposure to adenosine acting upon an A$_1$ receptor enhances an intracellular molecular event to a subsequent occupancy of that receptor, with a time course which matches the temporal requirements of preconditioning [7]; that is the intracellular event should decay over 60-120 min.

There are alternative explanations to the suggestion that occupancy of adenosine receptors is required both to elicit preconditioning and to provide the protective effect during the prolonged ischaemic insult [34]. One option is that the adenosine released by the conditioning period actually evokes its protective effects during the ensuing reperfusion phase. Then by administering the adenosine receptor antagonist after the transient conditioning ischaemia, the drug could prevent the development of the cardioprotective phenomenon during the subsequent period of reflow, independent of any effect during the prolonged ischaemic period. Varying the timing of administration of the adenosine antagonist (by allowing reperfusion to take place and introducing the antagonist either immediately before or just after induction of the prolonged ischaemia period) should address this potential interpretation.

Alternatively, it is well recognized that adenosine is cardioprotective even in naive (i.e. not preconditioned) hearts. Administration of either an adenosine receptor antagonist [37] or adenosine deaminase [38] to degrade endogenously produced adenosine, exacerbates ischaemia-reperfusion injury. Consequently, the finding that blockade of adenosine receptors in preconditioned hearts prior to the prolonged ischaemic event prevents the reduction in infarct size normally seen with preconditioning [34] might indicate the simultaneous occurrence of two separate (albeit related) events. That is, the hearts are preconditioned, but the ultimate measure of this, infarct size, is compromised by an independent suppression of the protective effects of the adenosine released during the prolonged ischaemia by the adenosine receptor antagonist.

ATP-sensitive K^+ channels are functionally linked to A_1 receptors

In 1983 Noma [39] reported that the previously undefined outward K^+ current activated during myocardial ischaemia was actually via a K^+ channel regulated by the intracellular concentration of ATP, and, to a lesser extent, ADP. He postulated that these ATP-sensitive K^+ channels may play a cardioprotective role during ischaemia or hypoxia. ATP-sensitive K^+ channel activity can be modulated by other factors produced during ischaemia apart from the decrease in ATP, including pH, lactate and certain nucleotide diphosphates. In 1990, Kirsch and colleagues [31] found that ATP-sensitive K^+ channels could also be opened by activation of adenosine A_1 receptors in rat ventricular myocytes, and that this process was coupled by a pertussis toxin-sensitive G protein. A functional link between adenosine and ATP-sensitive K^+ channels was provided by studies of Daut and co-workers [40] in which the ATP-sensitive K^+ channel blocker glibenclamide attenuated the coronary vasodilation induced by either hypoxia or adenosine in the isolated guinea-pig heart; a finding confirmed in vivo [41].

While the link between adenosine receptors and ATP-sensitive K^+ channels seems well established in the coronary circulation, adenosine-induced coronary vasodilation was previously attributed to activation of A_2 receptors rather than the A_1 receptors which are coupled to the ATP-sensitive K^+ channel. Also, it is not clear whether the K^+ channels in the coronary circulation are the same as those described in the myocytes. However, two recent reports [42, 43] suggest that the infarct reduction in naive hearts evoked by either adenosine combined with dipyridamole, or by R-PIA, can be attenuated by glibenclamide. These observations imply a functional link between the A_1 receptors and the ATP-sensitive K^+ channel in the myocytes (because they occurred in the absence of changes in collateral blood flow during ischaemia [42]), which result in cardioprotection.

The effects of myocardial preconditioning are comparable to those produced by activation of ATP-sensitive K^+ channels

In anaesthetized open-chest dogs Gross and Auchampach [44] reported that glibenclamide given either as a pretreatment before or immediately after the conditioning ischaemia, prevented the cardioprotection associated with preconditioning. Similar results were obtained by Grover and co-workers [43] who further demonstrated that the infarct-size limiting effect of a 15 min intracoronary pretreatment with R-PIA was abolished by glibenclamide. These observations support the proposal that activation of ATP-sensitive K^+ channels, probably as a result of A_1 receptor stimulation, evokes preconditioning. The finding that glibenclamide blocks the response either when given as a pretreatment before the conditioning ischaemia [44] or early on during the prolonged ischaemic insult, suggests that A_1-linked ATP-sensitive K^+ channel opening is both the inducer and the mediator of myocardial preconditioning.

In contrast to the foregoing studies, Thornton and Downey [45] were unable to block preconditioning with glibenclamide in the rabbit. Decreases in blood glucose levels were observed, consistent with enhanced insulin secretion secondary to ATP-sensitive K^+ channel blockade. Moreover, in the non-preconditioned rabbit, glibenclamide exacerbated ischaemia-reperfusion injury resulting in a larger infarct. Despite this enhancement of injury, glibenclamide-treated rabbits showed just as much benefit from preconditioning as untreated control animals [45]. Concern that glibenclamide-induced increases in insulin secretion or decreases in blood glucose might be responsible for ablation of the protective effects of preconditioning, rather than blockade of cardiac ATP-sensitive K^+ channels, caused Auchampach and colleagues [46] to examine another ATP-sensitive K^+ channel blocker, sodium 5- hydroxydacanoate (5-HD), in the setting of myocardial preconditioning in the dog. 5-HD is thought to block the ATP-sensitive K^+ channel only during ischaemia by competing with the ATP binding site, and does not affect pancreatic ATP-sensitive K^+ channels and hence, insulin secretion or blood glucose levels. Like glibenclamide, 5-HD was able to prevent ischaemic preconditioning in the dog [46]. Resolution of these opposite observations was initially, and rather unsatisfactorily, attributed to species differences in the mechanism of preconditioning in dogs and rabbits [45, 46]. Unsatisfactorily because usually teleological evidence for retention of a mechanism across different species is regarded as strong support for the importance of the phenomenon. Recently, Downey et al. [47] found that glibenclamide did block preconditioning in rabbits when the anaesthetic was switched from pentobarbitone to ketamine/xylazine. Why differences in anaesthetic are important in apparently unmasking the role of ATP-sensitive K^+ channels is not clear. Downey et al. [47] suggested that ATP-sensitive K^+ channels in combination with the anaesthetic agent might affect the interstitial adenosine levels during ischaemia, rather than ATP-sensitive K^+ channel opening representing the final pathway of protection.

Consistent with the proposed role of ATP-sensitive K^+ channel activation in myocardial preconditioning, ATP-sensitive K^+ channel openers such as aprikalim are cardioprotective in naive hearts resulting in a decrease in infarct size ([48]; see chapter 13). Similar results have been obtained with cromakalim and pinacidil when given directly into the coronary artery to circumvent the haemodynamic effects of the compounds which can offset their cardioprotective activities [49].

> "If there's no meaning in it", said the King, that saves a world of trouble, you know, as we needn't try to find any. And yet I don't know," he went on...." I seem to see some meaning in them after all." [1]

Further considerations regarding ATP-sensitive K^+ channels in preconditioning

Before the concept of ATP-sensitive K^+ channel activation-mediated preconditioning can be accepted, it is important to recognize the limitations of the evidence produced to date.

First, just because ATP-sensitive K⁺ channel openers reduce infarct size in naive hearts [48, 49] does not imply ATP-sensitive K⁺ channel opening is the mechanism of preconditioning. Preconditioning is a phenomenon in which a prolonged change is triggered by a transient event. The ATP-sensitive K⁺ channel openers such as aprikalim have an extended duration of activity, and when used to "mimic" preconditioning are still present during the period of ischaemia and reperfusion to protect the heart. Studies using pretreatment with an ATP-sensitive K⁺ channel opener with a very short duration of action, such that the activity of the drug has resolved by the time of the ischaemic insult, are required to determine if ATP-sensitive K⁺ channel openers can mimic or evoke preconditioning. Moreover, in naive hearts ATP-sensitive K⁺ channel openers decrease infarct size about 30-40% [48, 49], whereas preconditioning decreases infarct size by 70-80% [7, 46].

The finding that ATP-sensitive K⁺ channel openers are effective cardioprotective agents in naive hearts raises the issue of what is unique as regards ATP-sensitive K⁺ channel opening in preconditioning. If ATP levels fall and adenosine levels rise during ischaemia, resulting in activation of ATP-sensitive K⁺ channels, how does a transient period of ischaemia augment this phenomenon to result in such dramatic protection? For example, it can be speculated that the link between the A_1 receptor and the ATP-sensitive K⁺ channel is somehow improved by the conditioning ischaemia so that opening of ATP-sensitive K⁺ channels occurs more readily or with less ATP loss or at lower adenosine levels when a second bout of ischaemia is imposed; or that phosphorylation of some critical site was triggered by the first ischaemic period. However, to date no explanation involving ATP-sensitive K⁺ channel activation after a transient ischaemic event promoting subsequent ATP-sensitive K⁺ channel-mediated protection during a second prolonged ischaemic insult, consistent with the results obtained with glibenclamide [44] cited above, has been proposed. An important link is to define the temporal pattern of ATP-sensitive K⁺ channel activation to determine if it is consistent with the established time-course of the preconditioning phenomenon.

Another factor which complicates interpretation of studies with ATP-sensitive K⁺ channel blockers in myocardial preconditioning is that coronary artery occlusion and reperfusion provokes other phenomena, such as myocardial stunning, no-reflow, neutrophil activation or ventricular arrhythmias [4] which are sensitive to modulation of ATP-sensitive K⁺ channels [44]. It has been demonstrated, for example, that glibenclamide exacerbates the post-ischaemic contractile failure termed myocardial stunning [50]. ATP-sensitive K⁺ channel openers attenuate neutrophil activation and accumulation in the ischaemic-reperfused heart [48]. The extent to which glibenclamide induced modification of these concomitant events may contribute to the suppression of myocardial preconditioning is not clear. While it is established that myocardial stunning and preconditioning are separate phenomena, they are clearly inter-related because they are triggered by transient periods of ischaemia and reperfusion, and are both modified by agents which modulate adenosine or ATP-sensitive K⁺ channel activity.

Finally, many studies have relied upon blockade of a response by glibenclamide as proof of mediation via ATP-sensitive K⁺ channels. While patch-clamp studies have verified the selectivity of this compound for the ATP-sensitive K⁺ channel it may have additional properties which contribute to its pharmacological profile. For example, it was recently reported that glibenclamide inhibits contraction of vascular smooth muscle in vitro induced by prostaglandins E_2 and D_2 and F_α independent of an effect on ATP-sensitive K⁺ channels [51]. The extent to which other activities of glibenclamide contribute to its cardiac effects is not known. Certainly confirmation of the results obtained with glibenclamide in myocardial preconditioning with another ATP-sensitive K⁺ channel antagonist, 5-HD, strengthen the proposed role for ATP-sensitive K⁺ channels in that phenomenon.

It has been argued that the small decrease in ATP observed during the prolonged ischaemic insult in preconditioned hearts by itself might be insufficient to increase ATP-sensitive K⁺ channel opening enough to elicit protection. The potential for A_1 receptor activation to synergize with small changes in ATP in opening ATP-sensitive K⁺ channels to a level sufficient to induce cardioprotection has not been investigated to this author's knowledge. Indeed, de Weille [52] suggested that in systems other than the ß-cell, ATP regulation of the ATP-sensitive K⁺ channel might play only a minor role in the physiological functioning of the channel. Moreover, different channels with different effectors and functions probably exist, inappropriately labelled as a homogenous group of ATP-sensitive K⁺ channels [51]. Recognition of the subtype of ATP-sensitive K⁺ channel involved in preconditioning, and its regulation, might permit the development of new therapeutic agents to protect the heart from injury evoked by ischaemia and reperfusion.

"I could have done it in a much more complicated way,"
said the Red Queen, immensely proud. [1]

References

1. Carroll, Lewis: excerpts from "Alice in Wonderland" and "Through the Looking Glass"
2. Hearse DJ, Yellon DM (1984) Why are we still in doubt about infarct size limitation? The experimentalist's viewpoint. In: Hearse DJ, Yellon DM (eds) Therapeutic approaches to myocardial infarct size limitation. Raven Press, NewYork, pp 17-41
3. White HD, Norris RM, Brown MA, Takayama M, Maslowski A, Bass NM, Ormistron JA, Whitlock T (1987) Effect of intravenous streptokinase on left ventricular function and early survival after acute myocardial infarction. N Engl J Med 317: 850-855
4. Hearse DJ (1992) Myocardial injury during ischaemia and reperfusion: concepts and controversies. In: Yellon DM, Jennings RB (eds) Myocardial Protection: the pathophysiology of reperfusion and reperfusion injury. Raven Press, New York, pp 13-57
5. Reimer KA, Murry CE, Yamasawa I, Hill ML, Jennings RB (1986) Four brief periods of ischemia cause no cumulative ATP loss or necrosis. Am J Physiol 251: H1306-H1315
6. Murry CE, Jennings RB, Reimer KA (1986) Preconditioning with ischemia: a delay of lethal cell injury in ischemic myocardium. Circulation 74: 1124-1136
7. Downey JM, Liu GS, Thornton JD (1993): Adenosine and the anti-infarct effects of preconditioning. Cardiovasc Res 27: 3-8
8. Alkhulaifi AM, Pugsley WB, Yellon DM (1992) The influence of the time period between ischaemic preconditioning and prolonged ischaemia on myocardial protection. Circulation 86 (Suppl I): 1-342
9. Nao BS, McClanachan TB, Groh MA, Schott RJ, Gallagher KP (1990) The time limit of effective ischemic preconditioning in dogs. Circulation 82 (Suppl III): 271
10. Walker DM, Yellon DM (1992) Ischaemic preconditioning: from mechanisms to exploitation. Cardiovasc Res 26: 734-739
11. Shizukuda Y, Mallet RT, Lee SC, Downey HF (1992) Hypoxic preconditioning of ischemic canine myocardium. Cardiovasc Res 26: 534-542

12. Ovize M, Kloner RA, Hale SL, Przyklenk K (1992) Coronary cyclic flow variations "precondition" ischemic myocardium. Circulation 85: 779-789
13. Marber MS, Walker DM, Yellon DM, Walker JM (1992) Rapid atrial pacing fails to precondition the rabbit heart. J Mol Cell Cardiol 24 (Suppl I): S92
14. Shiki K, Hearse DJ (1987) Preconditioning of ischaemic myocardium: reperfusion-induced arrhythmias. Am J Physiol 253: H1470-H1476
15. Cohen MV, Liu GS, Downey JM (1991) Preconditioning causes improved wall motion as well as smaller infarcts after transient coronary occlusion in rabbits. Circulation 84: 341-349
16. Barakat O, Van Wylen DGL, Mentzer RM, Lasley RD (1991) Ischemic preconditioning improves postischemic recovery of function but shortens time to onset of ischemic contracture in isolated rat hearts. Circulation 84 (Suppl II): 433
17. Liu GS, Thornton J, Van Winkle DM, Stanley AWH, Olsson KA, Downey JM (1991) Protection against infarction afforded by preconditioning is mediated by A_1 adenosine receptors in rabbit heart. Circulation 84: 350-356
18. Belardinelli L, Linden J, Berne R (1989) The cardiac effects of adenosine. Prog Cardiovasc Dis 32: 73-97
19. Lasley RD, Rhee JW, Van Wylen DGL, Mentzer RM (1990) Adenosine A_1 receptor mediated protection of the globally ischaemic isolated rat heart. J Mol Cell Cardiol 22: 39-47
20. Tsuchida A, Miura T, Timura O (1991) Role of adenosine receptor activation in infarct size limitation by preconditioning in the heart. Circulation 84 (Suppl II): 191
21. Thornton J, Liu GS, Olsson RA, Downey JM (1992) Intravenous pretreatment with A_1-selective adenosine analogues protects the heart against infarction. Circulation 85: 659-665
22. Schwarz ER, Mohri M, Sack S, Arras M (1991) The role of adenosine and its A_1-receptor in ischaemic preconditioning. Circulation 84 (Suppl 2): II-191
23. Gruber HE, Hoffer ME, McAllister DR, Laikind PK, Lane TA, Schmid-Schoenbein GW, Engler RL (1989) Increased adenosine concentration in blood from ischemic myocardium by AICA riboside: effects on flow, granulocytes and injury. Circulation 80: 1400-1411
24. Mullane K (1993) Acadesine: the prototype adenosine regulating agent for reducing myocardial ischaemic injury. Cardiovasc Res 27: 43-47
25. Tsuchida A, Liu GS, Mullane K, Downey JM (1993) Acadesine lowers the temporal threshold for the myocardial infarct size limiting effect of preconditioning. Cardiovasc Res 27: 116-120
26. Linden J (1991) Structure and function of A_1 adenosine receptors. FASEB J 5: 2668-2676
27. Thornton JD, Liu GS and Downey JM (1993) Pretreatment with pertussis toxin blocks the protective effects of preconditioning: evidence for a G-protein mechanism. J Mol Cell Cardiol: in press
28. Brechler V, Pavoine C, Lotersztajn S, Garbarz E, Pecker F (1990) Activation of Na^+/Ca^{2+} exchange by adenosine in ewe heart sarcolemma is mediated by a pertussis toxin- sensitive G protein. J Biol Chem 265: 16851-16855
29. Kurachi Y, Nakajima T, Sugimoto T (1986) On the mechanism of activation of muscarinic K^+ channels by adenosine in isolated atrial cells: Involvement of GTP-binding proteins. Pflügers Arch 407: 264-274
30. MacDonald RL, Skerritt JH, Werz MA (1986) Adenosine agonists reduce voltage- dependent calcium conductance of mouse sensory neurons in cell cultures. J Physiol (Lond) 370: 75-90
31. Kirsch GE, Codina J, Birnbaumer L, Brown AM (1990) Coupling of ATP-sensitive K^+ channels to A_1 receptors by G proteins in rat ventricular myocytes. Am J Physiol 259: H820-H826
32. Jacobsen KA, van Galen PJM, Williams M (1992) Adenosine receptors: pharmacology, structure-activity relationships, and therapeutic potential. J Medic Chem 35: 407-422
33. Murphy E, London RE, Steenbergen C (1992) Some adenosine antagonist block glucose uptake. Circulation 86 (Suppl I): I-214
34. Thornton J, Downey J (1992) Blocking adenosine receptors between preconditioning and prolonged ischemia prevents protection. Circulation 86 (Suppl I): I-343
35. Kitakaze M, Hori M, Takashima S, Sato H, Kamada T (1991) Augmentation of adenosine pro-

duction during ischaemia as a possible mechanism of myocardial protection in ischaemic preconditioning. Circulation 84 (Suppl II): II-306
36. Dorheim T, Mentzer R, Van Wylen D (1991) Preconditioning reduces interstitial fluid purine metabolites during prolonged myocardial ischemia. Circulation 84 (Suppl 2): II-0760
37. Zhao ZQ, McGee S, Nakanishi K, Johnston WE, Vinten-Johansen J (1992) Receptor-mediated cardioprotection by endogenous adenosine is exerted during reperfusion after coronary occlusion in the rabbit. Circulation 86 (Suppl I): I-213
38. Bolling SF, Bove EL, Gallagher KP (1991) ATP precursor depletion and postischaemic myocardial recovery. J Surg Res 50: 629-633
39. Noma A (1983) ATP-regulated K channels in cardiac muscle. Nature 305: 147-148
40. Daut JW, Maier-Rudolph W, von Beckerath N, Mehrke G, Günther K, Goedel-Meinen L (1990) Hypoxic dilation of coronary arteries is mediated by ATP-sensitive potassium channels. Science 247: 1341-1344
41. Belloni FL, Hintze TH (1991) Glibenclamide attenuates adenosine-induced bradycardia and coronary vasodilation. Am J Physiol 261: H720-H727
42. Auchampach JA, Gross GJ (1992) Activation of ATP-dependent potassium (KATP) channels by adenosine: a possible mechanism for ischemic preconditioning in dogs. Circulation 86 (Suppl I): I-340
43. Grover GJ, Sleph PG, Dzwonczyk S (1992) Role of myocardial ATP-sensitive potassium channels in mediating preconditioning in the dog heart and their possible interaction with adenosine A_1-receptors. Circulation 86: 1310-1316
44. Gross GJ, Auchampach JA (1992) Blockade of ATP-sensitive potassium channels prevents myocardial preconditioning in dogs. Circ Res 70: 223-233
45. Thornton JD, Thornton CS, Sterling DL, Downey JM (1993) Blockade of ATP-sensitive potassium channels increases infarct size but does not prevent preconditioning in rabbit hearts. Circ Res 72: 44-49
46. Auchampach JA, Grover GJ, Gross GJ (1992) Blockade of ischaemic preconditioning in dogs by the novel ATP dependent potassium channel antagonist sodium 5- hydroxydecanoate. Cardiovasc Res 26: 1054-1062
47. Downey JM, Liu Y, Ytrehus K (1993) Adenosine and the anti-infarct effects of preconditioning. In: Przyklenk K, Kloner R, Yellon D, Klewer, Norwell MA (eds) Ischaemic preconditioning: the concept of endogenous cardioprotection. In press
48. Auchampach JA, Maruyama M, Cavero I, Gross G (1991) The new K^+ channel opener aprikalim (RP 52891) reduces experimental infarct size in dogs in the absence of haemodynamic changes. J Pharmacol Exp Ther 259: 961-967
49. Grover GJ, Dwonczyk S, Parham CS, Sleph PG (1990) The protective effects of cromakalim and pinacidil on reperfusion function and infarct size in isolated rat hearts and anaesthetized dogs. Cardiovasc Drugs Ther 15: 465-474
50. Auchampach JA, Maruyama M, Cavero I, Gross GJ (1992) Pharmacological evidence for a role of ATP-dependent potassium channels in myocardial stunning. Circulation 86: 311-319
51. Zhang H, Weir B, Cook D (1992) Antagonism of eicosanoid-induced contraction of rat aorta by sulphonylureas. Circulation 86 (Suppl I): I-834
52. de Weille JR (1992) Modulation of ATP sensitive potassium channels. Cardiovasc Res 26: 1017-1020

CHAPTER 16
Potassium channel openers in hypertension; a clinician's perspective

RJ Kovacs

Antihypertensives are the most commonly prescribed class of therapeutic agents for long-term use [1]. In the United States, one in six persons meets clinical criteria for systemic hypertension. K^+ channel openers represent but one of many classes of antihypertensive agents available to the clinician. Each new class of antihypertensives benefits some patient populations and complicates the care of others. K^+ channel openers will prove to be the same. The simple addition of another antihypertensive or class of antihypertensives to the pharmacopoeia offers little to the clinician who already finds a bewildering array of agents that, by and large, are effective and safe. Each new agent will be compared with its peers, and will be considered in light of the broad spectrum of patients who require antihypertensive therapy. In order to understand the role of K^+ channel openers in hypertension, it will be useful to understand the factors relevant to the decision to treat hypertension, the selection of an agent, and the assessment of a drug's utility in a given patient or patient population. This review will outline the current clinical approach to the treatment of hypertension, address the role of K^+ channel openers as they relate to existing pharmacological and non-pharmacological therapies and discuss the use of K^+ channel openers in particular patient populations. Possible drug interactions will be considered as well as the interaction of K^+ channel openers with pathophysiological cardiovascular states associated with hypertension. Much of the discussion of applicability, by necessity, will be speculative.

Data available on the K^+ channel openers suggest that these agents will be of clinical importance. We believe that the most important role for these agents will be in the treatment of patients with hypertension and other manifestations of cardiovascular disease, such as ischaemic heart disease, congestive heart failure, or cardiac arrhythmias.

Evaluation of the hypertensive patient is needed before drug selection

Initial evaluation

A basic knowledge of the approach to the patient with hypertension is fundamental to the understanding of drug selection for these patients. Initial evaluation usually follows screening for hypertension and measurement of elevated blood pressure on several measurements. Mass screenings are common in high-risk populations in the United States, including blacks and the elderly.

When sustained elevation of the blood pressure is documented, further evaluation includes a complete history and physical examination, with special attention to features of the patient that might indicate causes for hypertension remediable by therapy other

Table 1. Types of hypertension

1. Primary, essential, or idiopathic		95%
2. Secondary		5%
Due to	Renal disease renovascular disease	
	Endocrine disease	
	Exogenous hormones	
	Coarctation of the aorta	
	Pregnancy neurologic disease	
	Acute stress	
	Alcohol & drugs	

than antihypertensive drugs (secondary hypertension; see Table 1). The initial evaluation should also carefully address the possibility of co-morbid conditions, especially those conditions that may influence the selection of an antihypertensive [2]. The existence of co-morbid conditions is quite common, and is often an important factor in the selection of therapy (Table 2). Complications of hypertension existing at the time of the initial evaluation are also examined for, and influence the choice of therapy as well (Table 3). Routine laboratory screening for secondary causes of hypertension has largely lost favor because of the very low incidence of the various secondary diseases (Table 1). In numerous studies, the percentage of patients with primary (essential) hypertension is about 95% of all patients with hypertension. The yield of screening for all secondary causes of hypertension becomes quite small, unless clinical suspicion of such an aetiology is raised by the history and physical examination. Evaluation of renal function by serum creatinine measurement is probably an exception, as renal impairment is so common in this population and critically influences the choice of drug therapy.

Selection of an antihypertensive agent

The general principles of antihypertensive selection have undergone considerable change during the past 30 years, progressing from empirical therapy, to a rigid stepwise therapy scheme, and now to the notion of individualized therapy [3]. Stepwise therapy dictated the initiation of treatment with a diuretic, followed by a ß-adrenergic blocker, and finally a vasodilator if needed. Individualized therapy considers the patient's age (the elderly may respond better to diuretics and calcium antagonists), race (blacks respond less well to ß-blockers), and concomitant conditions (Table 2) in the selection of initial therapy. Vasodilator therapy is now considered a rational first-step therapy in many patients. The clinician has a broader choice than ever of agents for initial therapy. The existence of co-morbid conditions is often the most important factour in selecting therapy. As mentioned above, concomitant conditions are common (Table 2), and suggest a better tolerated agent. The simultaneous existence of angina and hypertension may favour the use of a ß-blocker or calcium antagonist. Post-myocardial infarction care may dictate use of a ß-blocker. The presence of congestive heart failure may influence the choice of an angiotensin converting enzyme inhibitor. Conversely, co-morbid conditions may preclude the use of certain agents, such as ß-blockers in congestive heart failure, bradyarrhythmias,

Table 2. Co-morbid conditions. From [2]

Diabetes	18.1%
Arthritis	40.5%
Chronic lung disease	7.4%
Angina	8.0%
Myocardial infarction	2.2%
Congestive heart failure	5.9%
None	37.3%

chronic lung disease, the use of angiotensin converting enzyme inhibitors in patients with renal insufficiency, or the use of diuretics in patients prone to volume depletion. Despite careful selection, 10-20% of patients will require discontinuation of an agent for one reason or another. The clinician must be prepared to use alternative agents.

The rational use of K^+ channel openers in patients with hypertension will therefore depend on many factors, including efficacy of the agents in lowering blood pressure, safety, tolerability, and interaction of the agents with drugs and co-morbid conditions common in the hypertensive population. It is the interplay between the K^+ channel openers and these co-existing conditions that will largely determine the niche that these agents will occupy. Although the addition of an effective, safe, and well-tolerated class of antihypertensives would be welcomed in the clinical arena, some of the theoretical advantages or disadvantages of K^+ channel openers discussed below will determine the ultimate utility of these drugs in treating cardiovascular diseases.

Among K^+ channel openers, diazoxide and minoxidil have had the longest clinical use as antihypertensives

Antihypertensives that activate K^+ channels in man behave clinically as other direct-acting vasodilators. Of the agents known to activate K^+ channels, diazoxide and minoxidil have had the longest clinical use. Diazoxide has been useful as a parenteral agent for the treatment of hypertensive crises, but is not used chronically. Minoxidil has been used primarily in refractory hypertension [3], especially in patients with renal dysfunction. More recently, pinacidil and cromakalim have been developed, with pinacidil having more extensive clinical data.

In its oral form, pinacidil given once or twice daily is an effective and well-tolerated antihypertensive. The potency of this particular K^+ channel opener is comparable to moderate doses of hydralazine [4], prazosin [5], or methyldopa [6]. Like many vasodilators, pinacidil has been combined with diuretic therapy to increase effectiveness and decrease the peripheral œdema seen in 15-30% of patients on the drug [7]. Likewise, the addition of ß-blockade is useful both in increasing the antihypertensive effect of the drug, and to control the reflex tachycardia associated with the direct vasodilation.

The utility of K^+ channel openers in patients refractory to other agents or combinations of agents remains unknown. Two factors would seem to favour the probability that these drugs would be useful. First, the K^+ channel openers act as direct-acting vasodila-

Table 3. Cardiovascular complications of hypertension

Haemorrhagic stroke
Nephrosclerosis
Aortic dissection
Congestive heart failure
Coronary heart disease
Peripheral vascular disease

tors, the class of agents most often relied upon in these patients. Diazoxide has been utilized in hypertensive crises for many years, and has proven to be a safe and effective agent. It should be noted that in the most severe hypertensive crises, clinicians generally prefer agents that can be given parenterally and titrated over short periods of time to respond to minute-to-minute fluctuations in blood pressure. The incidence of refractory hypertension in the general population of hypertensives is about 5%. The incidence of hypertensive crises or malignant hypertension is even smaller. A much larger clinical experience will be necessary to assess the role of K^+ channel openers in these subsets.

K^+ channel openers are generally well tolerated, with side effects similar to those of most direct-acting vasodilators: headache, œdema, dizziness, tachycardia, palpitations, and flushing. In addition, more non-specific side effects such as diarrhea, nausea, asthenia, and dyspnea can be seen as well. The overall incidence of side effects with pinacidil was 2 to 13% [8], and this figure probably estimates the overall incidence of side-effects for the class. Biochemical and haematological effects of these agents are relatively infrequent, but include deterioration of renal function, development of positive antinuclear antibodies, and elevated levels of uric acid. These agents (except for diazoxide) do not appear to alter glucose tolerance or insulin secretion. Without entertaining an extensive discussion of all the clinical side effects of these agents, it can be summarized that the K^+ channel openers used thus far in the clinic demonstrate an acceptable degree of tolerability and efficacy comparable to other classes of antihypertensives.

In the clinical setting where multiple agents exist for the treatment of hypertension, and 95% of patients are easily controlled, other more subtle factors may play a role in the clinical utility of K^+ channel openers. Short-term acceptance of the drug may be influenced by its ability to be easily titrated, avoiding steep drops in blood pressure that may produce intolerable side effects such as headache. Once or twice daily dosing is preferable. Some patients may prefer a transdermal delivery system. Cost must be mentioned as a factor in patient and physician acceptance of a therapy, and as cost-containment measures become more stringent, third parties who pay for the therapy will demand that cost of a new agent be considered in its overall application, especially if less expensive alternatives exist.

As mentioned above, it is often not the direct applicability of an agent to hypertension, but rather the interplay between drug, patient, and other common consequences, complications or co-morbid conditions, that determines the clinical utility of a class of drugs. It is in this area that the K^+ channel openers present the most interesting possibilities for both benefit and harm.

The use of K⁺ channel openers as antihypertensives should be evalued in the setting of co-morbid conditions

Ischaemic heart disease

Since ischaemic heart disease is a common clinical companion of hypertension, the consequences of K⁺ channel openers in the setting of ischaemia need to be viewed from several perspectives. First will be the consideration of therapy in the setting of an acute ischaemic event. Second will be the treatment of hypertension in the setting of chronic ischaemic heart disease. Clinical trials do not exist to answer these questions, so extrapolation from basic studies needs to be made.

Several lines of evidence suggest that K⁺ channel openers are anti-ischaemic [9], limit infarct size [10], and play a beneficial role in the recovery from ischaemia/reperfusion-induced myocardial stunning [9]. A more detailed discussion of these properties is contained in chapter 13, but one may speculate that the patient at risk for acute ischaemic events may benefit from the choice of a K⁺ channel activator as his chronic antihypertensive therapy. This remains speculative, however, and one can argue that pre-existing therapy with ß-blocker does not seem to provide a clinical advantage for these patients, although pretreatment of experimental animals with ß-blockers is capable of limiting infarct size as well.

In the setting of chronic ischaemic heart disease, the situation may be more problematic. There is no clinical evidence that K⁺ channel openers will be useful as anti-anginal agents because of direct coronary vasodilating properties. On the other hand, one might speculate that, as with other vasodilators, the reflex tachycardia produced by these agents in the absence of ß-blockage might exacerbate angina by increasing myocardial oxygen demand.

The diagnosis of ischaemic heart disease might be made more difficult by the administration of these agents because of the recognized effect of K⁺ channel openers such as pinacidil on the electrocardiogram [7]. Pinacidil produces T-wave changes in about 30% of patients, which are expected because of the recognized effect of these agents on repolarization of ventricular myocardium. T-waves are flattened or inverted. Such T-wave alterations will decrease the predictive accuracy of the exercise electrocardiogram for the diagnosis of ischaemic heart disease. In addition, these resting EKG abnormalities will also interfere with the ability of the resting electrocardiogram to diagnose other types of myocardial disease, such as left ventricular hypertrophy. Thus we may speculate that these agents may be less frequently used in patients in whom either resting or exercise electrocardiographic criteria are important for follow-up.

Cardiac arrhythmias

The interplay between hypertension, ischaemic heart disease, left ventricular hypertrophy, and cardiac arrhythmias makes consideration of this co-morbid condition important in the selection of an antihypertensive agent. Here, predictions of the clinical responses to K⁺ channel openers in what amounts to a very heterogeneous group of arrhythmias are very problematical. Cromakalim and pinacidil have been found to be both proarrhythmic and antiarrhythmic; depending on the experimental preparation. These mechanisms are reviewed in chapter 13, and extrapolation to the clinical setting of hypertension is not possible at this time. However, it will be crucial to determine proarrhythmia or antiarrhythmic effects in the various patient populations before these agents are widely used.

Two observations are relevant, however. First, clinical experience with antihypertensives known to be K^+ channel activators (diazoxide and minoxidil) has not been associated with a recognized excess of cardiac arrhythmias. Second, indirect evidence about the potential actions of K^+ channel openers can be derived from experiments utilizing K^+ channel blockers in the myocardium to investigate the role of these channels in arrhythmogenesis. Experimentally produced ischaemic arrhythmias in the dog heart can be suppressed by glibenclamide treatment, suggesting that K^+ channel activation may be potentially arrhythmogenic. It is useful to recall that in early trials of oral hypoglycaemics, an excess of sudden (possibly arrhythmic) deaths was noted [11]. The issue is clearly unresolved.

Diabetes

There is no evidence to suggest that concomitant use of K^+ channel activators in the patient with diabetes would produce adverse effects. The interaction between K^+ channel activators and the ATP-sensitive K^+ channel blockers routinely used to treat non-insulin-dependent diabetes mellitus has not been investigated. Hyperglycaemia has not been a reported side effect of these agents in clinical trials; but, again, from experience with the available agents, it should be noted that diazoxide, as well as being a potent antihypertensive, is also clinically used as a treatment for insulin-secreting tumours.

Lipid disorders

Adverse effects of antihypertensive agents on serum lipid values influence the clinical choice of these agents. Many hypertensive patients are at high risk for the complications of atherosclerosis. Agents with adverse effects on lipid metabolism will be avoided in these patients. Limited data are available on K^+ channel openers but the existing data suggest that K^+ channel openers should have a neutral or possibly beneficial effect on serum cholesterol and triglyceride levels and have an adverse effect on the levels of high density lipoproteins (HDL) [12, 13].

Drug interactions

Limited information is available on drug-drug interactions as they pertain to K^+ channel openers. Drugs which inhibit the cytochrome P450 system delay the elimination of pinacidil. Drugs which induce hepatic microsomal enzymes will increase the rate of elimination [14]. No clinically relevant drug interactions have been determined thus far, and complete data will not be available until the drugs are more commonly used.

K^+ channel openers will prove effective for the treatment of hypertension. As a class, it is unlikely that they will be widely used as monotherapy, because of the side-effects of peripheral vasodilators-specifically œdema and tachycardia.

The ultimate clinical utility of these agents will depend on their interaction with other cardiovascular and non-cardiovascular diseases, and other cardiovascular and non-cardiovascular therapies.

We would like to thank Debra Brooks for excellent secretarial assistance, and Dr. Paul E. Schmidt for reading the manuscript. Supported in part by a grant from the Showalter Trust.

References

1. Kaplan, NM (1992) Systemic Hypertension Therapy. In: Braunwald E (ed) A textbook of cardiovascular disease. WB Saunders, Philadelphia, pp 1296
2. Greenfield AL, Hays RD, et al. (1989) Functional status and well-being of patients with chronic conditions. Results from the Medical Outcomes Study. JAMA 262: 907-914
3. 1988 Joint National Committee (1988) The 1988 Report of the Joint National Committee on detection, evaluation, and treatment of high blood pressure. Arch Intern Med 148: 1023-1028
4. Abraham PA, Halsterson CE, Matzke GR, Keane WF (1987) Comparison of antihypertensive renal haemodynamic and humoral effects of pinacidil and hydralazine monotherapy. J Clin Hypertension 3: 439-451
5. Goldberg MR, Sushak CS, Rockhold FW, Thompson WL and the Pinacidil vs. Prazosin Multicenter Investigator Group (1988) Vasodilator monotherapy in the treatment of hypertension: comparative efficacy and safety of pinacidil, a potassium channel opener, and prazosin. Clin Pharmacol Ther 44: 78-92
6. Callaghan JT, Goldberg MR, Brunelle R (1988) Double blind comparator trials with pinacidil, a potassium channel opener. Drugs 36 (Suppl 7): 77-82
7. Goldberg MR, Offen WW (1988) Pinacidil with and without hydrochlorothiazide: dose response relationships from results of a 4 X 3 factorial design study. Drugs 36 (Suppl 7): 83-92
8. Friedel HA, Brogden RN (1990) Pinacidil. Drugs 39: 929-967
9. Grover GJ, McCullough JR, Henry DE, Conder ML, Sleph PG (1989) Anti-ischaemic effects of the potassium channel activators pinacidil and cromakalin and the reversal of these effects with the potassium channel blocker glyburide. J Pharmacol Exp Ther 251: 98-103
10. Auchampach JA, Maruyama M, Cavero I, Gross GJ (1991) The new K^+ channel opener Aprikalim reduces experimental infarct size in dogs in the absence of haemodynamic changes. J Pharmacol Exp Ther 59: 961-967
11. University Group Diabetes Program (1970) A study of the effects of hypoglycemic agents on vascular complications in patient with adult-onset diabetes. Diabetes 19: 747-830
12. Corden CN, Goldberg MR, Alaupovic PA, Price MD, Furste SS (1992) Lipid and apolipoprotein levels during therapy with pinacidil combined with hydrochlorothiazide. Eur J Clin Pharmacol 42: 65-70
13. Saku K, Ying H, Arakawa K (1990) Effects of pinacidil on serum lipid, lipoprotein, and apolipoprotein levels in patients with mild to moderate hypertension. Clin Ther 12: 132-138
14. Ahnfelt-Ronne I (1988) Pinacidil: preclinical investigations. Drugs 36 (Suppl 7): 4-9

CHAPTER 17
Potassium channel openers in coronary heart disease

A Berdeaux and JM Lablanche

Ischaemic heart disease remains one of the major scourges afflicting the population of industrialized nations. The difficulties encountered to prevent its occurrence and to reduce its risk factors highlight the necessity, at least for the coming decades, to continue to call on curative treatment, of which drug therapy plays a paramount role. The mere observation that most of the ischaemic heart disease patients are treated with several antianginal agents illustrates the fact that the ideal antianginal agent has not yet been found, and therefore new and original antianginal agents are still needed. Potassium channel openers may slot in that category.

Because of their remarkable potency in relaxing vascular smooth muscle and hence decreasing arterial blood pressure (see chapter 12), K+ channel openers were initially developed as a new class of antihypertensive agents. At that time, there was little evidence to suggest that these agents could exert a significant antianginal action since they were used at high doses inducing marked hypotension and tachycardia. In experimental models, such high doses lead to coronary steal and/or proarrhythmic effects during acute myocardial ischaemia [1-3].

However, recent experimental studies conducted in isolated perfused hearts submitted to global ischaemia reveal that K^+ channel openers exert cardioprotective effects without the confounding influence of changes in haemodynamic status. These effects have been further confirmed using in vivo models of myocardial stunning when the drugs where administered at non hypotensive doses. Furthermore, recent findings [4] also suggest that close analogies exist between the cardioprotection conferred by K^+ channel openers and ischaemic preconditioning leading to the concept that these agents may afford a permanent "chemical preconditioning".

Taken together, these data suggest that K^+ channel openers exhibit a new pharmacological profile and might represent an original alternative for the treatment of ischaemic heart disease since their properties differ from those reported for conventional antianginal drugs such as nitrates derivatives, ß-blockers and Ca^{2+} channel antagonists. With this in mind, the goal of the present review is to summarize current clinical information concerning the effects of K^+ channel openers on the myocardium made ischaemic by either fixed coronary stenosis or vasospasm. Most clinical experience with K^+ channel openers in coronary heart disease have been accumulated with nicorandil, a nicotinamide nitrate derivative which is not a pure K^+ channel opener but also has properties similar to those of the nitrates. However, experimental data obtained with more specific K^+ channel openers will also be discussed.

K+ channel openers improve the myocardial oxygen supply/demand balance

Activation of ATP-sensitive K+ channels by potassium openers in vascular smooth muscle results in membrane hyperpolarization and ultimately in inhibition of Ca^{2+} entry into these cells leading to vasodilation (see also chapter 12). These effects might be expected to be beneficial in situations resulting from a disequilibrium between oxygen supply and demand at the cardiac level as occurs during myocardial ischaemia. However, the cardiac and haemodynamic effects of K^+ channel openers in the whole organism not only depend on their intrinsic effects at the level of vascular and cardiac ATP-sensitive K^+ channels but also depend on compensatory neurohormonal mechanisms triggered, for instance, by activation of baroreceptor reflexes as a consequence of systemic hypotension. Thus, following oral or intravenous administration of K^+ channel openers, a dose-dependent hypotension is actually observed without development of marked negative inotropic and chronotropic effects, a net tachycardia even occurring when the drug is given acutely [5]. As previously mentioned, this explains why K^+ channel openers were initially suspected to induce coronary steal in patients with angina pectoris, since they were historically developed as antihypertensive agents and used at high doses inducing deleterious effects at the coronary level. However, when administered at non-excessive hypotensive doses, or at doses which do not induce excessive compensatory haemodynamic effects, K^+ channel openers always improve the oxygen supply/demand balance of the myocardium, and this remains a fundamental property of any drug aiming to exert a beneficial effect in a patient with angina pectoris. In this context, it is important to underline that among the vasodilator agents used in the treatment of angina pectoris, and when administered at equivalent hypotensive doses, nicorandil exerts unique haemodynamic effects by inducing a balanced decrease of both preload and afterload (nitrates decrease mostly preload whereas Ca^{2+} channel blockers decrease mostly afterload).

An illustration of this property comes from a clinical study where the effects of 0.1 and 0.2 mg/kg of i.v. nicorandil were compared with those of 10 µg/kg of nifedipine and 0.1 mg/kg of isosorbide dinitrate on systemic and coronary haemodynamics in 36 patients with prior myocardial infarction [6]. The results indicate that nicorandil and nifedipine induced a significant decrease in systemic vascular resistance and increase in cardiac index whereas isosorbide dinitrate decreased cardiac index and only slightly reduced vascular resistance. Pulmonary capillary wedge pressure decreased significantly with all compounds. Coltart and Signy [7] examined the effects of a 40 mg single dose in fifteen patients with normal left ventricular function, scheduled for cardiac catheterization. The study was open and used left- and right-side catheterization. Route of administration was oral for seven patients and sublingual for eight. Peak plasma concentrations were achieved after 30 to 45 min after oral or sublingual administration. A reduction in preload was apparent from a decrease in left ventricular end-diastolic pressure (from 7.4 ± 1.7 to -2.6 ± 1.5 mmHg), probably reflecting the venous vasodilatory activity of the drug as indicated by a significant increase in venous capacitance assessed by plethysmography. Total peripheral resistance decreased to a maximum of 19%, while systolic and diastolic blood pressure decreased by 34% and 21%, respectively. A transient increase in heart rate (from 71 ± 13 to 81 ± 4 beats/min) was observed. Cardiac contractility was transiently improved, as reflected by an increase in dP/dT max (from 1364 ± 73 to 1515 ± 99 mmHg/s).

Thus the nicorandil-induced decreases in preload and afterload reduce myocardial oxygen demand. This unloading effect of nicorandil, in conjunction with coronary vasodilation which increases myocardial oxygen supply, is beneficial in the prevention of

Table 1. Comparison of actions of different antiischaemic agents in humans

	Rest			Effort	Vasodilation			
	HR	SBP	DBP	HR X SBP	Systemic	coronaries Large Vessels	Small Vessels	LVEDP
ß-blockers	↓↓	↓	↓	↓	(-)	(-)	(-)	↑
Nifedipine	↑	↓↓	↓	=	+++	++	+++	=
Diltiazem	=↓	↓↓	↓	=↓	++	+	++	=
Nicorandil	=	=↓	=	=	+	+++	+	↓
Nitrates	=	=↓	=	=	=	+++	0	↓↓

effort-induced angina. For a similar decrease in preload, nitrates always exhibit a smaller decrease in afterload than nicorandil [8], clearly indicating that the haemodynamic profile of nicorandil is unique among the antianginal drugs presently used for their vasodilating properties. Table 1 compares the coronary and systemic effects of nicorandil with those of the conventional antianginal agents.

The unloading effects of nicorandil were also found to be useful, acutely, in heart failure. Cohen-solal et al. [9] found that single oral doses of nicorandil (40, 60 or 80 mg) in congestive heart failure patients (stage III or IV NYHA) induce, after 30 minutes, a 34% reduction in the pulmonary wedge pressure and a 55% increase in cardiac index. Systemic vascular resistance and systemic diastolic pressure decreased, by 36% and 15% respectively, while systemic systolic pressure and heart rate did not significantly change. These beneficial effects (reduction in pulmonary capillary wedge pressure and increase of cardiac output) were also observed by other teams, following administration of acute oral or intravenous doses of nicorandil [10, 11]. This favorable influence of nicorandil in patients with depressed ventricular function needs to be confirmed in long-term trials, but may be an advantage over beta-blockers and dihydropyridines which are not indicated in this subset of patients [12].

K+ channel openers have a beneficial effect on exercise induced angina

It is well known that myocardial ischaemia always occurs when there is an imbalance between myocardial oxygen supply and demand, although ischaemic episodes may either be asymptomatic or may produce typical anginal pain. Ischaemia may primarily result from the presence of fixed atherosclerotic stenoses that limit the delivery of oxygen to the myocardium during conditions where myocardial oxygen demand is increased, such as exercise. On the basis of their pharmacological profile, it is clear that K+ channel openers may favourably influence the two components of the myocardial oxygen supply/demand balance and it can be inferred that these agents should reduce myocardial ischaemia resulting from the development of fixed stenosis. Available clinical studies conducted with nicorandil actually provide some answers to these questions. Even though not a "pure" K^+ channel opener, this drug offers the opportunity to compare its antianginal profile with that of other "conventional" therapies.

Comparison with β-blockers

The efficacy of nicorandil was compared with atenolol in 37 patients using a placebo-controlled, parallel study design [14]. After six weeks of treatment, exercise tests were performed two hours after 20 mg of nicorandil or 100 mg of atenolol. The overall improvement was similar in both groups but the mechanism of action of the drugs inferred by the haemodynamic changes was different. Atenolol improved exercise time by reducing myocardial oxygen demand regardless of the work loads. Nicorandil did not reduce the double product; peak oxygen demand was therefore not altered by the drug. Resting blood pressure was reduced, suggesting a peripheral vasodilating action. The improvement in exercise capacity with nicorandil was thus mediated through a reduction in afterload and the small increase in peak exercise rate-pressure product may reflect a degree of coronary vasodilation. Similar observations were reported when propranolol was compared with nicorandil.

Comparison with calcium antagonists

The antianginal efficacy of nicorandil (20 mg, bid) was compared with nifedipine (20 mg bid) in a double blind, randomized, multicenter study with 58 patients [15]. After eight weeks of treatment, exercise was performed two hours after the drug intake. At rest, systolic blood pressure was reduced significantly with nicorandil, though less than with nifedipine, which increased heart rate and induced a more pronounced reduction of both systolic and diastolic blood pressure. The improvement in exercise duration and time to 1 mm ST depression was not significantly different between the two groups of patients. The incidence of side effects was identical in both groups with more frequent headache in the nicorandil group and more peripheral vasodilation in the nifedipine group.

Nicorandil (30 mg) was also compared with diltiazem (60 or 120 mg) and propranolol (40 mg) [16]. The increase in exercise duration was significant with all drugs. Nicorandil increased the peak double product more than did a low-dose of diltiazem but similarly to a high-dose. In another study involving 123 patients during 3 months [17], nicorandil (20 mg, bid) and diltiazem (60 mg, tid) administered in a double-blind manner, showed a similar efficacy and the peak double product remained unchanged.

Comparison with nitrates

Two studies were reported by Döring [18], the first comparing nicorandil (20 mg, bid) with isosorbide 5 mononitrate (20 mg, bid), and the second comparing nicorandil (same regimen) with isosorbide dinitrate (20 mg, tid). In both studies, involving a total of 129 patients, a significant improvement as compared with the baseline was demonstrated on exercise tests with all drugs but no significant differences were observed between the study groups. No tolerance developed during the study, regardless of the compound tested.

Thus the different clinical studies at present available with the K^+ channel opener, nico-

Table 2. Potential actions of K⁺ channel openers in coronary artery disease

	Favourable action	Unfavourable action
Arterioles	Increase in transmural blood flows	Steal in CAD (?)
Epicardial arteries	Spasmolytic	
Afterload	Decrease	Hypotension
Preload	Decrease (*)	–
Inotropism	No cardiac depression	–
Rhythm	Prevention of long QT arrhythmias (?)	Profibrillatory effect (?)
Cardioprotection	Prevention of myocardial stunning	–

(*) Only so far demonstrated with nicorandil, (?) Not so far demonstrated in humans

randil, all demonstrate a beneficial effect of the drug during exercise-induced angina. This effect is similar to that induced by ß-blockers, Ca^{2+} antagonists and nitrates, but the mechanism of action of nicorandil seems to be mainly related to its vasodilator properties which in some way encompass those induced by both nitrates and Ca^{2+} antagonists (Table 2).

K⁺ channel openers act directly on coronary vessels

Many studies in humans, using either quantitative coronary angiography to measure alterations in coronary diameter or intracoronary doppler techniques to measure alterations in coronary blood flow, have shown that vasomotor responses in large epicardial and small coronary resistance vessels are altered in patients with atheromatous coronary disease. In control subjects, sympathetic stimulation induced by the cold pressor test or by exercise causes dilation of epicardial vessels accompanied by an increase in coronary blood flow, indicating a concomitant dilation of conductive and resistive coronary vessels. In contrast, in patients with coronary atherosclerosis, sympathetic stimulation by the cold pressor test results in epicardial constriction and a decrease in coronary blood flow [19]. This alteration in vasomotor response has been attributed to the presence of a dysfunctional endothelium that is no longer able to synthesize or release adequate amounts of endothelium-derived relaxing factors [20, 21]. These observations demonstrate that variations in vasomotor tone, that could cause or contribute to the pathogenesis of myocardial ischaemia, may occur not only in epicardial coronary vessels but also at the level of resistance vessels and in the microvasculature where coronary blood flow is primarily regulated. Furthermore, these observations underline the importance of vasodilator agents in the treatment of myocardial ischaemia in patients who exhibit even pure exercise-induced angina.

The role of vasospasm in the genesis of episodes of myocardial ischaemia is of major importance in patients who have angina at rest. Classical coronary vasospasm, defined as reversible total or subtotal epicardial occlusion, occurs frequently in some subgroups of patients whose anginal symptoms occur predominantly or exclusively at rest, in patients who experience angina in the weeks after an acute myocardial infarction and in patients whose anginal symptoms recur after percutaneous transluminal coronary angioplasty. Despite the well-demonstrated efficacy of nitrates and Ca^{2+} antagonists in the treatment of such coronary vasospasm, some patients remain symptomatic, even after treatment with combinations of these drugs. Thus new more effective treatments for vasospasm are required and K^+ channel openers might represent an original alternative for such design.

K^+ channel openers preferentially relax small coronary vessels

Besides the well-known presence of abundant Ca^{2+}-activated K^+ channels in coronary smooth muscle cells, the presence of ATP-sensitive K^+ channels has been reported in both large epicardial and small resistive coronary arteries [5, 22](see chapters 6 and 7). Direct recording of the tension of isolated coronary arteries or measurement of coronary blood flow in animal preparations clearly demonstrated that glibenclamide, a blocker of ATP-sensitive K^+ channels (see chapter 10), produced a parallel shift rightward of the dose-response curves of K^+ channel openers such as pinacidil, cromakalim, nicorandil or aprikalim while not affecting the dose-response curve to nitroglycerin or Ca^{2+} antagonists [23, 24]. These K^+ channel openers are more potent as relaxing agents for small resistive coronary arteries (i.e. arteries of less than 100 µm in diameter) than for large conductance vessels (see also chapter 12), although a definitive conclusion on such differential vascular selectivity requires further investigations conducted in experimental models where the reactivity of coronary arteries with different vascular sizes can be studied simultaneously.

Presently, it is not definitively established whether or not the coronary vasodilator effect of K^+ channel openers depends on the presence of an intact endothelium. It has been demonstrated that cromakalim still dilates isolated denuded rabbit aorta but in conscious dogs, neither cromakalim nor pinacidil were able to dilate large epicardial arteries previously de-endothelialized by selective coronary angioplasty [25]. This question is of importance since coronary stenosis and spasms mainly occur in these conductive vessels and patients with atherosclerosis as well as patients who have undergone coronary angioplasty are not able to dilate these epicardial arteries following intracoronary injection of endothelium-dependent dilators (e.g. acetylcholine) or to increase blood flow secondary to arteriolar dilation (the so called "flow-dependent dilation") [19-21]. In this context, nicorandil, which also exhibits a marked endothelium-independent vasorelaxant activity on epicardial coronary arteries through its "nitrate-like" properties, may possesses some advantages over pure K^+ channel openers for the treatment of unstable angina where coronary spasms occur at a high frequency. However, recent experimental data suggest that coronary spasm mainly results from a decrease in K^+ conductance in large epicardial conductance vessels. Thus hyperpolarization of the smooth muscle by any K^+ channel opener could reduce Ca^{2+} entry through voltage-operated Ca^{2+} channels and indirectly relax the vessel or prevent coronary spasm regardless of the effect of such a drug on vascular endothelium.

K$^+$ channel openers increase regional myocardial blood flow

Experimental studies using the microsphere technique to measure residual collateral blood flows resulting from an acute or chronic coronary artery occlusion have demonstrated that nicorandil, as well as other "pure" K$^+$ channel openers such as EMD 52692, significantly increase collateral blood flow when systemic arterial pressure is not allowed to decrease under a threshold value or when the drug is directly delivered through the coronary artery [5]. When the drug is given at hypotensive doses, the ultimal transmural distribution of ischaemic myocardial blood flow between the endocardial and epicardial layers of the ischaemic region will directly depends on the severity of the stenosis and the decrease in coronary perfusion pressure [3]. In this regard, K$^+$ channel openers behave like any other potent coronary dilating agents and ultimately a coronary steal may occur when marked hypotensive doses accompanied by reflex tachycardia are administered [3]. However, it has to be kept in mind that K$^+$ channel openers exert a direct cardioprotective effect at doses which do not necessarily induce a decrease in coronary perfusion pressure through excessive arterial hypotension [4] and this property might be of considerable interest as compared to other antianginal drugs such as Ca^{2+} channel antagonists (see chapter 13). In addition, clinical studies carried out with nicorandil have not yielded any concern with regards to proischaemia, a term recently coined to designate the potential unwanted effects of some dihydropyridines, possibly linked to the steal phenomenon [26].

Nicorandil causes coronary vasodilation in humans

Two studies using quantitative coronary angiography have demonstrated, in humans, the vasodilator properties of nicorandil. Kishida [27] studied the effects of nicorandil (10 mg, sublingually) on coronary diameter in 26 patients. A significant epicardial vasodilatory response was well documented at 30 and 60 min following the administration of this drug. In the subgroup of patients with vasospastic angina, the vasodilatory effect of nicorandil was significantly more pronounced. In another study, nicorandil proved superior to nitrates. Suryapranata [28] compared the effects of nicorandil and isosorbide dinitrate administered through an intracoronary route in two groups of patients. The former received nicorandil (6 µg/kg) followed by isosorbide dinitrate (ISDN, 2 mg) whereas the sequence of administration was reversed in the latter group. They reported that the two medications had a similar vasodilatory effect on normal coronary arteries. However, in stenotic segments, nicorandil induced a significant increase from baseline of 20 % in obstruction diameters, whereas the increase was not significant with ISDN (+ 8 %). Moreover, nicorandil could induce an additional significant increase (+ 13%) in obstruction diameter when given after ISDN, while no such additional effect could be evidenced when ISDN was administered after nicorandil.

Nicorandil relieves coronary spasm in human

A potential role for nicorandil in the treatment of vasospastic angina has been suggested by two studies. Lablanche et al. [29] investigated the effects of nicorandil in 13 male patients in whom spontaneous or ergometrine-induced coronary spasm had been documented during coronary arteriography. All previous antianginal therapy, except aspirin (100 mg), was discontinued 48 hours before the study and a randomized, double-blind, cross-over design was used. Every day, over four consecutive days, each patient received orally either nicorandil (30 mg), nifedipine (10 mg), or a placebo (on two occasions). An ergometrine test was performed one hour after each drug intake. Ergometrine was administered intravenously in incremental doses to a maximum cumulative dose of 0.75 mg unless chest pain or electrocardiographic evidence of myocardial ischaemia occurred. A 12-lead electrocardiogram was recorded every minute and isosorbide dinitrate (3 mg, i.v.) was routinely administered after the highest dose of ergometrine, or earlier if pain or ST-segment change occurred. All patients exhibited reproducible chest pain and objective evidence of myocardial ischaemia when ergometrine tests were performed after administration of the placebo. After administration of nicorandil, the test was negative in nine patients and positive in three patients, although at a significantly higher cumulative dose of ergometrine than following the placebo. After nifedipine, the test became negative in five patients and the mean cumulative dose of ergometrine that produced spasm was significantly greater than after the placebo. The effect of nicorandil on ergometrine-induced coronary spasm was slightly, albeit not significantly, more pronounced than that of nifedipine, which is the reference drug for this test.

Aizawa et al. [30] documented angiographically the effect of nicorandil on spontaneous and ergometrine-induced coronary spasm. In five patients who exhibited spontaneous spasm during coronary arteriography, nicorandil (4 mg, i.v.) relieved the spasm in all five patients. In five other patients in whom spasm was induced by ergometrine, administration of nicorandil (2 mg, i.c. or 4 mg, i.v.) relieved the spasm after a delay of 30 to 140 s.

The effects of long-term treatment in patients with variant angina are particularly difficult to evaluate since there is a high spontaneous variation in symptoms in such patients. However, short term studies have documented the efficacy of oral treatment with nicorandil in patients with vasospastic angina. One such study examined the effects of oral nicorandil (20 or 40 mg/day over 3 days) after a placebo period of two days in 32 patients who exhibited episodes of ST-segment elevation during angina. Nicorandil administration was associated with a significant reduction in the frequency of episodes of anginal pain (from 3.6 to 0.7 episodes/day) and in the frequency of episodes of ST-segment elevation during holter monitoring [31].

Thus all the above clinical studies clearly demonstrate that nicorandil exerts a beneficial action in the treatment of vasospastic angina and this effect is at least as potent as that induced by Ca^{2+} antagonists.

K^+ channel openers exert cardioprotective effects

Considerable experimental evidence has recently been accumulated suggesting that K^+ channel openers exert direct and potent cardioprotective effects on the ischaemic myocardium. The arguments which support this assertion have been developed in chapter 13 but it is important to underline that K^+ channel opener-induced cardioprotection is promising since no comparable action has been previously reported with conventional antianginal drugs. Regardless of the animal model used, both in vitro and in vivo experiments have clearly shown that K⁺ channel openers exhibit cardioprotective effects on the ischaemic myocardium when they are administered at doses which do not alter the haemodynamic status of the preparation. The fact that this cardioprotective effect is stereospecific and that the ATP-dependent K⁺ channel blocker glibenclamide prevents the beneficial actions of different pure K⁺ channel openers: (i) on the release of lactate dehydrogenase (LDH) in isolated rat hearts subjected to global ischaemia [32]; (ii) on the post-ischaemic contractile dysfunction, and ultimately; (iii) on the infarct size [33], strongly suggest that these drugs act as cardioprotective agents by opening ATP-dependent K⁺ channels. Preservation of myocardial ATP levels by nicorandil after ischaemia and subsequent reperfusion in the canine heart as well as the isolated rat heart also support a metabolic basis for the beneficial effect of opening the ATP-sensitive K⁺ channels [34]. Furthermore, there is a general agreement in observing a wide separation between cardiodepression and cardioprotection when K⁺ channel openers were tested on isolated perfused ischaemic hearts, thus making them somewhat different to the Ca^{2+} antagonists studied in these models.

One of the key points presently under investigation is related to identification of the biological signal which triggers the opening of ATP-dependent K⁺ channels during myocardial ischaemia (see chapter 13). Current evidence suggests that the initial depletion in high energy phosphates is directly related to this action, but the concomitant release of adenosine, which also exerts potent cardioprotective effects on the ischaemic myocardium, has lead to the hypothesis that adenosine could induce the opening of K⁺ channels through activation of an A1 receptor [35].

In humans, the cardioprotective effect of K⁺ channel openers has not been extensively investigated since no relevant model is presently available. However, ischaemia induced by repetitive short periods of coronary occlusion can be studied during angioplasty. Saito et al. [36] investigated four groups of 10 patients in which nitroglycerin (0.2 mg), verapamil (5 mg), nicorandil (4 mg) and saline (5 ml) were administered after two control coronary occlusions. ST segment deviation was significantly reduced by verapamil and nicorandil but not by nitroglycerin. Side effects were observed after nitroglycerin and verapamil but not after nicorandil, which appeared safer and more efficacious than verapamil in this model of acute myocardial ischaemia. Further studies are required to evaluate the mechanism of action and the potential clinical implications of the cardioprotection induced by K⁺ channel openers.

Overall, potassium channel openers may well turn out to become, after calcium channel blockers, the major breakthrough for treating myocardial ischaemia. An ever increasing body of evidence suggests this class has the peculiarity to provide, at non hypotensive doses, direct myocardial protective effects, somewhat close to preconditioning, a feature that makes these compounds promising. Nicorandil is, at this point, the only such agent, available to clinicians or about to be, and proved to be efficacious and safe in ischaemic heart disease. Wider clinical experience with this new drug and others to come, from the same class, will tell us if potassium channel openers live up to their expectations, in ischaemic heart disease.

References

1. Imai N, Liang CS, Stone C, Sakamoto S, Hood WB Jr (1988) Comparative effects of nitroprusside and pinacidil on myocardial blood flow and infarct size in awake dogs with acute myocardial infarction. Circulation 77: 705-711
2. Sakamoto S, Liang CS, Stone CK, Hood WB Jr. (1989) Effects of pinacidil on myocardial blood flow and infarct size after acute left anterior descending coronary artery occlusion and reperfusion in awake dogs with and without a coexisting left circonflex coronary artery stenosis. J Cardiovasc Pharmacol 14: 747-755
3. Bache RJ, Dai X, Baran KW (1990) Effect of pinacidil on myocardial blood low in the presence of coronary artery stenosis. J Cardiovasc Pharmacol 15: 618-625
4. Auchampach JA, Maruyama M, Cavero I, Gross GJ (1992) Pharmacological evidence for a role of ATP-dependent potassium channels in myocardial stunning. Circulation 86: 311-319
5. Richer C, Pratz J, Mulder P, Mondot S, Giudicelli JF, Cavero I (1990) Cardiovascular and biological effects of K$^+$ channel openers, a class of drugs with vasorelaxant and cardioprotective properties. Life Sci 47: 1693-1705
6. Murakami M, Takeyama Y, Matsubana H, Hasegawa S, Nakamura M, Sekita S, Katagiri T (1989) Effects of intravenous injections of nicorandil on systemic and coronary haemodynamics in patients with old myocardial infarction: a comparison with nifedipine and ISDN. Eur Heart J 10 (Abstr): 426
7. Coltart DS, Signy M (1989) Acute hemodynamic effects of single-dose nicorandil in coronary artery disease. Am J Cardiol 63: 34J-39J
8. Krumenacher M, Roland E (1992) Clinical profile of nicorandil: an overview of its haemodynamic profile and therapeutic efficacy. J Cardiovasc Pharmacol 20 (Suppl 3): S93-S102
9. Cohen-Solal A, Jaeger P, Bouthier J, Juliard JM, Dahan M, Gourgon R (1989) Hemodynamic action of nicorandil in chronic congestive heart failure. Am J Cardiol 63: 44J-48J
10. Galie N, Varani E, Maiello L, Boriani G, Boschi S, Binetti G, Magnani B (1990) Usefulness of nicorandil in congestive heart failure. Am J Cardiol 65: 343-48
11. Giles TD, Pina IL, Quiroz Ac, Roffidal L, Zaleski R, Porter RS, Karalis DJ, Morhland JS, Wolf DL, Hearron AE, Sander GE (1992) Hemodynamic and neurohormonal responses to intravenous nicorandil in congestive heart failure in humans. J Cardiovasc Pharmacol 20: 572-578
12. Elkayam U, Amin J, Mehra A, Vasquez J, Weber L, Rahimtoola S (1990) A prospective, randomised, double-blind cross-over study to compare the efficacy and safety of chronic nifedipine therapy with that of isosorbide dinitrate and their combination in the treatment of chronic congestive heart failure. Circulation 82: 1954-1961
13. Why HJF, Richardson PJ (1993) Nicorandil as monotherapy in chronic stable angina pectoris: comparison with placebo. Eur Heart J (in press)
14. Hughes LO, Rose EL, Lahiri A, Raftery EB (1990) Comparison of nicorandil and atenolol in stable angina pectoris. Am J Cardiol 66: 679-682
15. Ulvenstam G, Diderholm E, Frithz G, Gudbrandsson T, Hedbäck B, Höglund C, Moeltad P, Perk J, Sverrisson JT (1992) Antianginal and anti-ischaemic efficacy of nicorandil compared with nifedipine in patients with angina pectoris and coronary heart disease: a double blind, randomized, multicenter study. J Cardiovasc Pharmacol 20 (Suppl 3): S67-S73
16. Kinoshita M, Hashimoto K, Ohbayashi Y, Inoue T, Tagushi H, Mitsunami K (1989) Comparison of antianginal activity of nicorandil, propranolol and diltiazem with reference to the antianginal mechanism. Am J Cardiol 63: 71J-74J
17. Guermonprez JL, Blin P, Peterlongo F (1993) A double blind comparison of the long term efficacy of nicorandil and diltiazem in stable angina pectoris. Eur Heart J (in press)
18. Döring G (1992) Antianginal and anti-ischaemic efficacy of nicorandil in comparison with isosorbide-5-mononitrate and isosorbide dinitrate: results from two multicenter, double blind, randomized studies with stable coronary heart disease patients. J Cardiovasc Pharmacol 20 (Suppl 3): S74-S81

19. Zeiher AM, Drexler H, Wollschläger H, Hanjörg J (1991) Endothelial dysfunction of the coronary microvasculature is associated with impaired coronary blood flow regulation in patients with early atherosclerosis. Circulation 84: 1984-1992
20. Berdeaux A, Drieu La Rochelle C, Gosgnach M, Roupie E, Richard V, Giudicelli JF (1991) La dilatation débit-dépendante. Aspects pharmacologiques et physiopathologiques. Arch Mal Cœur 84 (III): 67-72
21. Cooke JP, Rossitch E, Andon NA, Loscalzo J, Dzau VJ (1991) Flow activates an endothelial potassium channel to realease an endogenous nitrovasodilator. J Clin Invest 88: 1663-1671
22. Giudicelli JF, Drieu La Rochelle C, Berdeaux A (1990) Effects of cromakalim and pinacidil on large and small coronary arteries in conscious dogs. J Pharmacol Exp Ther 255: 836-842
23. Quast U, Cook NS (1989) In vitro and in vivo comparison of two K^+ channel operners, diazoxide and cromakalim, and their inhibition by glibenclamide . J Pharmacol Exp Ther 250: 261-271
24. Cavero I, Mondot S, Mestre M (1989) Vasorelaxant effects of cromakalim in rats are mediated by glibenclamide-sensitive potassium channels. J Pharmacol Exp Ther 248: 1261-1268
25. Drieu La Rochelle C, Richard V, Dubois-Randé JL, Roupie E, Giudicelli JF, Hittinger L, Berdeaux A (1992) Potassium channel openers dilate large epicardial coronary arteries in conscious dogs by an indirect, endothelium-dependent mechanism. J Pharmacol Exp Ther 263: 1091-1096
26. Waters D (1991) Proischemic complications of dihydropyridine calcium channel blockers. Circulation 84: 2598-2600
27. Kishada H, Hata N, Kusama Y (1990) Angiographic response to a vasodilating drug nicorandil in patients with coronary artery disease. Jpn Heart J 31: 135-143
28. Suryapranata M, Macleod D (1992) Nicorandil and cardiovascular performance in patients with coronary artery disease. J Cardiovasc Pharmacol 20 (Suppl 3): S45-S51
29. Lablanche JM, Bauters C, Leroy F, Bertrand ME (1992) Prevention of coronary arterial spasm by nicorandil: comparison with nifedipine. J Cardiovasc Pharmacol 20 (Suppl 3): S82-S85
30. Aizawa T, Ogasawara K, Nakamura F (1989) Effects of nicorandil on coronary spasm. Am J Cardiol 63: 75J-79J
31. Matsumura Y, Kodama M, Mishima M (1990) Effect of nicorandil on coronary vasospasm. Ther Res 11: 110-114
32. Grover GJ, Sleph PG, Dzwonczyk SS (1990) Pharmacologic profile of cromakalim in the treatment of myocardial ischemia in isolated rat hearts and anaesthetized dogs. J Cardiovasc Pharmacol 16: 853-864
33. Grover GJ, Dzwonczyk SS, Parham CS, Sleph PG (1990) The protective effect of cromakalim and pinacidil on reperfusion function and infarct size in isolated perfused rat hearts and anaesthetized dogs. Cardiovasc Drugs Ther 4: 465-474
34. Gross JG, Auchampach JA (1992) Blockade of ATP-sensitive potassium channels prevents myocardial preconditioning in dogs. Circ Res 70: 223-233
35. Liu GS, Thornton J, Van Winkle DM, Stanley AWH, Olsson RA, Downey JM (1991) Protection against infarction afforded by preconditioning is mediated by A1 adenosine receptors in rabbit heart. Circulation 84: 350-356
36. Saito S, Tamura Y, Moriuchi M, Mizumura T, Hibiya K, Honye J, Ando T, Kamata T, Tsuji M, Ozawa Y (1991) Comparative efficacy and safety of nitroglycerin, verapamil and nicorandil during coronary angioplasty. J Am Coll Cardiol 17 (Abstr): 377A

CHAPTER 18
The potential of potassium channel openers in peripheral vascular disease

NS Cook, JR Fozard and RP Hof

Studies in a variety of different animal models of peripheral vascular disease have shown K^+ channel openers to have a unique profile of action. In haemodynamic investigations with the benzopyran derivatives SDZ PCO-400 and cromakalim, a selective vasodilation of collateral vessels supplying ischaemic skeletal muscle has been shown, often occuring at doses below those affecting systemic blood pressure. Diversion of blood away from the ischaemic area towards the normally perfused tissue ("steal") does not appear to be a problem at these low doses, unlike previously studied vasodilators. In peripheral vascular disease models, K^+ channel openers have been shown to dramatically improve skeletal muscle energy metabolism under conditions resembling those seen in patients with intermittent claudication, and also to improve skeletal muscle performance in a fatiguability model under ischaemic conditions. These effects were manifest at sub-hypotensive doses and, at least in terms of energy metabolism, could be demonstrated 24 hours after an oral dose of SDZ PCO-400.

In the present chapter, possible mechanisms contributing to these interesting results are discussed, including actions on different K^+ channels, differences between collateral and normal arterial blood vessels, and the endothelium and nerve endings as possible target sites for K^+ channel openers which may contribute to these actions. Finally, we consider the therapeutic potential of these drugs, both in peripheral vascular disease and in other ischaemic disorders and emphasize the need for clinical trials to validate the true potential of K^+ channel openers in these indications.

Available drugs show poor efficacy in intermittent claudication

Occlusive vascular disease of the legs, predominantly due to arteriosclerosis, is a disabling disorder of high frequency and with a prevalence which increases with age. Cardiovascular mortality from ischaemic heart disease and stroke is increased in these patients, resulting in a shortened life expectancy [1]. Whilst amputation or major surgical intervention is not warranted in the majority of cases, therapeutic intervention to relieve the exercise-induced fatigue and associated pain ("intermittent claudication"), would be of great benefit to the patient. A number of drugs have been evaluated for this indication, two of the most recent being ketanserin and pentoxifylline. However, clinical trials with these agents have not been straightforward, in many cases yielding conflicting results [2]. The currently held view is that none of the available therapies offer significant benefit to the patient with peripheral vascular disease.

Non-selective vasodilation is contra-indicated in peripheral vascular disease [3]. Such an action is likely to dilate vessels supplying normally perfused tissue, therefore diverting blood away from the ischaemic area. A lowering of the peripheral perfusion pressure will also limit blood supply to the afflicted limb. In this regard, Ca^{2+} antagonists [4, 5] and other vasodilator drugs [6] were found in clinical trials not to improve the symptoms of claudication or to increase exercise capacity. In view of these considerations, the desirable profile of a vasodilator for the treatment of intermittent claudication would be one which selectively dilates the collateral vessels supplying the ischaemic region, at doses which do not reduce systemic blood pressure.

K^+ channel openers have a characteristic regional vasodilator profile

The K^+ channel openers are potent peripheral vasodilators in animals, with a characteristic regional haemodynamic profile. This is illustrated by haemodynamic experiments performed in anaesthetized cats with the benzopyran, SDZ PCO-400, which extends our previous investigations with cromakalim in rabbits [7-9]. Cumulative intravenous doses of 3 - 100 µg/kg SDZ PCO-400 were chosen, which caused a dose-dependent decrease in blood pressure and increase in cardiac output, with no change in heart rate (fig. 1A).

The regional vasodilatator profile of SDZ PCO-400 (fig. 1B) showed differences from that of various calcium antagonists in the same cat model [10, 11]. Equipotent blood pressure lowering doses of both types of drug dilated coronary vessels to a similar extent, but the cerebral circulation was less responsive to SDZ PCO-400 than to Ca^{2+} antagonists. Concerning renal conductance, SDZ PCO - 400 caused dilation at low doses (significant at 10 and 30 µg/kg). However, the 100 µg/kg dose lowered blood pressure below the autoregulatory threshold, which may well explain the reduction in renal conductance relative to the lower doses. Strong dose-dependent vasodilation over the whole SDZ PCO-400 dose-range was observed in the stomach and the small intestine, which is similar to results in rabbits with cromakalim and its (-)-enantiomer [7, 8].

The regional haemodynamic differences between Ca^{2+} antagonists and K^+ channel openers are particularly apparent with respect to skeletal muscle blood flow. All calcium antagonists and especially dihydropyridine derivatives, dilate skeletal muscle blood vessels [10, 11]. SDZ PCO-400 by contrast caused marked and significant vasoconstriction at the lower doses, which, however, changed into a significant vasodilation at the 100 µg/kg dose (fig. 1B). This biphasic behaviour is not cat specific: (-)-cromakalim evoked similar changes in skeletal muscle vascular conductance in rabbits [8]. A more detailed analysis of skeletal muscle flow follows in the next section. It is noteworthy, however, that skeletal muscle vasoconstriction was observed at relatively low doses, where blood pressure changes were modest and counterregulatory homeostatic mechanisms presumably activated only weakly. Such effects were not seen with calcium antagonists.

These observations on regional vascular activity of SDZ PCO-400 in cats, in particular the lack of a prominent vasodilation of skeletal muscle beds, agrees well with earlier observations of K^+ channel openers in rabbits [7, 8], cats [12], pigs [13] and rats [14, though see 15]. However, vasodilation has been observed in man following intra-arterial infusion of cromakalim in the arm, which suggests a contribution from skeletal muscle vasodilation [16]. With respect to peripheral vascular disease, a minimal effect of K^+ channel openers on the normal arterial resistance vessels of the skeletal muscle vascular bed would be desirable, such that a "steal" phenomena is unlikely to occur.

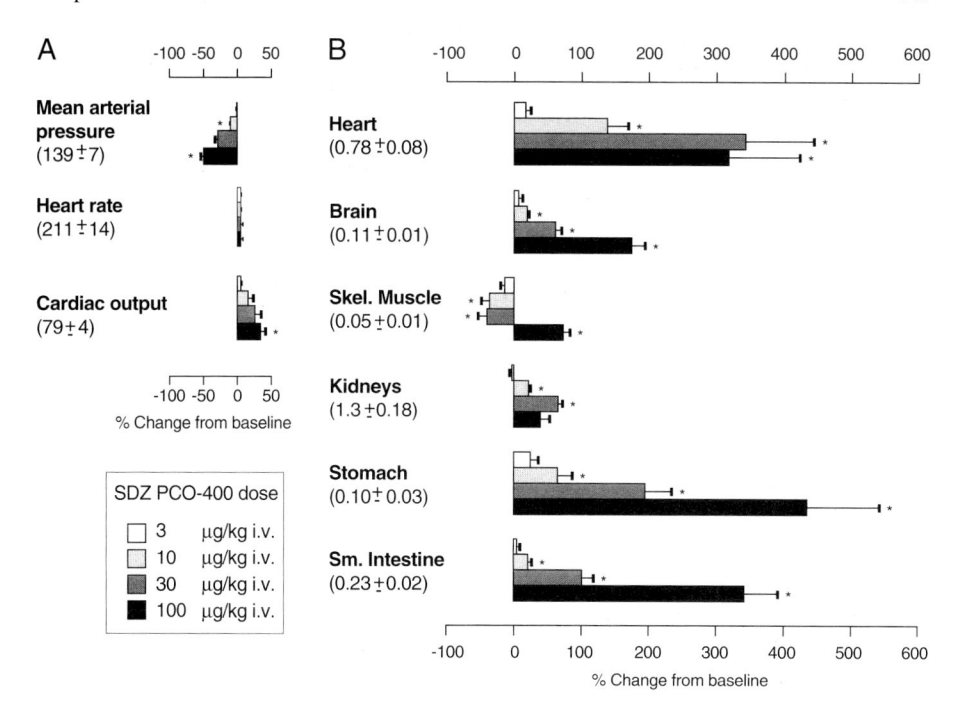

Fig. 1 A, B. Dose-dependent effects of SDZ PCO-400 on **A** systemic haemodynamics and **B** regional conductances in chloralose-urethane anaesthetized cats. Methods are essentially as described in [10] with regional flow assessed by injection of ~150,000 radioactive microspheres (^{125}I, ^{141}Ce, ^{51}Cr, ^{85}Sr, ^{46}Sc) at different times throughout the experiment, calibrated to give estimates of absolute flow according to the reference flow method [10,11]. Following baseline measurements, SDZ PCO-400 was infused into the jugular vein for 10 min at each dose, with further measurements at the end of each infusion period. Drug effects are shown as the mean % change (± SEM) from the pretreatment baseline values, which are indicated beneath the corresponding parameter in parentheses (mean arterial pressure, mmHg; heart rate, beats/min; cardiac output, ml/min/kg; vascular conductances, ml/min/mmHg per 100g of wet tissue). *Skel. Muscle,* skeletal muscle; *Sm Intestine,* small intestine. * denotes a significant change from baseline ($p < 0.05$, n = 5)

K⁺ channel openers improve collateral perfusion in acute skeletal muscle ischaemia

In an isolated rabbit ear model of acute arterial occlusion, (-)-cromakalim (0.5 and 3 μM) was found to significantly enhance perfusion through pre-existing collaterals [17]. This action was seen in the absence of changes in total perfusion pressure, suggesting a redistribution of flow by selective opening of collaterals at concentrations not producing general relaxation of the vessels of the whole ear preparation. In contrast, the Ca^{2+} antagonist, verapamil, caused a potent lowering of tone of the perfused ear, but did not influence collateral flow [17].

Our haemodynamic studies of SDZ PCO-400 in cats were extended to investigate effects on skeletal muscle blood flow following acute ligation of the femoral artery supplying the right hindlimb. The cat is well suited to acute investigations of this kind, since the cat hindlimb is known to have pre-existing collaterals, which open rapidly in response to arterial ligation [18]. Occlusion of the femoral artery high in the groin region, allows only flow through collateral vessels to reach the distal parts of the leg, whereas the proximal parts will also receive flow directly from other arteries.

In the normally perfused left leg, SDZ PCO-400 (3 - 30 µg/kg i.v.) caused a uniform and dose-related decrease of conductance to both the proximal (above the knee) and distal (below the knee) parts of the leg muscle (fig. 2). The decrease was fully reversed at the highest dose of 100 µg/kg, thereby generating a bell-shaped dose-response curve.

The baseline conductance of the distal (but not the proximal) part of the occluded right leg was significantly less than that of the normally perfused left leg, confirming that the femoral artery is the primary supplier of blood to the distal part of the leg muscle under normal conditions. Changes in conductance induced by SDZ PCO-400 to the proximal muscles of the occluded leg were similar to those of the healthy leg (fig. 2). This is to be expected, since much of this part of the leg obtains its blood supply from arteries other than the occluded femoral artery.

The fully collateral-dependent distal part of the leg showed a quite different effect of SDZ PCO-400. The lowest (3 µg/kg) dose was almost inactive, which is significantly different from the constrictor effect seen in normally perfused leg. At 10 and 30 µg/kg conductance increased, again significantly different from the constrictor effect of SDZ PCO-400 on the normal leg (fig. 2). However, at the highest dose the drug-induced changes from baseline were no longer different. Thus, in agreement with in vitro studies of collateral flow to the rabbit ear [17], these in vivo studies further demonstrate a blood flow redistribution by a K^+ channel opener due to selective dilation of pre-existing collateral vessels. Additionally, these studies highlight the importance of the "therapeutic window" for these effects; at higher doses where conductance of the normal leg is also increased, a steal phenomenon is likely to occur and the beneficial effect on flow redistribution lost.

K^+ channel openers selectively enhance blood supply to chronically ischaemic rat hindlimb

The pioneering studies with K^+ channel openers in peripheral vascular disease were performed by Angersbach and Nicholson [14] in a rat model of intermittent claudication. Chronic unilateral ligation of the femoral artery supplying the rat hind limb was performed, to create a situation resembling chronic occlusive arterial disease in man [19]. In this animal model, the K^+ channel openers cromakalim, pinacidil and nicorandil were shown to cause a significant increase in hypoxic muscle blood cell flux, whereas several different Ca^{2+} antagonists and the direct-acting vasodilator, hydralazine, were inactive. The K^+ channel openers, but not the Ca^{2+} antagonists or hydralazine, also increased the oxygen tension of the chronically ligated leg muscle [14].

We have carried out experiments in a model of occlusive arterial disease based on that of Angersbach and Nicholson [14] but where flow to the individual hindlimbs was monitored by pulsed Doppler flow probes placed around the respective iliac arteries [20]. Consistent with earlier findings [14,19], ligation of one femoral artery three weeks prior to the experiment results in a significant reduction in resting flow to the limb (unligated limb, 3.54 ± 0.43 kHz; ligated limb, 2.26 ± 0.23 kHz; $p < 0.05$). Infusion of

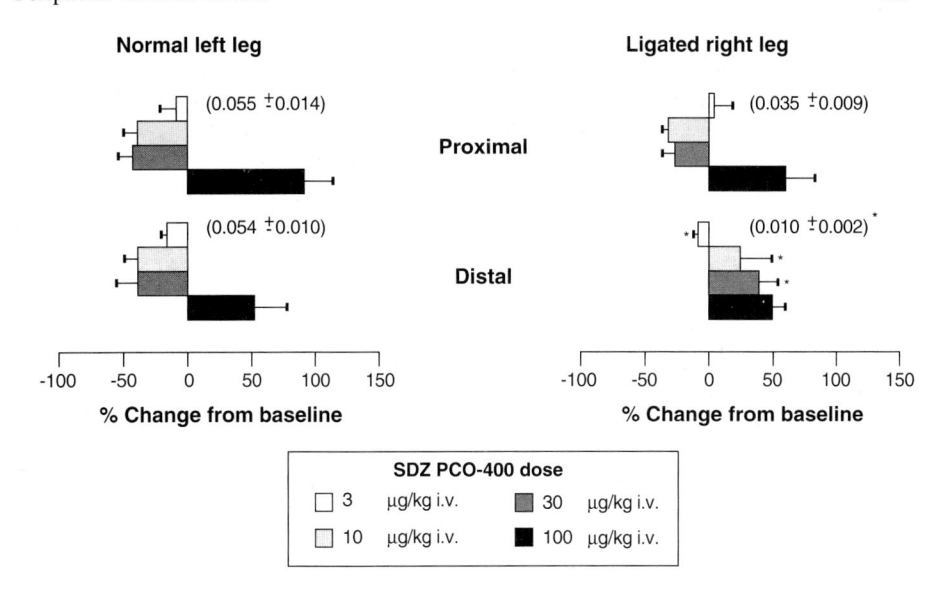

Fig. 2. Regional vasodilation (ml/min/mmHg per 100g wet tissue) in the hindlimb skeletal muscle of anaesthetized cats (same experiments as in fig. 1) in which the femoral artery supplying the right leg had been occluded in the upper groin area around 2 hours previously. SDZ PCO-400 was infused i.v. at the doses shown. See legend to fig. 1 for further details. Drug effects are shown as the mean % change in conductance (± SEM) to the proximal and distal parts (above/below knee, respectively) of the normally perfused left leg (left) and the occluded right leg (right). * indicates a significant difference between the corresponding region of the normally perfused leg and the occluded leg (p < 0.05, paired t-test, n = 5). Statistical computations were carried out on the absolute (and not the%) values, due to significant differences in the baseline values

SDZ PCO-400 intravenously at cumulative doses of 2, 4 and 8 µg/kg/min induced dose-related falls in mean arterial pressure and tachycardia (fig. 3A). SDZ PCO-400 did not markedly affect flow to the intact limb although conductance increased dose-dependently, reflecting the fall in blood pressure (fig. 3B, C). In contrast, blood flow and conductance in the ligated limb increased dose-dependently during infusion of 2 and 4 µg/kg/min of SDZ PCO-400, such that the initial difference in flow between the ligated and unligated limbs was eliminated (fig 3B, C). These data confirm, using a different K$^+$ channel opener, the findings of Angersbach and Nicholson [14], that these drugs dilate selectively the vessels of chronically ischaemic skeletal muscle in the rat.

Energy metabolism in ischaemic skeletal muscle is improved by K$^+$ channel openers

The exercise-induced fatigue which characterizes intermittent claudication has been investigated in man in terms of the biochemical abnormalities relating to skeletal muscle energy metabolism. Keller et al. [21] used ^{31}P-NMR spectroscopy to demonstrate that the recovery of leg muscle phosphocreatine (PCr) levels following a period of ischaemic exercise was

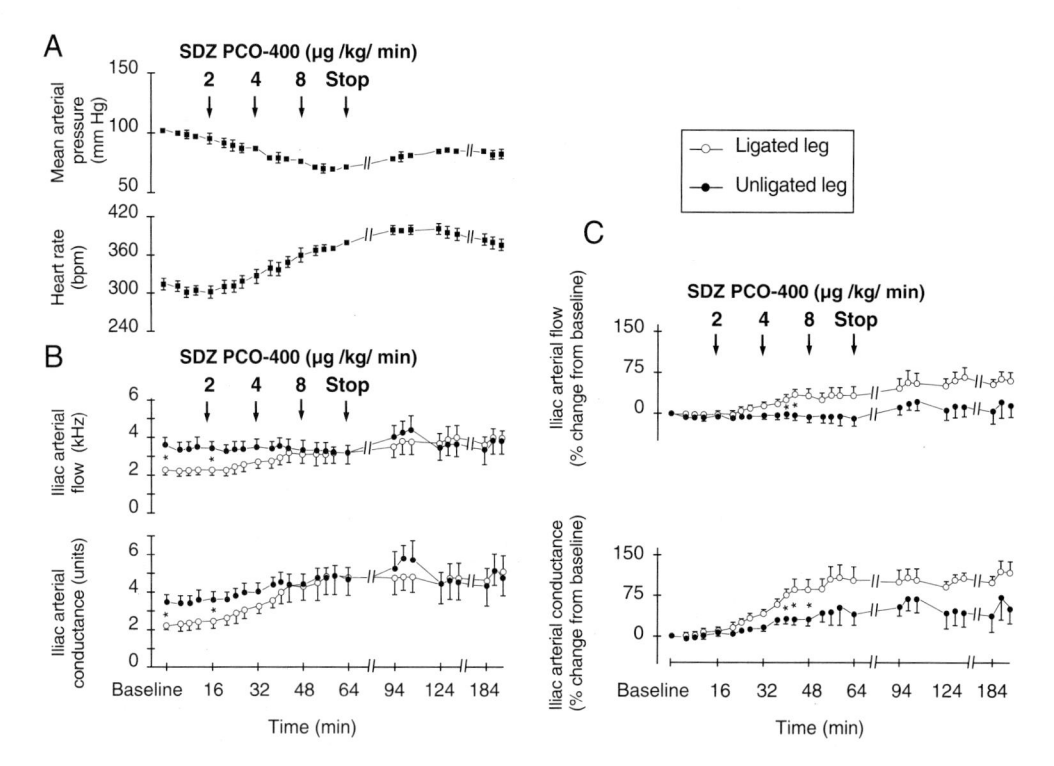

Fig. 3 A-C. Effect of SDZ PCO-400 on **A** mean arterial pressure and heart rate and **B** iliac arterial blood flow and conductance in anaesthetized Sprange-Dawley rats in which one femoral artery was ligated three weeks prior to the experiment. The changes in iliac arterial flow and conductance, as a % change from baseline, are shown in **C**. SDZ PCO-400 was infused (0.1 ml/min) intravenously at cumulative doses of 2, 4 and 8 µg/kg/min, starting at the time-points indicated. Points represent mean values (± SEM) from 5 individual experiments. For technical reasons, flow and conductance values at the later time-points were recorded from just 4 (48 - 94 min) or 3 (124 min - end) animals, respectively. * $p < 0.05$ that the values at a particular-time point differ significantly. Further details of the method can be found in [20]

significantly slower in patients with intermittent claudication than in normal individuals (see fig. 4A). Based on these clinical observations, we have developed a rat model of peripheral vascular disease [22]. ^{31}P-NMR spectroscopy was used to monitor PCr levels of the hindlimb during tourniquet-induced ischaemia and subsequent recovery upon release of the tourniquet. Comparable to the clinical situation [21], PCr recovery rate from acute ischaemia was significantly slower in those rats where the femoral artery had been ligated two weeks previously, as compared with sham-operated animals (fig. 4B) [22].

Two K$^+$ channel openers, SDZ PCO-400 and cromakalim, and the dihydropyridine Ca^{2+} antagonist nitrendipine, have been investigated for their effects in this rat NMR model of peripheral vascular disease. ^{31}P spectra were recorded during two periods of ischaemia and recovery. The PCr recovery time from the first period of ischaemia (before drug administration) was very similar for the different animal groups (see [22]), but marked differences were seen in the second recovery time (in the presence of drug). This is illustrated in fig. 5, showing a significant increase in the recovery rate from the second

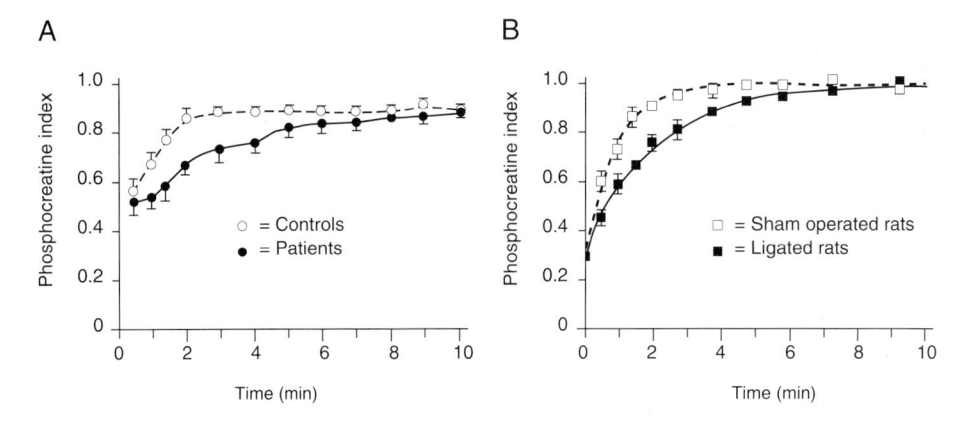

Fig. 4. A Clinical NMR study showing the recovery of calf muscle phosphocreatine, following a 3 min period of ischaemic exercise in 11 patients with angiographic and symptomatic evidence of occlusive arterial disease of the legs and in 7 age-matched normal volunteers. The half-time for PCr recovery was 37 ± 6 s for the control group and 203 ± 74 s for the patient group. ATP levels did not decrease significantly in either group during the ischaemic exercise. Redrawn with permission from [21]. **B** Recovery of the hindlimb phosphocreatine peak from a 35 min period of ischaemia (under resting conditions) in rats in which the right femoral artery had been ligated 2 weeks prior to the experiment ("ligated") or in sham-operated rats ("sham"). The half-time for PCr recovery was 37 ± 1 s (sham) and 109 ± 6 s (ligated). Means $\pm$ SEM, n = 4 rats per group. ATP levels remain unaltered throughout. Redrawn from [22] with the permission of Blackwell Scientific Publications Inc

period of ischaemia in the group of rats treated with SDZ PCO-400 (10 µg/kg i.v.), as compared with the control rats.

Similar experiments have been performed with different i.v. doses of SDZ PCO-400, cromakalim and nitrendipine. The results, together with the concomitant effects on blood pressure, are summarized in fig. 6. Effects on PCr recovery are expressed as the ratio (in %) of the recovery-(2) / recovery-(1). In control animals, the ratio was $148 \pm 13\%$, indicating a slower recovery from the second ischaemia than from the first. A significant reduction in the PCr recovery ratio was seen with doses of 1 - 30 µg/kg SDZ-PCO 400 and with 30 µg/kg cromakalim. At these doses, neither drug significantly lowered blood pressure. At higher doses where blood pressure was significantly reduced, the beneficial effect of the K^+ channel openers on PCr recovery was lost, and, in the case of 300 µg/kg cromakalim, a marked slowing of the recovery of PCr from ischaemia was evident (fig. 6). Thus both SDZ PCO-400 and cromakalim exhibited "U"-shaped dose-response curves.

Nitrendipine (10 - 300 µg/kg i.v.) caused only a minor improvement in PCr recovery rate, which did not reach significance (fig. 6), in accordance with the lack of clinical efficacy of Ca^{2+} antagonists in intermittent claudication [4, 5]. Predictably, at a dose causing prominent falls in blood pressure, a marked slowing of the recovery from ischaemia was seen in the presence of nitrendipine.

We have also evaluated the effects of oral administration of SDZ-PCO 400 in this rat model, with ischaemia and recovery in the presence of drug being evaluated 24 hours after drug administration [22]. Under these conditions, the recovery-(2) / recovery-(1) ratio was $110 \pm 14\%$ in the control group and $62 \pm 11\%$ in the group treated with SDZ PCO-400 (0.3 mg/kg p.o., 24 hours previously). This suggests a long-lasting bene-

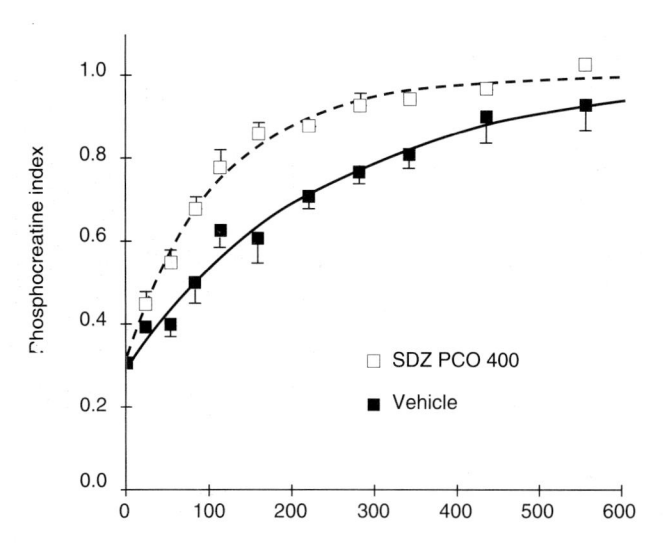

Fig. 5. Effect of SDZ PCO-400 (10 µg/kg i.v.) or vehicle upon the recovery of rat leg muscle phosphocreatine from a 35 min period of ischaemia, induced with a tourniquet, in rats in which the femoral artery supplying the leg had been ligated 3 weeks previously. Half-times for PCr recovery were 78 ± 8 s (SDZ PCO-400, n = 4) and 112 ± 5 s (vehicle control, n = 7). Recovery times from a first ischaemic period (in the absence of drug/vehicle) were normalized to 150 s for the two groups, in order to allow comparison of the second recovery in the presence of drug/vehicle (for further details, see [22])

ficial effect of oral SDZ PCO-400 on skeletal muscle energy metabolism in this animal model of peripheral vascular disease.

K⁺ channel openers reduce fatigue in ischaemic skeletal muscle

Positive effects of K⁺ channel openers on both blood flow to ischaemic tissue and on energy metabolism of skeletal muscle are complemented by results in a fatiguability model, where skeletal muscle performance under acute ischaemic conditions was measured. Following acute ligation of the femoral artery, the reduction in twitch height in response to sciatic nerve stimulation was investigated [23]. In rats treated with cromakalim (1 - 100 µg/kg i.v.), the reduction in twitch height was inhibited by 45% when compared with control rats [23]. Interestingly, the 1 µg/kg dose, which did not affect systemic blood pressure or total limb blood flow, was equi-active with the higher doses, whilst at a dose of 300 µg/kg (causing a prominent fall in blood pressure) the beneficial effect was lost. Consistent with the findings reported above in the NMR model, selective redistribution of blood flow within the ischaemic region is a likely explanation for these results, although a direct effect of K⁺ channel openers on muscle energy metabolism cannot be excluded.

Do collateral vessels have distinct high affinity sites for K⁺ channel openers?

The unique efficacy of K⁺ channel openers in different peripheral ischaemia models and

Fig. 6. (Upper) Ratio between the recovery times before and after drug treatment in vehicle- treated control rats, and in rats treated with varying intravenous doses of SDZ PCO-400, cromakalim or nitrendipine. Results are expressed as the ratio in % of the recovery in the presence of drug / the initial recovery. A significant difference from the control group (non-paired t test) is indicated, * p < 0.02. Means ± SEM, n = 3 to 7 rats per dose for the drug-treated rats, n = 19 for the control group. (Lower) Change in blood pressure from baseline in the presence of the different drug doses, expressed as the area under the curve (AUC, in arbitrary units) of the Δ blood pressure vs time graph, for the period 30 to 60 min after drug administration, ie. corresponding to the time-period during which the recovery from the second period of ischaemia took place. Taken from [22] with permision of Blackwell Scientific Publications Inc

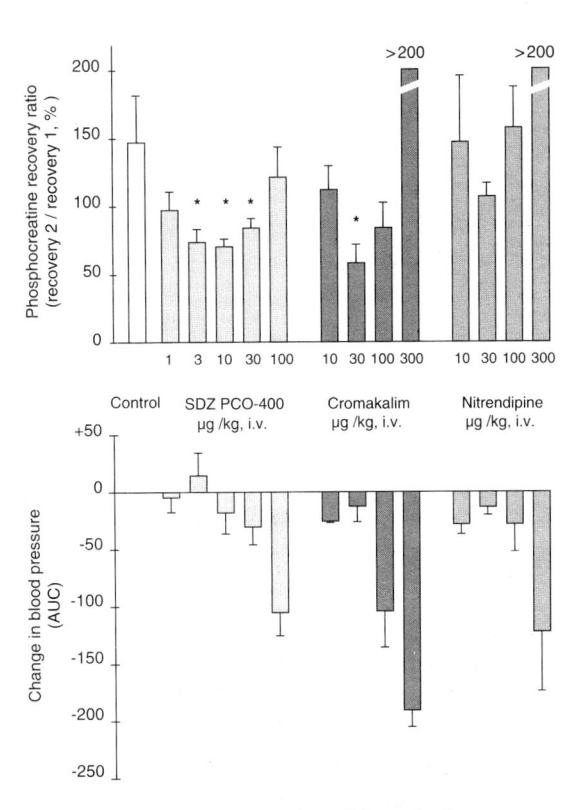

the mechanistic basis for this activity merit further consideration. Of particular note are the findings in models investigating muscle high energy phosphate metabolism [22] and fatiguability [23], where beneficial effects of K^+ channel openers were manifest at doses below those lowering blood pressure. Indeed, strongly hypotensive doses of both K^+ channel openers and nitrendipine had an adverse effect on the recovery of leg muscle phosphocreatine from a period of ischaemia. This supports the reasoning that lowering of the perfusion pressure supplying the already compromised limb, through non-selective vasodilation, would be detrimental to the treatment of intermittent claudication [3].

Surprisingly, in our blood flow studies in the rat hindlimb (fig. 3), the beneficial effects of SDZ PCO-400 on flow to the ligated limb persisted in the face of systemic cardiovascular changes. This differed, however, from the effects of K^+ channel openers in the cat and rabbit on the conductance of normally perfused skeletal muscle: The "U" shaped dose-responce curves seen with both SDZ PCO-400 (fig. 1) and (-)-cromakalim [8] in fact provide an additional mechanistic explanation for why the optimal effects of K^+ channel openers in peripheral vascular disease models may be manifest at low (sub-hypotensive) doses. At the higher doses, dilation of the arterioles supplying the non-ischaemic tissue would cause "steal" to occur. For the same reason, Ca^{2+} antagonists are unsuited to the treatment of intermittent claudication, since they dilate equally both healthy and diseased tissue.

The clear biphasic nature of the skeletal muscle conductance changes in cats and rabbits induced by SDZ PCO-400 and (-)-cromakalim (but interestingly not by (+)-cromakalim; see [8]), raises the question that more than one site of action of the K^+ channel openers may be responsible for these different vascular actions. At a microvascular level, (-)-cromakalim has been shown in vitro to dilate arterioles of different sizes in striated muscle of spontaneously hypertensive rats, with the most pronounced effect being on the small precapillary arterioles (< 35 µM diameter) [24]. Vessels of 35 - 120 µM diameter, which are considered important for blood pressure control, were less sensitive. Furthermore, the in vivo vasodilator effect of cromakalim and pinacidil on large coronary arteries of the dog, which is in part flow dependent [25], involves an endothelium-dependent component, whereas the relaxation of small coronary arteries is endothelium independent [26].

Some of the earliest studies with K^+ channel openers on the portal vein indicated that a single site of action may not be sufficient to define all the effects of these drugs on vascular smooth muscle [27]. Additional electrophysiological and ion flux studies subsequently confirmed that K^+ channel openers can open a variety of different K^+ channels, albeit at different drug concentrations [27-29]. Structurally different K^+ channel openers also show differences in their selectivity for different K^+ channels, the differences being particularly apparent between cromakalim-like agents, diazoxide and minoxidil [27, 29, 30]. These in vitro studies defining multiple sites of action for K^+ channel openers, may now have in vivo correlates in vascular ischaemia models.

The recent investigations with the perfused rabbit ear [17] are the first to look at K^+ channel opener effects on collateral vessels in vitro, and, together with the present studies in cats (fig. 2) and rats (fig. 3), suggest that collateral vessels are more sensitive to K^+ channel openers than are normal arterioles. Such an enhanced sensitivity could be explained either by a quantitative or a qualitative difference in the K^+ channel distribution between the different vascular beds, or metabolic differences, especially with respect to ATP-sensitive K^+ channels between collaterals and other arterioles. The presently available evidence is insufficient for us to decide between these possibilities.

A supersensitivity of coronary collateral vessels to vasoconstrictor agents (eg. vasopressin) has been reported [31]. In large arteries, cromakalim has been shown to elicit potent antivasoconstrictor effects over and above that seen with a Ca^{2+} antagonist [32], which may be an important factor in their ability to dilate collaterals. Collateral vessels are also known to have a greater dependence on intracellular Ca^{2+} stores than normal arteries [33], which could also be a reason why K^+ channel openers are more effective than Ca^{2+} antagonists at dilating these vessels: K^+ channel openers have been shown to inhibit intracellular Ca^{2+} movement and agonist-induced intracellular Ca^{2+}-dependent responses [29]. An additional point is that the endothelial lining of newly formed collateral vessels may differ in function from that found elsewhere in established blood vessels [31, 34]. Since a component of the in vivo vasodilator activity of K^+ channel openers has been shown to be endothelial (and probably flow) dependent [26], endothelial cell disparity may also contribute to the difference in sensitivity to K^+ channel openers between collateral and other vessels.

Are ATP-sensitive K^+ channels involved?

The target of K^+ channel opener drugs which to date has received most attention is an ATP-sensitive K^+ channel sensitive to inhibition by glibenclamide [35]. In vivo hypotensive effects of cromakalim and other K^+ channel openers can be inhibited by glibenclamide [30]. ATP-sensitive K^+ channels, of which a physiologically and pharmacologically diverse family are now known to exist, provide a link between the metabolic state of the cell and its excitability [35]. Such a channel would provide an attractive candidate for explaining the efficacy of K$^+$ channel openers in skeletal muscle ischaemia. As has been demonstrated in myocardial ischaemia [36, 37], a reduction in the intracellular ATP/ADP ratio and an extracellular acidosis would both favour ATP-sensitive K$^+$ channel opening. In peripheral vascular disease, a shift in open probability of ATP-sensitive K$^+$ channels during ischaemia might favour channel opening by the K$^+$ channel openers (as shown in cardiac cells; [38]), thus giving rise to their seemingly greater affinity under such conditions.

For the moment though, this remains purely speculation. Although the normal arteriolar vasodilator effect of K$^+$ channel openers in vitro and in vivo can be blocked by glibenclamide [30, 35], similar experiments to demonstrate a glibenclamide-sensitivity of K$^+$ channel opener effects on collateral vessels, or in animal models of peripheral vascular disease, have yet to be described.

Effects of K$^+$ channel openers on neural tissue may contribute to the selective vasodilation

Collateral vessels are innervated and, particularly in rapidly growing or newly formed vessels, adventitial nerve endings which become incorporated into the medial tissue may lead to overreactivity [31]. This may have a bearing on the effect of K$^+$ channel openers in peripheral vascular disease, since it has been shown that regional vasoconstrictor responses to spinal cord stimulation are inhibited by K$^+$ channel openers to varying degrees in different vascular beds [39]. It should be noted that the peripheral vascular disease models where low doses of K$^+$ channel openers showed beneficial effects, all involved in vivo experiments with anaesthetized animals having an intact nervous supply to the tissue in question.

With respect to the respiratory system, the inhibitory effect of sub-bronchodilator doses of K$^+$ channel openers on hyper-reactive airway smooth muscle, seems to be due in part to an inhibitiory effect at the level of neural transmission: bilateral vagotomy significantly inhibited the ability of SDZ PCO-400 to attenuate antigen-induced increases in airways resistance [9, 40]. This is in accordance with evidence that K$^+$ channel openers can act presynaptically to block non-adrenergic non-cholinergic responses in vitro and that they inhibit electrically-induced responses of airways smooth muscle in preference to responses evoked by exogenous stimuli [41]. A similar neural action might therefore underlie selective effects of K$^+$ channel openers on certain vascular tissues. Since only limited success has so far been achieved in trying to identify K$^+$ channel openers with selectivity for a particular smooth muscle type, nerve-selective K$^+$ channel openers may offer an alternative approach towards the development of drugs targetted to a particular therapeutic disorder.

Can results in peripheral vascular disease be extrapolated to other ischaemic disorders?

Long before the first results with K$^+$ channel openers in peripheral vascular disease were obtained, it was recognized that nicorandil, albeit not a selective K$^+$ channel opener [27], could increase collateral blood supply during acute coronary ischaemia in dogs [42].

Recently, a number of studies have addressed the potential of K^+ channel openers as cardioprotective agents during coronary ischaemia, showing beneficial effects on infarct size reduction or recovery of mechanical function [38]. Glibenclamide, in contrast, increased infarct size [43]. Furthermore, cardioprotection as a consequence of preconditioning prior to a coronary ischaemic insult (which may be mediated via endogenous adenosine release), may also involve the opening of glibenclamide-sensitive K^+ channels [38, 44].

Despite evidence suggesting a more widespread therapeutic utility of K^+ channel openers in different ischaemic conditions, the doses required to elicit these effects and the haemodynamic activity of the respective drugs in both diseased and healthy tissues, have also to be taken into account. In contrast to the situation in skeletal muscle ischaemia, a possible drawback concerning the coronary vasculature is the increases in myocardial vascular conductance seen with K^+ channel openers in normal tissue (see fig. 1B). Thus, unlike the situation in skeletal muscle, a steal phenomenon may be a problem with K^+ channel openers in coronary ischaemia. However, a sub-hypotensive dose of aprikalim significantly reduced infarct size in the dog heart [43]. This effect, which was sensitive to glibenclamide, appeared to be independent of either regional flow redistribution within the myocardium or changes in collateral blood flow, suggesting other mechanisms were involved.

The potential utility of K^+ channel openers in cerebral ischaemia remains an open question since no studies of these compounds in stroke models have so far been published. However, the brain has an extensive collateral network linking the main cerebral arteries. These collaterals appear to be important in limiting infarction following cerebral ischaemia [45] and could therefore serve as an interesting therapeutic target for selective vasodilators. In addition, both SDZ PCO-400 (fig. 2) and cromakalim [7] caused smaller increases in cerebral conductance than were seen with Ca^{2+} antagonists [9-11], suggesting that diversion of blood towards the normally perfused brain regions (steal) would be less of a problem with K^+ channel openers in cerebral ischaemia.

Future of K^+ channel openers in the treatment of intermittent claudication

Recent results suggest that K^+ channel openers have selective effects at low doses in different animal models of peripheral vascular disease, a clinical disorder where new and efficacious drugs are greatly needed. An ability to dilate collaterals at sub-hypotensive doses and show beneficial effects on skeletal muscle blood flow, energy stores and performance, has been demonstrated with several structurally different K^+ channel openers, suggesting this to be a class effect. However, the biphasic dose-response curves seen in several instances (see, for example, figs. 2 and 6), with potentially adverse effects occuring at higher doses, indicate the importance of dosing correctly to achieve the desired therapeutic effect.

In clinical studies, the breadth of this therapeutic window, which could differ from one compound to the next (particularly if different K^+ channels are responsible for the effects in collateral and arterial vessels), may well determine the ultimate fate of individual compounds in peripheral vascular disease. Pharmacokinetic differences may also distinguish K^+ channel openers from one another. In this regard, SDZ PCO-400 has been shown to beneficially influence chronically ischaemic skeletal muscle energy levels in rats, 24 hours after an oral dose of 0.3 mg/kg [22]. Long-acting, orally active drugs would certainly be desirable for the treatment of a chronic disorder of this kind.

How great is the potential of K^+ channel openers in peripheral vascular disease? As vasodilators, their haemodynamic profile would seem optimal for this indication. However, there remains the open question in patients of how much vasodilator reserve is available for a drug to exploit, over and above the normal autoregulatory response to chronic limb ischaemia. Unfortunately, animal testing is unlikely to provide an answer to this question. Furthermore, previous testing of drugs in animal models of intermittent claudication has shown a notoriously bad predictability of the subsequent clinical outcome [2, 6]. Whilst the K^+ channel openers have been profiled in a variety of new peripheral vascular disease models, which are designed to be optimally representative of the clinical disease, caution must be applied in extrapolating these results to the human situation. Future studies should aim to establish whether the effects of K^+ channel openers are maintained during prolonged therapy, or even if there is any regression of the disease process during chronic administration. Most importantly, clinical trials are now required, to see if these interesting results in animals can be reproduced in the clinic, thereby justifying the potential of K^+ channel openers in peripheral vascular disease.

We wish to acknowledge Akiko Hof, Marie-Louise Part and Charles Pally for their expert technical work, Markus Rudin with whom the NMR experiments were performed and Ulrich Quast for advice and many stimulating discussions.

References

1. Dormandy JA, Mahir MS (1986) The natural history of peripheral atheromatous disease of the legs. In: Greenhalgh RM, Jamieson CW, Nicolaides AN (eds) Vascular surgery - Issues in current practise. Grune and Stratton, Orlando, pp 3-17
2. Creager MA (1989) Can claudication be treated medically? J Vasc Med Biol 1: 269-270
3. Coffman JD (1988) Pathophysiology of obstructive arterial disease. Herz 13: 343-350
4. Creager MA, Roddy MA (1990) The effect of nifedipine on calf blood flow and exercise capacity in patients with intermittent claudication. J Vasc Med Biol 2: 94-99
5. Lorentsen E, Landmark K (1983) The acute effects of nifedipine on calf and forefoot blood flow in patients with peripheral arterial insufficiency. Angiol 34: 46-52
6. Coffman JD, Mannick JA (1972) Failure of vasodilator drugs in arteriosclerosis obliterans. Ann Intern Med 76: 35-39
7. Cook NS, Hof RP (1988) Cardiovascular effects of apamin and BRL 34915 in rats and rabbits. Br J Pharmacol 93: 121-131
8. Hof RP, Quast U, Cook NS, Blarer S (1988) Mechanism of action, and systemic and regional haemodynamics of the potassium channel activator BRL34915 and its enantiomers. Circ Res 62: 679-686
9. Cook NS, Chapman I (1992) The therapeutic potential of potassium channel openers in peripheral vascular disease and asthma. Cardiovasc Drugs Ther: in press
10. Hof RP (1983) Calcium antagonists and the peripheral circulation: differences and similarities between PY108-068, nicardipine, verapamil and diltiazem. Br J Pharmacol 78: 375-394
11. Hof RP, Hof A, Scholtysik G, Menninger K (1984) Effects of the new calcium antagonist PN200-110 on the myocardium and the regional peripheral circulation in anaesthetized cats and dogs. J Cardiovasc Pharmacol 6: 407-416
12. Buckingham RE, Clapham JC, Hamilton TC, Longman SD, Norton J, Poyser RH (1986) BRL 34915, a novel antihypertensive agent: comparison of effects on blood pressure and other hae-

modynamic parameters with those of nifedipine in animal models. J Cardiovasc Pharmacol 8: 798-804
13. Sassen LM, Duncker DJ, Gho BC, Diekmann HW, Verdouw PD (1990) Haemodynamic profile of the potassium channel activator EMD 52692 in anaesthetized pigs. Br J Pharmacol 101: 605-614
14. Angersbach D, Nicholson CD (1988) Enhancement of muscle blood cell flux and pO2 by cromakalim (BRL 34915) and other compounds enhancing membrane K^+ conductance, but not by Ca^{2+} antagonists or hydralazine, in an animal model of occlusive arterial disease. Naunyn-Schmied Arch Pharmacol 337: 341-346.
15. Shoji T, Aki Y, Fukui K, Tamaki T, Iwao H, Abe Y (1990) Effects of cromakalim, a potassium channel opener, on regional blood flow in conscious spontaneously hypertensive rats. Eur J Pharmacol 186: 119-123
16. Webb DJ, Benjamin N, Vallance P (1989) The potassium channel opening drug cromakalim produces arterioselective vasodilation in the upper limbs of healthy volunteers. Br J Clin Pharmacol 27: 757-761
17. Randall MD, Griffith TM (1992) Effects of BRL38227, sodium nitroprusside and verapamil on collateral perfusion following acute arterial occlusion in the rabbit isolated ear. Br J Pharmacol 106: 315-323
18. Schaub RG, Meyers KM, Sande RD, Hamilton G (1976) Inhibition of feline collateral development following experimental thrombotic occlusion. Circ Res 39: 736-743
19. Angersbach D, Jukna JJ, Nicholson CD, Ochlich P, Wilke R (1987) The effect of acute and chronic femoral artery ligation on rat calf muscle oxygen tension, blood flow, metabolism and function. Int J Microcirc Clin Exp 7: 15-30
20. Wright CE, Fozard J (1990) Differences in regional vascular sensitivity to endothelin-1 between spontaneously hypertensive and normotensive Wistar-Kyoto rats. Br J Pharmacol 100: 107-113
21. Keller U, Oberhensli R, Huber P, Widmer LK, Aue WP, Hassink RI, Müller S, Seelig J (1985) Phosphocreatine content and intracellular pH of calf muscle measured by phosphorous NMR spectroscopy in occlusive arterial disease of the legs. Eur J Clin Invest 15: 382-388
22. Cook NS, Rudin M, Pally C, Blarer S, Quast U (1992) Effects of the potassium channel openers SDZ-PCO 400 and cromakalim in an in vivo rat model of occlusive arterial disease assessed by ^{31}P-NMR spectroscopy. J Vasc Med Biol 4: 14-22
23. Hatton R, Heys C, Todd MH, Downing OA, Wilson KA (1991) Cromakalim reverses ischaemic depression of skeletal muscle contractions in rat hind limb. Br J Pharmacol 102: 280
24. Struijker Boudier HAJ, Messing MWJ, van Essen H (1992) Preferential small arteriolar vasodilation by the potassium channel opener, BRL 38227, in conscious spontaneously hypertensive rats. Eur J Pharmacol 218: 191-193
25. Giudicelli JF, Drieu la Rochelle C, Berdeaux A (1990) Effects of cromakalim and pinacidil on large epicardial and small coronary arteries in conscious dogs. J Pharmacol Exp Ther 255: 836-842
26. Drieu la Rochelle C, Roupie E, Hittinger L, Dubois-Rand JL, Richard V, Giudicelli JF, Berdeaux A (1991) Dilation of large coronary arteries induced by potassium channel openers in conscious dogs is abolished after coronary endothelium removal. Circulation 84 (Suppl II): II-185
27. Cook NS, Quast U (1990) Potassium channel pharmacology. In: Cook NS (ed) Potassium channels: structure, classification, function and therapeutic potential. Ellis Horwood, Chichester, pp 181-255
28. Kajioka S, Nakashima M, Kitamura K, Kuriyama H (1991) Mechanisms of vasodilation induced by potassium-channel activators. Clin Sci 81: 129-139
29. Quast U, Bray KM, Baumlin Y, Dosogne J (1992) Potassium channel openers: pharmacology and therapeutic prospects. In: Angeli P, Gulini U, Quaglia W (eds) Trends in receptor res. Elsevier, Amsterdam, pp 309-332

30. Quast U, Cook NS (1989) In vitro and in vivo comparison of two K^+ channel openers, diazoxide and cromakalim, and their inhibition by glibenclamide. J Pharmacol Exp Ther 250: 261-271
31. Schaper W (1991) Development and role of coronary collaterals. Trends Cardiovasc Med 1: 256-261
32. Cook NS, Weir SW, Danzeisen MC (1988) Anti-vasoconstrictor effect of the K^+ channel opener cromakalim on the rabbit aorta - comparison with the calcium antagonist isradipine. Br J Pharmacol 95: 741-752
33. Williams PB (1982) Effect of calcium concentration on collateral arteries from the hindlimb of the dog. J Pharmacol Exp Ther 223: 409-415
34. Engelberg H (1989) Endothelium in health and disease. Sem Thrombos Haemostas 15: 178-183
35. Quast U, Cook NS (1989) Moving together: K^+ channel openers and ATP-sensitive K^+ channels. Trends Pharmacol Sci 10: 431-435
36. Coetzee WA (1992) ATP-sensitive potassium channels and myocardial ischaemia: why do they open? Cardiovasc Drugs Therap 6: 201-208
37. Coetzee WA (1992) Stimulation of current through K(ATP) channels of guinea pig ventricular myocytes by extracellular pH. J Mol Cell Cardiol 24 (Suppl I): S 109
38. Escande D, Cavero I (1992) K^+ channel openers and "natural" cardioprotection. Trends Pharmacol Sci 13: 269-272
39. Richer C, Mulder P, Doussau MP, Gautier P, Giudicelli JF (1990) Systemic and regional haemodynamic interactions between K^+ channel openers and the sympathetic nervous system in the pithed SHR. Br J Pharmacol 100: 557-563
40. Chapman ID, Kristersson A, Mathelin G, Schaeublin E, Mazzoni L, Boubekeur K, Murphy N, Morley J (1992) Effects of a potassium channel opener (SDZ PCO 400) on guinea-pig and human pulmonary airways. Br J Pharmacol 106: 423-429
41. Ichinose M, Barnes PJ (1990) A potassium channel activator modulates both excitatory non-cholinergic and cholinergic neurotransmission in guinea-pig airways. J Pharmacol Exp Ther 252: 1207-1212
42. Lamping KA, Gross GJ (1984) Comparative effects of a new nicotinamide nitrate derivative, nicorandil (SG 75), with nifedipine and nitroglycerin on true collateral blood flow following an acute coronary occlusion in dogs. J Cardiovasc Pharmacol 6: 601-608
43. Auchampach JA, Maruyama M, Cavero I, Gross GJ (1991) The new K^+ channel opener Aprikalim (RP52891) reduces experimental infarct size in dogs in the absence of haemodynamic changes. J Pharmacol Exp Ther 259: 961-967
44. Gross GJ, Auchampach JA (1992) Blockade of ATP-sensitive potassium channels prevents myocardial preconditioning in dogs. Circ Res 70: 223-233
45. Coyle P (1987) Spatial relations of dorsal anastomoses and lesion border after middle cerebral artery occlusion. Stroke 18: 1133-1140

Index

A

A₁ receptor 76, 238, 275-278, 280-282, 301
A₂ receptor 157, 221, 276
Acadesine 275, 282
Acetylcholine 4, 21-22, 24, 55, 58-59, 62-63, 67-68, 74, 89-91, 103, 115, 138, 157, 208, 212, 221, 225, 240, 298
A channels 3-5, 7, 11, 13, 15-16, 18-19, 23, 36, 49, 55-56, 155, 315
Action potential duration 34, 53-54, 56-58, 61, 63-64, 66, 69-70, 75-81, 83, 85-87, 90-91, 172-173, 175, 177, 180-181, 184, 187-188, 190, 225, 229, 236-237, 241, 243, 251-253, 256, 259, 261, 263, 265-266
Action potential shortening 76, 81-82, 90, 158, 172, 180, 226-227, 229, 236-237, 239, 243
Activation 4, 11, 17-18, 24, 26, 29, 32, 46-47, 50, 53, 55-60, 62-63, 65, 67, 75-76, 78, 81-82, 85, 87, 89-90, 95-96, 98-106, 109-110, 112, 114, 116, 121, 130, 134, 138, 142, 145, 152-153, 155-157, 171-174, 188, 195, 197, 200, 206, 208, 214-215, 218, 221, 223, 225-226, 236-237, 240-243, 261, 263, 271, 275-276, 278-283, 290, 293-294, 301
Acute myocardial infarction 247-248, 250, 258, 269, 273, 281, 298, 302
Acute skeletal muscle ischaemia 307
Adenosine 21, 24, 51, 56, 62, 69, 76-77, 90, 103-107, 109, 111-117, 119-124, 130-132, 139, 208, 214, 225, 237-238, 240, 243, 273-283, 301, 303, 316
Adenosine diphosphate (ADP) 21, 49, 84, 103-104, 106, 109, 112, 115, 130, 163-165, 170, 173, 188, 235-238, 278, 315
Adenylate cyclase 74, 76-77

Adrenalin 56, 67, 127-129, 222
Afterdepolarizations 64-67, 272
Afterload 294-297
Almokalant 59, 181, 184-185, 190, 259, 264
α-adrenergic agonists 69, 78
Alternate splicing 26
Alternate transcription 26
Alveoli 141
Alzheimer's disease 176, 189
Amantadine 178
Ambasilide 181, 259, 261, 264-265, 271
Amiodarone 59-60, 88, 91, 180, 186, 190, 247-248, 251-252, 254-267, 269-272
Angina pectoris 195, 217, 286-287, 289, 294-295, 297-300, 302
Angiotensin II 99, 109, 114, 200, 208, 211-212, 215, 217
Anti-ectopic agents 256
Antiarrhythmic agents 19, 35, 53, 67, 69, 88, 172, 175, 177, 190, 248, 251, 253, 255, 257, 262-263, 271
  Class IA 177-178, 180, 248-249, 254
  Class IB 176-178, 248
  Class IC 177-178, 186, 248, 254, 261
  Class II 88, 178, 180, 248
  Class III 19, 35, 59, 63, 67, 69, 88-89, 91, 172, 176-181, 183-191, 196, 239, 248-249, 251-254, 256, 258-268, 270-272
  Class IV 88, 178, 180, 248, 263
Antihistamine drugs 175
Antihypertensive drugs 190, 194-195, 209-210, 214-215, 217, 220, 222, 241, 285-291, 293-294
Anti-ischaemic effects 225, 241, 291
Antivasoconstrictor effects 200, 202, 210, 314

Apamin 21, 98, 114, 144, 176, 206, 317
Aprikalim 194-196, 198-201, 207-209, 212-216, 220, 225, 227-228, 231-234, 238-240, 242, 279-280, 283, 291, 298, 316, 319
Arachidonate-activated K^+ channels 18, 21, 42
Arachidonic acid 21, 24
Arterial smooth muscle cells 95, 104-105, 109, 121, 151, 153-154, 157-158, 222
Arterioles 20, 96, 100-101, 105, 107-108, 110, 114, 118-120, 123-124, 128, 130, 132, 134-135, 137, 139, 214, 219, 297, 313-314
Arteriosclerosis 305, 317
Astemizole 175, 179
Atenolol 214, 296, 302
Atherosclerosis 218, 290, 297-298, 303
ATP 21, 26, 41, 46-50, 55, 72, 74, 79-80, 82-87, 90, 100, 104, 109, 112, 115-116, 130, 133, 143-144, 147-148, 152-155, 163-164, 167-174, 188, 196, 205-206, 215, 222, 227-228, 236-238, 241, 273, 278-281, 283, 301, 311, 315, 319
ATP depletion 41, 47-48, 222
ATP sensitivity 47, 236
ATP-sensitive K^+ channels 18, 20-21, 41-42, 44-51, 56, 166-173, 177-178, 188, 190, 194, 205-206, 215, 219, 222, 226-229, 235-238, 241, 243, 274, 276-282, 290, 293-294, 298, 301, 314-315, 319
ATPase 5, 139
Atrial cells 19, 24, 42, 54, 56, 58, 89, 91, 190, 252, 266, 275, 282
Atrial fibrillation 186-187, 249, 252, 257, 268, 270
Atrial flutter 184, 262, 269
Atrio-ventricular conduction 225
Atrio-ventricular node 53, 56, 62, 187, 256, 259
Autoregulation 114, 138-139, 195

B

Ba^{2+} 41, 43-44, 97-98, 100-103, 176, 266

Background current .. 41, 59, 62-63, 68
Ball-and-chain 28
BAY K8644 142
Benzopyran derivatives 183, 194-195, 201, 305
Bepridil 179
ß-blocker 88, 249-253, 256, 258-260, 265, 267, 270, 286, 289, 293, 295-297
ß-agonist 74, 76-77
ß-endosulphine 164
ß-receptor 74, 88, 129, 247, 252-253, 266
$ß_2$-receptor 128, 272
Bimakalim 195, 211-213, 226, 232, 234, 239, 242
Blood pressure 1, 95-96, 104-105, 127-129, 138, 206-217, 222, 230, 233-234, 239, 285, 287-288, 291, 293-294, 296, 305-306, 309, 311-314
Bradycardia 179, 183-184, 257, 262, 267, 283
Bradycardic effect 63-64, 249
Bradykinin 111, 116-119, 122

C

C-type inactivation 28
Ca^{2+}-activated K^+ channels 18, 20-21, 26, 42, 95, 97-102, 105, 113-118, 121, 144, 152-154, 158, 298
Ca^{2+} channels 6-7, 9, 11, 20, 22-24, 33, 47, 57, 58, 62, 65, 69, 71-72, 74, 96, 109, 114, 118, 137, 142, 155, 178-180, 185, 193, 196, 200, 215, 251, 263-264, 275-276, 293-294, 299
Ca^{2+} overload 59, 82, 114, 237-239
Ca^{2+} pump 72
Calcitonin gene-related peptide ... 102, 106, 208
cAMP 58, 74, 241
Carbon dioxide 123-124
Cardioprotective effect .. 230, 299, 301
Catecholamines 114, 126-127, 129, 134, 215
cDNA 25, 32, 34-35, 37
Celikalim 195, 203, 233, 242
Cell-attached .. 13, 42-43, 48, 169, 228

Cerebral arteries 101, 103-104, 106, 201, 221, 316
Cerebral ischaemia 316
cGMP 138
Charybdotoxin 20, 29, 31-33, 36, 97-100, 102, 114, 144, 153, 176, 206
Cholesterol 290
Chronic occlusive arterial disease 308
Chronotropic effects 21, 63-64, 129, 294
Ciclazindol 196, 205, 220
Clofilium 65-66, 181, 183, 185, 190, 239, 263-265, 271
Clonidine 178
CN⁻ 48, 79-80, 82, 220-221, 291
Collateral blood flow 230, 233, 273, 278, 299, 307, 316, 318-319
Conduction velocity 187, 252, 259, 262
Congestive heart failure 218, 256, 270, 285-288, 295, 302
Constant field equation 9
Contractility 1, 50, 58, 69-91, 74, 79, 91, 96, 107, 123-124, 126-127, 130, 136, 143, 146, 155, 157, 184, 200-201, 213-215, 221, 225, 227, 230, 232, 236-237, 280, 283, 294
Contracture 80-82, 237, 282
Coronary arteries 103-104, 106-107, 109, 111-112, 114, 120-121, 132, 139, 202, 214, 221, 223, 283, 298-299, 303, 314, 318
Coronary blood flow 107-109, 111, 113-115, 117, 119-122, 129, 213-214, 217, 223, 226, 230, 233, 297-298, 303
Coronary endothelial cells 116-119, 121-122
Coronary microvasculature 107, 112, 303
Coronary resistance arteries 107, 111-112
Coronary steal 214, 293-294, 299
Coronary stenosis .. 230, 233, 293, 298
Coronary vasodilation 132, 213, 238, 275, 278, 283, 296, 299
Coronary vasospasm 202, 298, 303
Cromakalim 83-84, 90, 102, 104, 106, 109, 111, 143, 145-147, 149, 154-157, 165-166, 190, 194-196, 198-216, 219-223, 225-226, 228-229, 231-233, 235-243, 279, 283, 287, 289, 298, 303, 305-306, 308, 310-314, 316, 318-319
Cs^+ 41, 43, 61, 176, 223, 242-243, 283, 291, 302-303
Current-voltage relation 16-17, 20, 41, 43, 45, 61
Cytoprotection 228-229, 235, 240

D

Deactivation 263
Delayed rectifiers 11, 15, 17, 19, 23, 26
Dendrotoxin 31-33, 37, 100, 176
Desethylamiodarone 256, 266-267, 272
Diabetes mellitus 290
Diacylglycerol 116
Diastole 44, 57, 62, 185, 226
Diastolic membrane potential 60
Diazoxide 97, 101-102, 104, 143, 146-147, 149, 156-157, 165-166, 171-172, 174, 194-195, 203-204, 207, 215-216, 219, 222, 225, 287-288, 290, 303, 314, 319
Dinitrophenol (DNP) 48, 206, 236
Dipyridamole 112-113, 278
Dofetilide . 59, 181, 183-185, 188, 190 259, 261, 264, 267, 271
DPI 201-106 .... 88, 91, 184
dV/dt 49, 227

E

E-4031 .... 89, 91, 259-261, 264-265, 267, 272
Early afterdepolarizations .. 64-67, 272
ECG .. 61, 65, 175, 179, 184, 230, 254, 259, 262, 289, 300
Effective refractory period 54, 189, 252, 257, 259, 261
8-phenyl-theophylline 112
8-sulphophenyltheophylline 274
E_K 8-11, 16, 20, 22-23, 41, 44, 80, 96, 101, 134, 156-157, 193, 199-202, 205

Electrochemical driving force 9-10, 12, 14
Electrochemical gradient 5, 7, 199
$E_{Na/Ca}$ 73
Encainide 247-249, 264-265, 268
Endogenous vasodilators 101-105, 138
Endothelin 99, 114, 125, 145, 157, 200-201, 211, 221, 318
Endothelium 103, 106-108, 112-122, 125, 137-138, 147, 150-151, 157, 197-198, 297-299, 305, 314, 318-319
Endothelium-dependent relaxing factor 111-112, 114, 118-119, 124-126, 138, 150, 157
Endothelium-derived hyperpolarizing factor 103, 105, 137-138, 150-151, 157
Endotoxic hypotension 104
Energy metabolism 41, 49, 107, 109, 112, 114, 120, 218, 305, 309, 312
Excitability 41, 43, 53-54, 67, 197, 200, 315
Excitation-contraction coupling 71-72, 89
Exercise 123-139, 141, 230, 289, 295-297, 306, 309, 311, 317
Exercise hyperaemia 123, 125, 130-132, 134-138
Exercise-induced angina 297-298
Extracellular K^+ accumulation 234-235

F
Fatigue 145, 305, 309, 312
Fibrillation 173, 186-187, 226, 230, 239, 244, 247-249, 251-254, 257, 260, 262, 268-271
Final repolarization .. 6, 53, 59-61, 182
5-hydroxydecanoate 178, 190, 196-197, 220, 229
Flecainide 177, 247-249, 253, 261, 263-265, 268, 271
Flow redistribution 308, 316
4-aminopyridine 32, 78-79, 97, 99-100, 143, 149-153, 155-157, 175-176, 179, 189

Frequency-dependent block 59

G
G-protein 22, 62, 91, 115, 225, 237-238, 275, 282
Gap junctions 50, 116-117, 119
Gating 4, 22, 25, 28-29, 31, 34, 36, 43-44, 46-47, 50, 109, 170, 205
Glibenclamide 21, 49, 82-84, 97-98, 100-105, 109, 111-113, 115, 117, 121, 132-133, 144-149, 154-157, 161-162, 164-165, 167-168, 171-174, 177-178, 190, 193, 196, 198-200, 202-210, 213-214, 216, 219-223, 226, 229, 231-232, 234-238, 241, 243, 278-281, 283, 290, 291, 298, 301, 303, 314-316, 319
Glibornuride 196, 199
Glipizide 196, 199, 204
Gliquidone 162, 168, 177
Glisoxepide 196, 199
Global ischaemia 228, 239-241, 293, 301
Glucose-sensors 168
Glycogen 131, 137
Glycosides 21, 87
Glycosylation 34
Goldman-Hodgkin-Katz equation 9-11
GTP-binding protein 5, 21-22
Guanylate cyclase .. 195, 201, 203, 220

H
H5 sequence 27, 29-30
Haloperidol 179
Heart failure 218, 256-257, 270, 285-288, 295, 302
Hibernation 232, 242
Histamine 115-116, 123-124, 137-138, 179, 200, 208
Holter monitoring 268, 300
Hydrophobicity analysis 25
Hyperaemia 104, 107, 112, 123-125, 128, 130-132, 134-139
Hypercapnia 141, 152, 158
Hyperosmolarity 123-124, 137
Hyperpolarizing vasodilators 101, 104-105, 145, 222

Hypertension 95, 122, 143-144, 156, 179, 195, 214, 216-217, 219, 221, 285-291
Hyperthyroidism 252, 266
Hypertrichosis 195
Hypertrophic cardiomyopathy 257, 270
Hypocalcaemia 186
Hypoglycaemia 161, 172, 194
Hypokalaemia 43, 61, 64, 179, 186, 239, 262
Hypomagnesaemia 64, 179, 262
Hypothalamic neurones 167-168, 170-171, 174
Hypotension 95, 104-106, 143, 207-208, 213, 215, 293-294, 297, 299
Hypothyroidism 186-188, 252, 265-267, 272
Hypoxia 21, 49, 79, 90, 101, 103-105, 107, 109-113, 120, 132, 141-151, 155-158, 235-237, 243, 274, 278
Hypoxic pulmonary vasoconstriction 142, 144-146, 148-151, 153-157
Hypoxic vasodilation 107, 113, 121

I
Iberiotoxin 97-99, 102
Ibutilide 181, 190, 259
i_f 1, 4, 8, 10-11, 22-24, 32, 43, 49, 57, 62-63, 70, 73, 81-82, 84-85, 114, 130, 136, 144, 149, 155-157, 161, 165, 171-173, 199-200, 205-206, 217-218, 227, 233, 247, 266-267, 273, 279-280, 286, 288, 300-301, 316-317
i_{K-ACh} 18, 21, 56, 62, 74-75, 263
i_{K-Na} 18
i_{Kr} 58, 181, 183
i_{Ks} 58
Iloprost 103
Imidazoline derivatives 196
Inactivation 11, 18, 26, 28, 30-32, 36, 44, 50, 53, 56-57, 65, 67, 78, 91, 100, 184, 275
Inactivation ball 28, 30
Indomethacin 103
Infarct size 223, 232-235, 242, 274, 278-283, 289, 291, 301-303, 316, 319
Initial repolarization 54, 56, 182
Inositol triphosphate 196
Inotropy 21, 67, 69, 74-79, 88-91, 129, 183-184, 191, 294
Insulin secretion 161-164, 174, 194, 196, 215-216, 279, 288
Intermittent claudication 305-306, 308-311, 313, 316-317
Intracellular acidification 49, 104
Intracellular Ca^{2+} 1, 3, 5, 8, 16, 20-22, 26, 28, 30-31, 41-46, 48-49, 51, 60, 62, 69-72, 79, 84, 86-87, 90-91, 97-101, 104-105, 108-110, 112-119, 122, 132, 134-135, 141, 143-145, 147, 152-154, 167-170, 172, 185, 188, 193, 196-197, 199-200, 206, 215, 227-228, 235-238, 241, 243, 275-278, 314-315
Intracellular Mg^{2+} 1, 3, 5, 8, 16, 20-22, 26, 28, 30-31, 41-46, 48-49, 51, 60, 69-72, 79, 84, 86-87, 108-110, 112-119, 122, 132, 134-135, 141, 143-145, 147, 152-153, 167-169
Intravascular pressure 96, 98-99, 109, 124-126, 138
Inward rectifier 18-20, 41-48, 50, 55, 59-61, 63-64, 95, 97-98, 100-101, 105, 134, 180-185, 187, 190, 227, 263-265
Ischaemia 8, 21, 49, 53, 55-56, 67, 69, 79, 82-85, 87, 90, 95, 107, 109, 111, 120, 132, 158, 172-173, 188, 214, 225, 228-233, 235-243, 251, 263, 268, 273-275, 277-281, 283, 289, 293-295, 297-298, 300-301, 307, 310-317, 319
Ischaemic heart disease 225, 252, 285, 289, 293, 301, 305
Ischaemic preconditioning 242, 273-283, 293, 301, 303, 316, 319
Ischaemic skeletal muscle 218, 305, 309, 312, 316
Isosorbide dinitrate 242, 294, 296, 299-300, 302
Isotope flux techniques 202
Isradipine 200, 220, 319

K

K+ channel openers 23, 66, 68, 90, 96, 102, 109, 143, 161, 165-166, 168, 171, 173-174, 193-223, 225-233, 235-244, 279-280, 283, 285, 287-291, 293, 295, 297, 299, 301, 303, 305, 317-318
K+ permeability 21, 24, 41-42, 48, 134
$K_{Ca, ATP}$ channels 153, 155
Ketanserin 179, 305
KRN2391 195

L

L-type Ca^{2+} channel 47, 65, 71-72, 74, 193, 200
Lactate 123-124, 135-137, 139, 273, 278, 301
Left ventricular hypertrophy 289
Levcromakalim 97, 102, 104, 143, 154, 165, 195, 206, 212, 214-215, 222, 235
Lidocaine 54, 67, 177-178, 241, 248, 250, 264-265, 268-269
Ligand-gated channels 4-5, 12
Ligustrazine 178
Linogliride 178
Local anaesthetics 127
Long QT syndrome 269

M

M channels 18
Mast cell degranulating peptide 32-33
Maximum diastolic potential 63
Meglitinide (HB 699) 162, 164, 167, 170, 174
Melperone 179
Membrane hyperpolarization 21, 96, 102, 106, 114, 151, 294
Membrane potential measurements 202
Membrane-spanning segments 26-27, 30
Metabolic inhibitors 48
Metabolic stress 168, 174
Methoxamine 78-79, 210-211
Methylsulphonamide 183
Mexiletine 248, 250, 269
Mg^{2+} block 20, 43-45, 50
MgATP 46-47, 166, 168

Minoxidil 97, 102, 104, 143, 157, 165-166, 194-196, 202-204, 219-222, 225, 287, 290, 314
Modulated-receptor 185
Monophasic action potential 65, 184, 237-238
Moricizine 248-249, 268
mRNA 25, 34
MS-551 .... 259, 264
Muscarinic K+ channels 18, 21, 24, 41-42, 44-45, 62-63, 74-76, 188, 227, 275, 282
Muscarinic receptors 21, 208
Muscle energy metabolism 305, 309, 312
Myo-endothelial regulatory unit .. 119-120
Myocardial infarction ... 218, 225, 233, 239, 242, 247-248, 250, 258, 267-271, 273, 281, 287, 294, 298, 302
Myocardial stunning 231, 242, 280, 283, 289, 293, 302
Myogenic response 114, 124-126, 138
Myogenic tone 95, 98, 114, 144

N

N-acetylprocainamide 261-262
N-type inactivation 28
Na+ -activated K+ channels 86-87, 227
Na+ channels 6, 10, 23, 35, 53, 88, 177-178, 180, 184, 250-251
Na/Ca exchange 50, 58, 69, 70, 72-73, 196, 275
Na/H exchange 135
Na/K pump 61, 87, 134-135
Nernst equation 8, 10, 193, 199
Neuropeptides 102
Neurotransmitter 1, 5, 11, 21-22, 130, 163, 172, 212
Nicorandil 83, 97, 102, 165, 195, 202, 205, 207, 213, 216, 220-221, 225-228, 230-232, 234-235, 239, 241-243, 293-303, 308, 315, 319
Nicotinic acetylcholine receptor 22
Nifedipine 142, 209-212, 221-222, 231-232, 234, 242, 294-296, 300, 302-303, 317-319

Nisoldipine 142
Nitrates 220, 293-299
Nitric oxide (NO) 8-9, 44, 47,
  58, 60, 63, 76, 78-79, 91, 98, 102-
  103, 106, 114, 118, 121, 124, 128,
  130, 134, 137-138, 140, 145-147,
  150-151, 155, 163-165, 169, 177,
  188, 195, 198-199, 205-206, 212,
  214, 216, 225, 227, 233, 239, 247-
  251, 253, 255-259, 263-266, 273-
  275, 277, 279-281, 289-290,
  296-297, 299-301, 306, 308, 316
Nitroglycerin 214, 220, 231-232,
  242, 298, 301, 303, 319
Non-selective cation channels ... 4, 166
Noradrenaline 24, 88, 99, 105,
  109, 127-129, 132, 196, 200, 202,
  210-212, 221
Noxiustoxin 176
Nucleotide diphosphates   47, 165, 278
Nucleotide triphosphates 47, 165

O

Occlusive vascular disease 305
Oil-gap method 44
On-cell recording 13
Oocyte 27-28, 32-33
Open probability   14-19, 24, 49, 83-84,
  86, 91, 97, 112, 152-153, 196, 276,
  315
Open-channel block 44
Oscillatory depolarizations 66
Osmolarity 137
Ouabain 42, 61, 186
Outside-out 13
Oxygen-derived free radicals 239

P

P1075 .... 66, 195, 204-205
Pacemaker potential 11, 22-24
Parasympathetic stimulation 62
Patch-clamp 6, 11-13, 24, 41-42,
  44, 95, 110, 116-117, 147, 151-153,
  166, 168, 175, 185, 205, 227, 236,
  238, 262, 280
Pco_2 .... 135-136, 141
PD 115, 119........................... 274, 276
Pentobarbitone 178, 206-207,
  209, 213, 216, 279

Pentoxifylline 305
Percutaneous transluminal coronary
  angioplasty 298
Peripheral resistance 1, 95-99,
  101, 103, 105, 231, 294
Peripheral vascular disease .. 218, 288
  305-313, 315-317
Perivascular nerve cells 107
Perivascular space 108, 112, 130
Permeabilized patch 13
Pertussis toxin .. 79, 137, 275-276, 282
pH 1, 21, 123, 133, 135-137, 152, 156,
  158, 188, 235, 238, 243, 278, 318-
  319
Phencyclidine 179-180, 190
Phenothiazines 179
Phentolamine 196, 208
Phosphatidylcholine 21
Phosphocreatine 309, 311-313, 318
Phospholipase C 115, 276
Phosphorylation 27, 34, 47, 50, 58, 89,
  99, 105, 116, 153, 165-166, 277,
  280
Picoamperes 15
Picosiemens 16-17
Pinacidil 83-85, 97, 102, 104, 106, 143,
  156-157, 165-166, 194-195, 201-
  205, 213-216, 219-221, 223, 225-
  226, 228-229, 233, 239-242, 244,
  279, 283, 287-291, 298, 302-303,
  308, 314, 318
Po_2 .... 110, 114, 135-136,
  141-142, 148, 155-156, 318
Polymerase chain reaction 276
Post-ischaemic contractile failure   274,
  280
Post-ischaemic vasodilation 107
Preload 294-295, 297
Pressure-heart rate product 234
Proarrhythmic effect 186, 253,
  262-263, 290
Procainamide 191, 248-249, 268
Prolonged cardiac repolarization .. 262
Propanolol 178
Prostacyclin 103, 114
Prostaglandin 118, 123-124,
  137, 140, 145
Protein kinase A 58
Protein kinase C 116

Protons 235-236
Pulmonary capillary wedge
　pressure 294-295
Pulmonary circulation 141-143
Pulmonary hypertension 143-144, 156
Pulmonary œdema 143
Pulmonary vascular tone 143-145, 156
Pulmonary vasculature 141, 143,
　145, 147, 149, 151, 153, 155, 157
Purkinje fibres 19, 50, 54, 56,
　65-67, 78, 91, 190, 226, 239, 241,
　243-244, 259, 261, 265, 269, 271
PVCs 247, 249, 253-254, 256-257, 262

Q
QT interval 64, 175, 179,
　186, 189-190, 251, 269
Quantitative coronary angiography
　297, 299
Quaternary ammonium ions 175
Quinidine 64, 91, 177-178, 180,
　187, 191, 239, 248-250, 257, 261-
　262, 264-265, 268, 271-272
Quinine 178

R
Radioligand binding assay 195, 204
Rate of diastolic depolarization 62-63
Rb^+ 18, 37, 41, 43-44, 50, 67-68,
　90, 221, 243, 271, 281
Re-entry 59, 239, 248, 266
Reactive hyperaemia 104, 107, 112
Recovery from inactivation 28, 36, 53,
　67
Rectification 12, 19-20, 24, 41, 43, 45,
　50, 59-60, 67, 74, 87, 91, 101, 205
Refractoriness 53-55, 57, 59, 61,
　63, 65, 67, 177, 188, 225, 235, 239,
　247-248, 251, 253, 260, 263, 265-
　266, 268
Refractory hypertension 287-288
Regional myocardial blood flow ... 299
Relaxation 96, 99, 101, 106,
　110, 124, 138, 145-149, 155, 157,
　185, 193-194, 198-199, 202-203,
　221, 225, 307, 314
Renal blood flow 212-213
Renin-angiotensin system 125,
　208, 215

Reperfusion damage 82-83,
　90, 242, 273, 281
Repolarizing currents 69, 182,
　262-263, 271
Resistance arteries 96, 98, 100,
　105-107, 111-112, 114
Resting potential (V_m) 7, 9-10,
　12, 16, 20, 41-43, 45-47, 49, 51, 53,
　56, 58-59, 61, 66, 73, 101, 109, 157,
　199, 201-202, 206, 222, 226
Reversal potential 9, 11, 22-23, 41-42,
　44, 48, 73
Reverse frequency-dependence ... 185,
　187
Rigor 80
Risotilide 181, 264-265
RP 49356 .... 97, 102, 143,
　156, 195, 201, 204-205, 214-216,
　220, 227-228, 241

S
S4 segment 28-29
Sarcoplasmic reticulum 71-73,
　185, 191, 197
SDZ PCO-400 .... 305-313, 315-316
Sematilide 181, 259, 261, 267, 272
Sepsis 95
Septic shock 104-105
Serotonin 99, 109, 179, 190, 200
Shaker 25-37
Sigma receptor 179
Single channel conductance 12, 15-
　16, 18, 20, 42, 45, 87
Sino-atrial 22-24, 41,
　53, 56, 60, 62, 64, 68, 74, 226
Sinus rhythm 185-186,
　249, 252, 257, 268
Site-directed mutagenesis 25, 30-31
Skeletal muscle 9, 34, 99, 101, 103,
　104, 109, 123-124, 127-135, 137-
　140, 155, 163, 212, 218, 225, 305-
　307, 309, 312-313, 315-316, 318
Smooth muscle 7, 10, 19, 21, 95-
　116, 118-119, 121-126, 130, 132,
　134-135, 138, 141, 144, 149-158,
　162, 166, 168, 171, 193-194, 196,
　198, 200, 203, 205-206, 215, 217,
　221-222, 225, 227, 280, 293, 298,
　314-315

Sodium 5-Hydroxydecanoate 190, 196-197, 220
Somatostatin 56, 62
Sotalol 88-89, 91, 178, 180-181, 190, 247, 249, 251-255, 257-262, 264-265, 267-268, 270-272
Sparteine 178, 196
Spike-dome 54, 57-58, 62
ST-segment 300
Stenosis 214, 223, 230, 233, 242, 293, 295, 298-299, 302
Stretch-operated channels 55
Stunning 218, 230-232, 238-239, 242, 280, 283, 289, 293, 297, 302
Substantia nigra 163, 168, 172
Sucrose gap 41
Sudden death 189, 247, 249-251, 253, 255, 257-258, 262-263, 265, 268, 270
Sulphonylurea 21, 82, 101-102, 104, 111, 121, 148-149, 161-171, 173-174, 178, 190, 199, 204, 208, 216, 220, 222, 243
Supraventricular tachycardias 62
Sympathetic nerve 23-24, 63, 105, 124, 126-130, 132, 139, 207, 211-212, 214, 218-219, 222, 247, 251-253, 297, 319
Syncope 262
Systemic hypertension 285, 291
Systole 6, 57, 185

T

T-wave 61, 289
Tachyarrhythmia 64-65, 186
Tacrine 176
Tail current 18
Tedisamil 58, 63-64, 181, 183, 190, 196, 220, 264-265
Terfenadine 175, 179, 189
Terikalant 181-186, 190, 264
Terminal arterioles 107-108, 110, 114, 119-120
Tetrabutylammonium 81, 175
Tetraethylammonium 22, 29-32, 36, 44, 88, 97-100, 102, 121, 143, 149-153, 175, 181
Thiopentane 178
Thrombin 115

Thrombolytic therapy 225, 273
Thromboxane ... 99, 145, 157, 199, 221
Thyroid hormone ... 252, 265, 267, 272
Tolbutamide 21, 81-84, 97, 101-102, 104-105, 148-149, 157, 161, 164, 167-174, 177, 196, 219, 226, 232, 236
Torsades de pointe 59, 65, 67, 175, 179, 183, 185-188, 239, 253, 259-260, 262-263, 265, 267
Transient outward current 19, 55-56, 77-79, 90, 175, 181-183, 185, 190, 271
Transmural pressure 95, 114
Triglyceride 290
2-deoxyglucose 79-80, 82, 227

U

U-37,883A 196, 220

V

Vagal stimulation 23-24, 56, 63, 187, 251, 269
Vascular membrane potential 100
Vascular resistance 95, 110, 123-124, 127, 149, 157, 206, 211, 213-214, 216-218, 234, 294-295
Vascular tone 95-96, 104-106, 109-110, 113-114, 125, 143-145, 156, 203, 206, 218
Vasoactive intestinal peptide 102
Vasoconstriction 95, 97-100, 105, 115, 125, 127-130, 139, 142-146, 148-151, 153-157, 200-201, 210-211, 215, 217, 306, 314-315
Vasodilation 95-96, 102-107, 109, 111-113, 115, 119, 121, 124, 126-138, 145, 147, 150, 156-157, 194-195, 199, 206, 208, 213-219, 221-223, 225, 238, 244, 253, 275, 278, 283, 286-287, 291, 294-299, 305-306, 308-309, 313-315, 317-318
Vasopressin 127, 200, 208, 211-212, 314
Vasospasm 95, 105, 202, 293, 298, 303
Vasospastic angina 299-300
Ventricular fibrillation 173, 226, 247 248, 251-254, 260, 262, 268-271

Ventricular premature beats 186
Ventricular tachycardia 135, 247-248, 251, 253, 256-257, 262, 268, 270
Verapamil 142, 178-179, 221, 252, 266, 301, 303, 307, 317-318
Voltage sensor 5, 31
Voltage-clamp 11-12, 15, 17, 41, 44, 46, 61, 70, 75-81, 145, 149, 187, 190, 205, 271
Voltage-gated channels 4-5, 12, 23

W
Whole-cell currents 13-15, 17-18, 24

X
Xenopus oocyte system 33

Y
Y-27,152 .... 195, 209, 211, 214-215, 220, 223

Impressions DUMAS
42100 SAINT-ÉTIENNE
Dépôt légal : 3ᵉ trimestre 1993
N° d'imprimeur : 31513
Imprimé en France